Migration is a key process in the population dynamics of many insects, including some of the most damaging pests. Multidisciplinary research over the last three decades has produced a series of important new insights. This book reviews current understanding of the ecological, behavioural, physiological and genetic bases of insect migration. The first part describes migration systems in Europe, Asia, Africa, Australia and America, with an emphasis on the role of regional weather and climate. The second part considers how insects are adapted for migration; it covers aerodynamics and energetics, the integration of migration in insect life cycles, environmental and genetic regulation of migratory potential and the evolutionary implications of habitat heterogeneity and variability. The third part addresses the application of this knowledge to operational pest forecasting. The book concludes with a comprehensive overview of insect migration, written from an evolutionary perspective.

# INSECT MIGRATION:

tracking resources through space and time

Participants in the symposium *Insect Migration: Physical Factors and Physiological Mechanisms* at the XIX International Congress of Entomology, Beijing, 29–30 June 1992. Left to right, *back row*: G.M. Tatchell, C.-S. Ma, J. Colvin, K. Wilson, T.P. Robinson, J.D. Knight, T. Watanabe, X.-N. Cheng, K. Hirai, J.-H. Lee, K.-B. Uhm, M. Lecoq, G. McDonald, J.I. Magor, Y.-J. Sun, T. Wada, K. Sogawa. *Front row*: S.J. Johnson, R.-L. Chen, P.C. Gregg, R. Kisimoto, M. Ishii, V.A. Drake, X.-X. Zhang, A.G. Gatehouse, R.A. Farrow, K.J. Walden, J.N. McNeil, D.E. Pedgley.

# INSECT MIGRATION:

tracking resources through space and time

*Edited by*

## V.A. DRAKE
*University of New South Wales, Australia*

*and*

## A.G. GATEHOUSE
*University of Wales, Bangor*

**CAMBRIDGE**
UNIVERSITY PRESS

Published by the Press Syndicate of the University of Cambridge
The Pitt Building, Trumpington Street, Cambridge CB2 1RP
40 West 20th Street, New York, NY 10011–4211, USA
10 Stamford Road, Oakleigh, Melbourne 3166, Australia

First published 1995

Printed in Great Britain at
the University Press, Cambridge

*A catalogue record for this book is available from the British Library*

*Library of Congress cataloguing in publication data*

Insect migration : tracking resources through space and time / edited
by V. A. Drake and A. G. Gatehouse.
p.   cm.
Includes bibliographical references and index.
ISBN 0 521 44000 9 (hardback)
1. Insects – Migration.   2. Insect pests – Migration.   I. Drake.
V. A.   II. Gatehouse, A. G.
QL496.2.147   1995
595.7052'5 – dc20   94–49434 CIP

ISBN 0 521 44000 9 hardback

RO

This book is dedicated to
the memory of
John S. Kennedy FRS
1912–1993

# Contents

ix

*Note added in proof:*
Professor R.-L. Chen died 28/3/95. The author for correspondence for
chapter 4 is now Ms Y.-J. Sun.

# Contributors

Prof. X.-Z. Bao
*Institute of Plant Protection, Jilin Academy of Agricultural Sciences, Gongzhuling, Jilin Province, 136100, China*

Prof. R.-L. Chen
*Institute of Plant Protection, Jilin Academy of Agricultural Sciences, Gongzhuling, Jilin Province, 136100, China*

Prof. X.-N. Cheng
*Department of Plant Protection, Nanjing Agricultural University, Nanjing 210014, China*

Dr J. Colvin
*Natural Resources Institute, Central Avenue, Chatham Maritime, Chatham, Kent ME4 4TB, UK*

Dr M. Cusson
*Laurentian Forestry Centre, 1055 PEPS, PO Box 3800, Ste Foy, PQ, G1V 4C7, Canada*

Dr R.K. Day
*International Institute for Biological Control, PO Box 30148, Nairobi, Kenya*

Dr J. Delisle
*Laurentian Forestry Centre, 1055 PEPS, PO Box 3800, Ste Foy, PQ, G1V 4C7, Canada*

Dr V.A. Drake
*Department of Physics, University College, University of New South Wales, Australian Defence Force Academy, Canberra, Australian Capital Territory 2600, Australia*

Dr R. Dudley
*Department of Zoology, University of Texas, Austin, TX 78712, USA*

Dr R.A. Farrow
*Division of Entomology, CSIRO, GPO Box 1700, Canberra, Australian Capital Territory 2601, Australia*

Dr G.P. Fitt
*Division of Entomology, CSIRO, PO Box 59, Narrabri, New South Wales 2390, Australia*

Dr A.G. Gatehouse
*School of Biological Sciences, University of Wales, Bangor, Gwynedd LL57 2UW, UK*

Dr P.C. Gregg
*Department of Agronomy and Soil Science, University of New England, Armidale, New South Wales 2351, Australia*

Dr K. Hirai
*National Agricultural Research Centre, Tsukuba, Ibaraki 305, Japan*

Dr S.J. Johnson
*Department of Entomology, Louisiana Agricultural Experiment Station, Louisiana State University Agricultural Center, Baton Rouge, LA 70803, USA*

Prof. R. Kisimoto
*Faculty of Bioresources, Mie University, 1515 Kamihama, Tsu 514, Japan*

Dr J.D. Knight
*Silwood Centre for Pest Management, Department of Biology, Imperial College at Silwood Park, Ascot, Berks. SL5 7PY, UK*

Dr M. Lecoq
*CIRAD-PRIFAS, BP 5035, 34032 Montpellier, CEDEX 1, France*

Dr J.-H. Lee
*Division of Applied Entomology, Department of Agricultural Biology, College of Agriculture and Life Sciences, Seoul National University, Suwon, 441–744, Korea*

Dr J.I. Magor
*Natural Resources Institute, Central Avenue, Chatham Maritime, Chatham, Kent ME4 4TB, UK*

Dr G. McDonald
*Plant Sciences and Biotechnology, Agriculture Victoria, La Trobe University, Bundoora, Victoria 3083, Australia*

Prof. J.N. McNeil
*Département de biologie, Université Laval, Cité universitaire, Québec, PQ, G1K 7P4, Canada*

Dr D.A.H. Murray
*Department of Primary Industries, Toowoomba, Queensland 4350, Australia*

Dr I. Orchard
*Department of Zoology, University of Toronto, Ramsay Wright Zoological Laboratories, 25 Harbord Street, Toronto, Ontario M5S 1A1, Canada*

Mr D.E. Pedgley
*Natural Resources Institute, Central Avenue, Chatham Maritime, Kent ME4 4TB, UK*

Dr D.R. Reynolds
*Natural Resources Institute, Central Avenue, Chatham Maritime, Kent ME4 4TB, UK*

Dr T.P. Robinson
*c/o The Foreign and Commonwealth Office (Lusaka), King Charles Street, London SW1A 2AH, UK*

Dr K. Sogawa
*Kyushu National Agricultural Experiment Station, Nishigoshi, Kumamoto 861–11, Japan*

Ms Y.-J. Sun
*Institute of Plant Protection, Jilin Academy of Agricultural Sciences, Gongzhuling, Jilin Province, 136100, China*

Dr G.M. Tatchell
*Rothamsted Experimental Station, Harpenden, Herts. AL5 2JQ, UK*

Prof. S.S. Tobe
*Department of Zoology, University of Toronto, Ramsay Wright Zoological Laboratories, 25 Harbord Street, Toronto, Ontario M5S 1A1, Canada*

Dr K.-B. Uhm
*Department of Entomology, Agricultural Science Institute, Rural Development Administration, Suwon, 441–707, Korea*

Mr K.J. Walden
*Western Australian Department of Agriculture, PO Box 110, Geraldton, Western Australia 6530, Australia*

Dr H.-K. Wang
*Department of Plant Protection, Nanjing Agricultural University, Nanjing 210014, China*

Ms S.-Y. Wang
*Institute of Plant Protection, Jilin Academy of Agricultural Sciences, Gongzhuling, Jilin Province, 136100, China*

Dr T. Watanabe
*Laboratory of Pest Management Systems, Kyushu National Agricultural Experiment Station, Nishigoshi, Kumamoto, 861–11, Japan*

Dr K. Wilson
*Department of Zoology, University of Cambridge, Downing Street, Cambridge CB2 3EJ, UK*

Dr M.P. Zalucki
*Department of Entomology, University of Queensland, Brisbane, Queensland 4072, Australia*

Dr B.-P. Zhai
*Institute of Plant Protection, Jilin Academy of Agricultural Sciences, Gongzhuling, Jilin Province, 136100, China*

Prof. X.-X. Zhang
*Department of Plant Protection, Nanjing Agricultural University, Nanjing 210014, China*

Prof. B.-H. Zhou
*Department of Plant Protection, Nanjing Agricultural University, Nanjing 210014, China*

# Preface

The spatial dimension of ecology, until recently rather unfashionable, is currently the subject of a surge of interest that appears to have arisen independently in a number of fields. Spatial heterogeneity and the spatiotemporal processes of dispersal and migration, and invasion and colonisation, are now receiving attention not only from 'island biogeographers' but also from population ecologists and geneticists, evolutionary biologists and applied ecologists. A recent editorial in the journal *Ecology* declared spatial dynamics to be 'the final frontier for ecological theory' (Kareiva, 1994). Geographers have established a new subdiscipline, 'landscape ecology', to integrate the biological, economic and cultural aspects of environmental heterogeneity. Technical advances in remote sensing (of both habitats and migratory movements), geographic information systems, geostatistics, biotelemetry, meteorology and climatology have spurred and sustained all of these approaches.

Central to any consideration of spatial dynamics in ecology must be the processes by which organisms move, or are moved, from one habitat location to another. Migration has been studied in many taxa, but work on insects has been particularly intense and productive, primarily because several of the world's most damaging pests are migratory, but also because insects' short life cycles make the inter-relations of migration with life history and the seasonal cycle especially evident and accessible to study.

Progress in understanding insect migration, and indeed animal migration in general, has been hampered for many years by confusion, in both concepts and terminology, about what constitutes 'migration'. This obstacle was finally removed by the late Professor J. S. Kennedy FRS, who first made clear the distinction between the behavioural and ecological meanings of this term. For many species, changes in the spatial distribution of populations are the result of no more than the day-to-day movements associated with foraging, together with the occasional and unpredictable adventitious displacements that occur when station-keeping behaviour is overwhelmed by disruptive forces such as high winds, storms, floods, or human intervention. However, in migratory species, specialised behaviour has evolved that enables individuals to leave the current habitat patch and find and colonise other patches that have arisen elsewhere. Kennedy (1985) proposed the following definition of this specialised behaviour:

> Migratory behaviour is persistent and straightened-out movement effected by the animal's own locomotory exertions or by its active embarkation on a vehicle. It depends on some temporary inhibition of station-keeping responses but promotes their eventual disinhibition and recurrence.

This behavioural characterisation of migration, and the distinction between the process of migration and its ecological consequences that it has fostered, form one of the cornerstones of the conceptual framework underlying contemporary approaches to migration studies.

Professor Kennedy died in 1993. We dedicate this volume to his memory, in recognition of his pivotal contribution to the emergence, over the last 40 years, of the study of insect migration from its origins in natural history to its present status as a modern, interdisciplinary branch of biology, incorporating biogeographic, ecological, behavioural, physiological and genetic dimensions.

This book had its origins in a symposium entitled *Insect Migration: Physical Factors and Physiological Mechanisms* held during the 1992 XIX International Congress of Entomology (ICE) in Beijing (Frontispiece). The symposium, organised by Alistair Drake, Gavin Gatehouse and Professor X.-X. Zhang (Nanjing Agricultural University), focussed on three areas in which significant progress had been achieved during the 1980s: the relation of migration to weather and climate, physiological adaptations for migration, and forecasting of migrant pests. After the symposium, the participants were commissioned to develop the material they had presented orally into the broader treatments that appear in this book. A slight geographical bias towards East Asia and Australia has been deliberately retained, to counterbalance a tendency in previous reviews and books on this topic to focus on Africa and North America. (It is worth noting that South America is unrepresented in this or earlier reviews: little seems to be known of insect migration in that continent.)

The book concludes with an overview chapter, in which we attempt a synthesis of current knowledge of insect migration within the framework of a single, comprehensive conceptual model.

We would like to thank the XIX ICE Programme Committee (Chairman Professor Y.-Q. Liu) and Professor X.-X. Zhang for their support of the Beijing Symposium, and all our chapter authors for responding positively and creatively to our proposals on the scope of their contributions and to our sometimes vigorous and critical editing. We are most grateful to H. Dingle, R.A. Farrow, P.C. Gregg, G. McDonald, D.E. Pedgley, D.R. Reynolds and K. Wilson for reading and commenting so constructively and promptly on various chapter drafts.

Alistair Drake, Canberra, Australia

Gavin Gatehouse, Bangor, North Wales, UK

October 1994

### References

Kareiva, P. (1994). Space: the final frontier for ecological theory. *Ecology*, **75**, 1.

Kennedy, J.S. (1985). Migration, behavioural and ecological. In *Migration: Mechanisms and Adaptive Significance*, ed. M.A. Rankin, pp. 5–26. *Contributions in Marine Science*, vol. 27 (Suppl.). Port Aransas, Texas: Marine Science Institute, The University of Texas at Austin.

# Part one

## Insect migration in relation to weather and climate

# 1

# Long-range insect migration in relation to climate and weather: Africa and Europe

## D. E. PEDGLEY, D. R. REYNOLDS AND G. M. TATCHELL

### Introduction

Although there are many records in the literature of individual insect migrations within Africa and Europe (see e.g. Williams, 1958; Johnson, 1969; Cayrol, 1972; Rainey, 1976; Pedgley, 1982; Dingle, 1985), there have been few attempts to review seasonal redistribution on a continental scale (but see Farrow (1990) for the migration of Acridoidea). Here we summarise the (now very extensive) available information on long-range, windborne movements by winged insects in the various climatic zones of Africa and Europe, and show how these migrations are related to the weather and to seasonal changes of climate and prevailing wind.

Some insect species migrate only short distances (a few hundred metres to a few kilometres) between habitat patches (see e.g. Solbreck, 1985), but we are concerned here with migration on scales covering countries and continents. These distances are so great, and the flying speeds of most insects are so slow (usually less than 3 m s$^{-1}$), that most migrations are dependent on the help of an external energy source: the wind. Persistent flight in a wind of, for example, 10 m s$^{-1}$ can easily result in migrations of 300–400 km in one day or night. In many (but by no means all) species, these migrations are made during the few days before the onset of breeding (Johnson, 1969; see also Chapter 10, this volume), and the movements are often dominated by a single weather system (Pedgley, 1982; Drake & Farrow, 1988).

Some large day-flying insects, particularly butterflies, appear to have directional migrations, exerting some control over their direction of displacement by flying largely within their 'flight boundary layer' (Taylor, 1974), or at least keeping within it when winds aloft prevent progress towards a preferred heading. In northern temperate areas some species orientate approximately northwards in spring and summer and southwards in autumn (Williams, 1930, 1958; Baker, 1968, 1978; Walker & Riordan, 1981; Walker, 1985, 1991). We have generally excluded these species from this review, as at least some of their movements may be relatively independent of the wind.

## Migration, wind systems and climate

The annual pattern of insect migration within and between climatic zones depends on the changing favourability of breeding habitats (due mainly to seasonal patterns of rainfall and, at higher latitudes, temperature) and to the changing availability of transporting winds. The main zones of natural vegetation (which are indicative of the climatic zones) are shown in Fig. 1.1 for the African–European region. These are:

> *Rainforest.* Rainy for much of the year. Centred on the equator, but does not extend into eastern Africa.
>
> *Savanna.* A 'wet/dry' climate with highly seasonal 'monsoon' rains. A transition zone between rainforest and steppe.
>
> *Desert and steppe* (including dry tropical scrub). Scarce and erratic rains associated with weather systems from tropical and temperate latitudes. Extends south across the Horn of Africa, because (for reasons that are still not well understood) the monsoon there brings little rain, and north into Central Asia, where distance from the oceans reduces rainfall.
>
> *Mediterranean vegetation.* Rain in winter only. A transition zone between desert and temperate latitudes.
>
> *Broadleaf and mixed forest.* Rainy all year. Climate milder in the west than the east.
>
> *Northern coniferous forest* ('*taiga*'). Rainfall modest. Winters very cold.

The seasonal winds associated with these climatic zones (and to a considerable extent giving rise to the rainfall patterns) are illustrated in Fig. 1.2. In the tropics, easterly trade winds dominate: northeasterlies in the northern hemisphere and southeasterlies in the southern, meeting at the Intertropical Convergence Zone (ITCZ). The ITCZ lies near the equator but moves into the summer hemisphere over land, even as far as latitude 25°. When that happens, the trade wind crossing the equator turns into an equatorial westerly (or 'monsoon'): the southeasterly trades, for example, become the southwesterly monsoon (Fig. 1.2). Rains equatorward of the ITCZ produce the wet seasons of the savanna

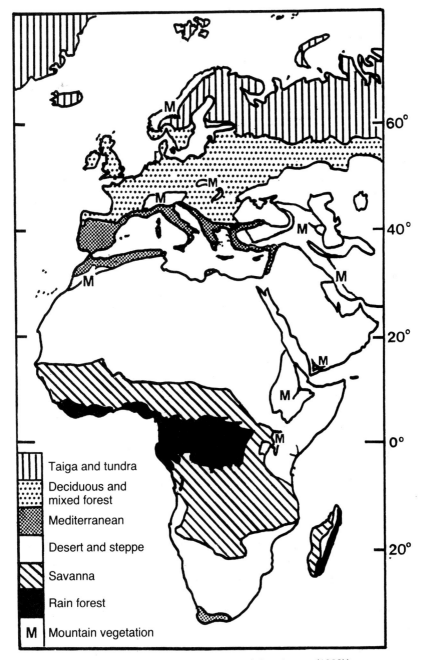

Taiga and tundra

Deciduous and
mixed forest

Mediterranean

Desert and steppe

Savanna

Rain forest

M | Mountain vegetation

Fig. 1.1. Natural-vegetation zones (after Anon. (1980)).

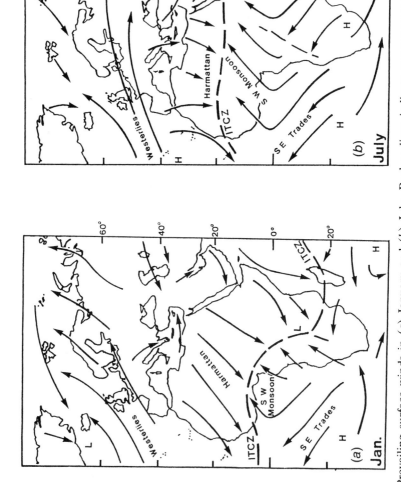

Fig. 1.2. Prevailing surface winds in (a) January and (b) July. Broken lines indicate convergence zones. ITCZ, Intertropical Convergence Zone; H, high pressure; L, low pressure.

zone. Windborne insects can be carried long distances on the trade and monsoon winds, and may become concentrated at the ITCZ. Superimposed on these seasonal changes are ephemeral deviations, mostly lasting up to a few days, caused by transient and often mobile weather systems. Such systems are often weak, but they can take migrating insects in directions different from those to be expected from the dominant winds.

Poleward of the tropics are the subtropics, which are dominated by large, slow-moving anticyclones centred over the oceans at about latitude 30° (with little seasonal change). The persistently descending air associated with these anticyclones allows little cloud to form and produces a desert climate. The low- and high-latitude fringes of this zone are affected by weather systems from tropical and temperate latitudes, respectively.

In the temperate zone, westerly winds are dominant in both hemispheres (Fig. 1.2), but they are much more variable than in the tropics, because of the incessant formation, movement and decay of a succession of cyclones and anticyclones, each lasting from a few days to a week or two. In winter, however, anticyclones can be very persistent over the cold interior of Asia. Because of the variability of winds in the temperate zone, windborne insects trace out rather complex trajectories there (see Johnson, 1969, p. 477).

Distortions of this simple global pattern are induced by the configuration of continents and oceans, and by highland areas. Highlands tend to be cool and wet; hence they may allow breeding in otherwise arid zones where it is impossible in adjacent lowlands. They also distort the wind patterns, partly through mechanical deflection. Winds flowing over highlands are cooled by the expansion resulting from lifting. Both the deflection and cooling can affect windborne migration, sometimes acting as barriers.

In the following sections we give examples of windborne migration from each climatic zone, beginning with the equatorial forest and moving towards the poles.

### Equatorial rainforest

There are very few records of long-distance migration within the rainforest zone. This may partly reflect a lack of observers and the difficulties of observation, but it is generally believed that rainforest insects are less mobile than, for example, tropical savanna species, because seasonal changes are less marked in the forest habitats (Southwood, 1962; Johnson

& Bowden, 1973). A comparison of the range of movement of species of the *Simulium damnosum* complex (Diptera: Simuliidae) in West Africa tends to support this: the forest species moved relatively short distances (e.g. 5 km in *Simulium yahense*, whose larvae inhabit permanently flowing small rivers) compared to the savanna species (e.g. 500 km in *Simulium sirbanum*, the larvae of which live in waterways that do not flow for part of the year) (Garms & Walsh, 1987).

It appears that there is very little information about insect movement above the canopy of the tropical rainforests of Africa. This topic would seem to merit further investigation.

### Savanna and steppe

#### West Africa

The steppe ('sahel') and savanna zones stretch across Africa from the Atlantic to the mountains of Ethiopia. Many insect species inhabiting this region are adapted to make seasonal long-distance migrations on the prevailing wind systems. As the ITCZ advances northwards during the northern hemisphere spring (Fig. 1.2), rain falls at progressively higher latitudes and long-distance windborne migrants are able to take advantage of renewed growth in the vegetation (Fig. 1.3a,b). Correspondingly, as the ITCZ retreats southwards and northeasterly winds (the 'harmattan') are re-established, breeding sites dry out and populations move southwestwards (Fig. 1.3c,d).

This seasonal reversal of windborne migration has been demonstrated for several species of grasshoppers, bugs, moths and flies, and it is likely to occur in other taxa. Of the many species of acridoid grasshoppers that make seasonal windborne migrations (see e.g. Lecoq, 1978; Duranton *et al.*, 1979; Fishpool & Popov, 1984; Reynolds & Riley, 1988; Farrow, 1990; Riley & Reynolds, 1990), the one that has been studied most extensively is the Senegalese Grasshopper *Oedaleus senegalensis*. Observations of changes in its distribution have given rise to a detailed descriptive model (Launois, 1979; Steedman, 1990; see also Chapter 19, this volume). The southward movements of *O. senegalensis* in September–November are better documented than the northward ones, because the populations undertaking them are usually larger, and they have in addition been observed directly by radar (Riley & Reynolds, 1979, 1983). Not all grasshopper movements are towards the south or southwest in autumn, however, as a long-distance migration of

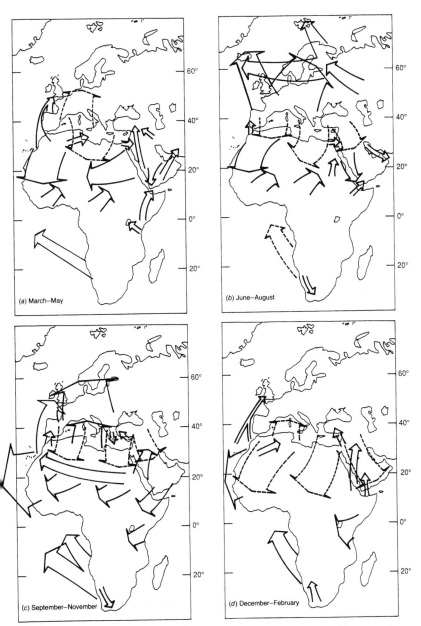

Fig. 1.3. Known and conjectured windborne insect migrations. Known migrations (continuous lines) are based on studies (see text); conjectured migrations (broken lines) are based on wind systems across sources of flying adults. Only principal migration directions are shown; other directions will also occur, especially in regions where winds are variable.

*Diabolocatantops axillaris*, *O.senegalensis* and other species towards the northeast and east has been observed in October, when the insects were to the south of the ITCZ (Reynolds & Riley, 1988).

Further evidence that many species of sahelian acridoids are capable of long-distance migration comes from captures on ships ~100–200 km off the coast of Western Sahara, Mauritania and Senegal, particularly between September and December during periods of northeasterly or easterly winds (Fig. 1.3c,d). The most commonly caught species include *O. senegalensis* (Ritchie, 1978), *D. axillaris* (Reynolds & Riley, 1988), *Anacridium* spp., the Migratory Locust *Locusta migratoria migratoriodes*, and *Ornithacris* spp. (Natural Resources Institute (NRI) archives, unpublished data). Other Orthoptera, such as gryllids (Ragge, 1972) and tettigoniids (particularly *Ruspolia differens*; NRI archives), and Lepidoptera are also caught at sea at this time (see summary by Bowden & Johnson, 1976). All these records of flight endurance pale, however, when compared to the crossing of the Atlantic from West Africa to the West Indies by swarms of the Desert Locust *Schistocerca gregaria gregaria* in October 1988 – a migration that seemingly required 4–6 days of continuous flight (Ritchie & Pedgley, 1989).

Solitaries of *S. g. gregaria* do not move southwestwards (following the retreat of the ITCZ) along with other acridoid species, for reasons that are unclear. In some years, however, *swarms* of this species do so move, only to return the next year (the so-called 'southern circuit'), but without breeding until they reach suitable habitats at the northern fringe of the monsoon rains (Pedgley, 1981). Swarms of *L. m. migratorioides* appear to make similar seasonal movements during plague years (when this locust can be widespread over West Africa). More usually, however, solitaries of this species make more complex movements that are associated with dry-season breeding in the inland delta of the River Niger in Mali as the floods recede. The locusts migrate northeastwards along the delta against the prevailing winds at that time of year (November–January) (Farrow, 1990).

Long-distance movements have been deduced for blood-feeding flies in West Africa. Waves of blackflies (*Simulium* spp.), particularly the savanna species *S. damnosum sensu strictu* and *S. sirbanum*, migrate on the monsoon winds. Some individuals travel at least 500 km in 4–5 weeks (Garms & Walsh, 1987; Baker *et al.*, 1990). Although most migrations studied have been south of the ITCZ, there is some evidence of southward flights on the north side (Magor & Rosenberg, 1980; Garms, Cheke & Sachs, 1991). It is likely that some other blood-feeding species make similar long-distance windborne movements, judging by the apparent spread of insect vectors of Rift Valley Fever virus from Sudan to southern Egypt during a transient surge of monsoon winds that extended unusually far to the north (Sellers, Pedgley & Tucker, 1982).

The latitudinal extent of successful movements is affected by availability of host plants and by climate. For example, the northern limit of spread of the pyrrhocorid bug *Dysdercus voelkeri* is determined by the distribution of its malvaceous host plants, while its southern limit at this time (August) is determined by the northern boundary of the heavy rains (which increase mortality due to fungal diseases and the drowning of nymphs) (Duviard, 1977). It is easy to envisage how successive generations of this species progressively colonise the more northerly summer distribution zone, but it is less clear how later generations move south, because the bugs cannot survive for long in the hot and dry harmattan winds.

## East Africa

In East Africa, the ITCZ separates northeasterly and southeasterly trade-wind airflows that are more alike in temperature and humidity than are the harmattan and monsoon in West Africa. Moreover, the ITCZ moves to well south of the equator during the southern summer (Fig. 1.2*a*). There are two rainy seasons (except at the northern and southern extremes of its movement): the Short Rains when the ITCZ moves southwards (in the northern autumn), and the Long Rains when it moves northwards (in the northern spring). In contrast to the mountains of West Africa, those of East Africa are so extensive that they have a strong influence on the pattern of winds and on the distribution of rainfall, and are therefore likely to affect the movement and distribution of migrant species.

Two migrant species have been much studied in East Africa: *S. g. gregaria*, and the African Armyworm *Spodoptera exempta*, a noctuid moth widespread in the grasslands of sub-Saharan Africa. Swarms of *S. g. gregaria* derived from Short Rains breeding in the Horn of Africa, and from somewhat earlier monsoon breeding in Ethiopia and Yemen, move southwestwards on the northeasterly trades into Kenya and Tanzania by the end of the year. Correspondingly, swarms from Long Rains breeding in East Africa move north in the middle of the year to the Horn of Africa, Ethiopia and Yemen on the then-prevailing southerly winds.

Rains associated with the ITCZ allow low-density populations of *S. exempta* to build up, from about October or November, with outbreaks anywhere from Kenya to Zimbabwe (Haggis, 1984). The flying moths are known to be concentrated by mesoscale wind convergence (Pedgley *et al.*, 1982; Tucker & Pedgley, 1983) and this, followed by synchronous egg laying, leads to outbreaks of 'gregarious' larvae. The adults developing from these infestations are obligate migrants, flying at heights up to several hundred metres above ground and being displaced downwind (Riley, Reynolds & Farmery, 1983; Rose *et al.*, 1985). Under the influence of dominantly easterly and northeasterly winds, moths spread westwards and southwestwards across eastern and central Africa, with new generations at about monthly intervals (Tucker, 1994). Brief spells of northerly winds associated with temperate-latitude weather systems of the southern hemisphere presumably take moths southwards into subtropical latitudes, but there do not seem to be any studies of such movements. From about April, as winds in equatorial latitudes turn more southeasterly, and later southerly, moths are taken northwards into Ethiopia and Yemen (Brown, Betts & Rainey, 1969; Haggis, 1984; Tucker, 1994) – a movement similar to that described above for *S. g. gregaria* swarms following Long Rains breeding. *S. exempta* movements out of breeding areas in Ethiopia and Yemen occur in September and October (i.e. earlier than those of *S. g. gregaria*), when winds are still southeasterly or easterly. Consequently, reinvasion of East Africa seems to be rare. This has been confirmed by backtracks from the first outbreaks in East Africa which point to sources in eastern Kenya or Tanzania, not Ethiopia (Tucker, 1984, 1994; Pedgley *et al.*, 1989).

### Subtropical desert

Studies of long-distance windborne migrations in the great subtropical deserts of Africa have been almost wholly confined to the Desert Locust: *S. g. gregaria* in the Sahara (and the deserts of Arabia and the Middle East), and *S. g. flaviventris* in Namibia, Botswana and western South Africa. Following the classic work of Z. Waloff and R. C. Rainey, the changes in distribution of the day-flying swarms of the northern subspecies during plague periods represent the best studied migration pattern of any insect (Johnson, 1969; Pedgley, 1981). To survive in its arid or semi-arid habitat, *S. g. gregaria* needs to find areas where rain has recently fallen, because the eggs must absorb water from the soil in order to develop. Rain is also necessary for the growth of the plants on which the insects feed. Breeding is mostly on the erratic rains of the desert fringes with long across-desert migrations between breeding sites. The locusts have consequently evolved behaviour that tends to move them towards rainfall, i.e. the strategy of downwind displacement (Rainey, 1963). The movement of a coherent swarm differs from that of an isolated insect: a swarm moves as a whole, owing to its complex internal structure, and it usually moves more slowly than the wind, because part is often settled at any one time. Nevertheless, seasonal swarm movements can often be used to indicate the directions, if not the speeds, of windborne migrations of other, less studied insect species. An extensive account of the effects of weather on migration of *S. g. gregaria*, along with many case studies, is presented by Pedgley (1981). The following is a brief outline.

Swarms produced on the monsoon rains, in a narrow latitudinal zone from Mauritania in the west to southern Arabia in the east, move downwind from about October (Fig. 1.3*c*). Those in the east go southwestwards into East Africa (but some go north towards the Middle East); those in the west go westwards and northwestwards into northwestern Africa. The northward components, across or even against the dominant winds at that time of year, are caused by the longer flights on days with warm southerlies, particularly ahead of cold fronts (see also Gerbier, 1965; Shaw, 1965). These fronts move from west to east and are most noticeable during the winter half of the year. Movements then at higher latitudes are restricted by low temperatures and by breeding, and directions are variable, because of the influences of the southern fringes of temperate-latitude mobile cyclones and anticyclones (Fig. 1.3*d*). Breeding on winter and spring rains in northwestern Africa and the Middle East is followed by downwind movement of the new-generation swarms southward into the monsoon breeding areas (Fig. 1.3*a,b*), although tracks may be tortuous if influenced by several wind systems. Breeding at this time of year also takes place in countries around the southern Red Sea, from where new-generation swarms move usually southwards and westwards, but also northwards ahead of cold fronts.

It is possible that other species migrate across the deserts in similar patterns to *S. g. gregaria* swarms, but their success will depend on flight duration and ability to survive both the high temperatures and the lack of food sources. Such movements are poorly documented, but some evidence is provided by the reappearance of several moth species at oases in Algeria, upper Egypt, and the Middle East (Cayrol, 1972; Zaazou *et al.*, 1973; Wiltshire, 1946), implying migration, presumably from sources to the north. Winter reappearances in northern Sudan (Schmutterer, 1969) by the well-known migrants *Agrotis ipsilon* (Noctuidae) and *Plutella xylostella* (Yponomeutidae) imply even longer southward movements.

Comparable northward trans-Sahara movements are likely to be confined to the winter and spring, when transient spells of warm southerly winds are most frequent. An example is perhaps provided by the specimen of the noctuid moth *Helicoverpa armigera* seen in southern Algeria (Gatter & Gatter, 1976), apparently on southwesterly winds ahead of a cold front. Such movements may be on a broad scale but involving populations at low densities. Similar movements have been inferred for blood-feeding midges (*Culicoides* spp.) carrying African Horse Sickness virus across the Middle East: the spread of disease from southwestern Iran across Iraq and Syria to Turkey and Cyprus during spring coincided with successive spells of southeasterly winds ahead of cold fronts (Sellers, Pedgley & Tucker, 1977). In this case, the wind direction was modified to southeasterly by the mountains of Iran. Although this direction is against the dominant northwesterly winds of the area, it can be attributed to the longer flights on days (or nights) with warmer southeasterly winds.

The role of Saharan highlands in providing possible refuges for migrant insects is unknown, but the rainy and cool Ethiopian highlands (in contrast to the hot and dry plains at comparable latitudes in Sudan and along the Red Sea coast) may provide a link for migrants between the Mediterranean or the Middle East and the highlands of eastern and southern Africa. This may have been the route by which the Russian Wheat Aphid *Diuraphis noxia* extended its range from the Middle East to southern Africa. Such a link may also account for the presence of some moth species both north and south of the Sahara, e.g. the noctuids *A. ipsilon*, *Agrotis segetum*, *Mythimna loreyi* and *Trichoplusia ni*. Geographical features that provide a conduit for the north/south migration of some species may impose a barrier against the east/west movements of others. For example, restriction of movement from the arid Horn of Africa to similar belts at the same latitude to the west, by the Ethiopian highlands, has apparently given rise to endemic species in the Somali peninsula, even in relatively mobile insects such as *Oedaleus* grasshoppers (Ritchie, 1981).

One of the few examples of windborne migration in the subtropical zone of southern Africa is the movement of swarms of *S. g. flaviventris* (Waloff & Pedgley, 1986). This locust can breed only during the single (monsoon) rainy season (between October and April) and there seem to be two generations during this period. Adults of the summer generation appear in April–May. Swarms of these adults leaving the outbreak areas in southern Namibia and adjoining areas of Botswana and South Africa tend to move southeast or south during the southern autumn-to-spring period, i.e. between May and November; during this time the locusts remain immature. These movements appear to be downwind on northerlies or northwesterlies (Fig. 1.3*b,c*), particularly in spells of warm prefrontal airflows ahead of eastward-moving depressions, although for much of the period temperatures are probably too low for flight. Adults of the spring generation appear in January and February, following breeding over a large area of Cape Province. The resulting swarms and scattered populations move north on southerly winds (Fig. 1.3*d*) caused by the development of a semi-permanent cyclonic centre over northern Namibia and southern Angola.

Evidence of the very long flight endurance of some southwestern African species comes from records of insects caught in the South Atlantic Ocean (summarised by Bowden & Johnson (1976)). Ships' records (in the NRI archives) of Orthoptera in the South Atlantic mainly concern the tettigoniid *R. differens* caught between ~4° and ~12° S from November to January while the winds were from the southeast (Fig. 1.3*c,d*). It is likely that these insects originated in grasslands in southwestern Africa. Similar explanations have been given for an invasion of Ascension Island by a leafhopper, *Balclutha* sp. (Cicadellidae) (Ghauri, 1983), and for the presence of a 'swarm' of the butterfly *Cynthia cardui* seen 1600 km west of St Helena (Bowden & Johnson, 1976). In the latter case the butterflies were 3000 km away from the most likely source in southwestern Africa.

### Mediterranean

There have been few studies of migration within the Mediterranean basin compared with countries in central and northern Europe. In Cyprus, it has been shown that occasional *H. armigera* recorded in winter are associated with spells of southerly and southeasterly winds that brought the same species simultaneously to neighbouring countries (Israel, Lebanon, Turkey) (Pedgley, 1986). Also in Cyprus, it has been shown that outbreaks of two animal diseases, African Horse Sickness and Bluetongue, were probably caused by windborne midges (*Culicoides* spp.) bringing viruses from known sources in eastern Turkey and Syria (Sellers *et al.*, 1977, 1979). In the first disease outbreak there was a spell of a few days with strong northerly winds interrupting the usual summer westerlies. In the second, the interruption was caused by easterly and northeasterly winds on the northern side of a weak cyclonic circulation that drifted slowly

slowly eastwards to the south of Cyprus over a three-day period. Movement on that occasion would have been at altitudes above about 0.5 km because winds near sea level were light and variable. In both instances the movements would have taken less than a day. Two months after the second occasion, Bluetongue appeared in western Turkey, where the outbreak can be linked to a spell of strong southeasterly winds ahead of a vigorous cyclone moving northeastwards from Libya (Sellers & Pedgley, 1985): an example of the greater variability of wind directions as winter approaches (Fig. 1.3c). Movement almost certainly took place in a single night and may have been ended by rain accompanying the cyclone.

At the western end of the zone, similar movements of virus-carrying insects have been inferred from Morocco to Spain (Sellers *et al.*, 1977) and to Portugal (Sellers, Pedgley & Tucker, 1978). The first can be narrowed down to a single day when strong southerly to southwesterly winds were blowing ahead of a cold front approaching from the west (Fig. 1.3c). The second can be narrowed down even further: to a few hours on one night during a brief and unusual (for summer) spell of southerly winds on the eastern side of a cyclonic circulation that formed off the coast of northern Morocco and then intensified as it moved northwards off the west coast of Portugal (Fig. 1.3b).

The effects on migration of the variability of wind directions in this zone, demonstrated by the preceding examples, is further illustrated by catches of various species of moths at a desert light trap in Israel (Pedgley & Yathom, 1993). Catches peak when the wind blows from the Nile Delta and the Levant; these source regions apparently generate vast populations of moths, with different species predominating in different seasons. One species trapped, the Egyptian Cotton Leafworm *Spodoptera littoralis* (Noctuidae), is usually considered to be rather sedentary, but there is also evidence from Egypt that flights exceeding 50 km downwind of the Delta have occurred (Nasr, Tucker & Campion, 1984). An example of a longer movement is the sudden occurrence of *A. ipsilon* at lights in Israel during a brief spell of warm southwesterly winds, part of a cyclone moving northeastwards from the Delta (Odiyo, 1975). Because winds in this region are frequently northerly, moths must often stream into the deserts, with consequent heavy losses.

Long-distance movements northward from the Mediterranean or southern Europe to central or northern Europe often imply the crossing of high mountains. Indeed, reports from Alpine passes in Switzerland and France show that many of the moth species mentioned can migrate over snow and in temperatures below freezing (Aubert, Goeldlin &

Lyon, 1969; Aubert, Aubert & Pury, 1973; Poitout *et al.*, 1974; Rezbanyai, 1980; Hachler, 1989). Flight in mountain fog greatly increases the catch of migrants (Aubert *et al.*, 1973), presumably because they fly close to the ground. The presence of dead specimens on Alpine and Pyrenean snowfields (Labonne, 1978; Liston & Leslie, 1982) emphasises the occurrence of long-distance migrations (Edwards, 1987).

Southward moth movements into the Mediterranean region, perhaps most clear-cut in the autumn when numbers have built up over Europe, are poorly documented. A sudden influx of *H. armigera* in Crete occurred on 12 September 1977 following the passage of a cold front and the onset of northerly winds (Malicky, 1981). Similar increases of several species, notably *Utetheisa pulchella* (Arctiidae), were likewise associated with the passage of cold fronts over the southern Caspian during August and September 1961 (Sutton, 1966).

### Northern temperate zone

Throughout the year, this zone is characterised by a great variability of wind direction, and therefore windborne insects can be expected to move in any direction, as long as flight is not inhibited by low temperatures. Such movements are easiest to detect when insects are entering areas where breeding has not been possible in previous months, particularly in winter. Several studies in Britain have related the arrival of moths on warm southerly winds – such species as *Spodoptera exigua* (Hurst, 1969a; Johnson, 1969) and *H. armigera* (Pedgley, 1985). Such winds occur in spells of a few days on the western sides of large anticyclones centred over eastern Europe, or to the east of large cyclones approaching western Europe from the Atlantic.

With *S. exigua*, backtracking suggested that sources early in the year were in the Mediterranean zone of Morocco (Fig. 1.3*a*), but from May onwards they were mostly in northwestern Spain (Fig. 1.3*b*). Many other species, some of them rare in Britain, arrive with similar weather patterns in all but the coldest months of the year, and it is likely that they travel similar distances. Indeed, one such long-distance movement was confirmed when a specimen of *Nomophila noctuella* (Pyralidae) caught in March was found to be carrying a radioactive particle that was almost certainly picked up close to the site of an atomic weapons test in Algeria nearly a month earlier (Kettlewell & Heard, 1961). Shorter movements are illustrated by the arrivals in southeastern England of the aphid *Myzus persicae* in large numbers in July 1947 (Hurst, 1969b) and July 1982

(Heathcote, 1984): in both cases, backtracking suggested sources in the Netherlands or Germany, 200–300 km away. In 1982, suction trap records (Taylor, 1986; Tatchell, 1991) indicated that the migration continued for three days with flights penetrating across the whole of England, but it is not clear whether this was made up of repeated flights by the same individuals on successive days or whether there were further immigrants from mainland Europe (Taylor, 1985).

Arrivals in Britain are often a mix of species. For example, during one ten-day spell of unseasonably warm southerly and southwesterly winds in October 1988, *H. armigera* was recorded on the first day, followed by *S. exigua* and *Palpita unionalis* (Pyralidae) on the next (Bretherton & Chalmers-Hunt, 1989). The last species, an olive leafworm, suggests a Mediterranean origin for at least some of the arrivals. Other subsequent arrivals included not only the well-known migrant *Mythimna unipuncta* (Noctuidae) but also *Ostrinia nubilalis* (Noctuidae), for which there is little evidence of migratory ability. At the end of the invasion, as winds veered to the south-southwest, an influx of *S. g. gregaria* arrived ahead of a cold front, having probably taken at least two days to cross the sea from Morocco, where many swarms had arrived recently. Generally, in the poleward-blowing winds ahead of a temperate-latitude cold front it is those winds immediately ahead of the front, along with the insects carried by them, that are most likely to have come from the furthest south, as is illustrated by this example.

Not all arrivals in northwestern Europe are from the south. A massive invasion of *P. xylostella* on easterly winds reached northeastern Britain on 28 June 1958, crossing Scotland to the Faroe Islands and Iceland, as well as to the mid-Atlantic at 59° N 20° W. Moths had been seen at the end of June in Finland, Estonia, Sweden, Denmark and Norway, and backtracks suggest a source in the west of the former USSR (Shaw, 1962). Other long-distance windborne migrations have been recorded into Finland: *P. xylostella* in June 1978 (Markkula, 1979), when moths reached both the Kola Peninsula of Russia (68° N; Kutsenin, 1980) and even Spitzbergen (78° N; Lokki, Malmstrom & Suomalainen, 1978); and *S. exigua* in August 1964 (Mikkola & Salmensuu, 1965), when moths reached Denmark a week later (Mikkola, 1970). These movements came from sources in western and southern Russia on southeasterly winds blowing around large anticyclones centred over northwestern Russia, and ahead of cold fronts approaching from the west. Other taxa that have been found to invade Scandinavia, after crossing the Baltic Sea on warm southerly winds ahead of cold fronts, are the Colorado Beetle,

*Leptinotarsa decemlineata* (Chrysomelidae) (Wiktelius, 1981) and the aphid *M. persicae*, carrying Sugar Beet Yellowing virus (Wiktelius, 1977, 1984).

Within the former USSR, moths of the pyralid leafworm *Margaritia* (*Loxostege*) *sticticalis* sometimes migrate in huge numbers following springtime mass emergence from overwintered larvae. As with *S. exempta*, take-off starts soon after dusk and flight continues at heights of 50 m or more, sometimes through the night (Mel'nichenko, 1936). Weather maps were used for the first time with this species to help understand long-distance windborne migration, covering several hundred kilometres in one or two nights. Individuals sometimes reach northern Europe; e.g. Finland in 1977 (Mikkola, 1986), and England in 1970 (Bretherton, 1982) and 1989 (Bretherton & Chalmers-Hunt, 1990).

## Discussion

This review has shown that windborne insect movements over hundreds of kilometres occur throughout Africa and Europe, except perhaps in the equatorial forest zone. Such migrations allow insects to move between habitats separated by distances too great to be traversed by their own intrinsic locomotor powers, and to exploit newly available resources such as fresh vegetation arising from seasonal rain or warmth. Movements have been studied on both daily and seasonal time scales: the former in relation to individual transient wind systems; the latter in relation to seasonal wind changes.

Because of the relative day-to-day constancy in the monsoon and trade winds of the tropics, windborne migrations in the savanna and steppe zones of Africa tend to occur in consistent directions, and the seasonal reversal of the winds can be exploited for movements between complementary breeding (or breeding and diapausing) areas. Moreover, as the poleward spread of the monsoon winds is associated with the advance of the rains, windborne migrations are particularly advantageous, with insects moving into areas where vast seasonal resources are becoming available (e.g. new growth of sahelian grasses for *O. senegalensis* or newly flowing waterways for savanna species of *Simulium*). In autumn, equatorward movements predominate. Some populations may move far enough to reach areas where late rains will allow breeding, while others move less far and persist in dry-season 'refuges'. Thus, windborne movements in the sahel and savanna belts are particularly adaptive, as favourable habitats occur throughout the year in a predominantly downwind direction. These migration 'circuits', and the life-history strategies that have evolved with them, are clearly effective despite the occasional transport of large populations of migrants into unsuitable areas (e.g. the Atlantic Ocean or the Sahara).

Similar remarks apply to *S. g. gregaria*, which survives in its specialised subtropical desert habitat by downwind movement around several enormous inter-connected seasonal circuits (Uvarov, 1977). In this species, the movement of whole populations (identified by biogeographical analysis) is consistent with the changing windfields, and downwind displacement generally represents an optimal strategy, because the insects travel persistently until they reach areas where rain has fallen. Movements are not always on dominant winds, however, because cold northerlies sometimes immobilise locust swarms, and flight then occurs during spells of warmer, southerly airflows.

Movements poleward from the tropics or subtropics in spring and summer allow the seasonal colonisation of temperate areas by highly mobile species. Benefits include new vegetative growth developing in response to rising temperatures, and, for the first immigrants at least, low numbers of natural enemies and competitors. Also, survival is possible in species which otherwise would be killed by the aridity and extreme heat of summer in the subtropics. Examples of migration into the temperate zone have been mentioned (e.g. for *A. ipsilon*, *H. armigera* and *S. exigua*) but the process is less well documented than in North America (Chapter 2, this volume) or East Asia (Chapters 3 and 4, this volume). This may be partly because there have been few studies of such movements in northern Africa and southern Europe, but long-distance migrations of this type may be less adaptive and thus less common than in other continents due to the combined hazards of the Sahara and the Mediterranean Sea. In addition, we note that the Sahara represents a relatively impoverished source when compared to the subtropical source areas of the species that spread annually into northern temperate areas of China and North America. This illustrates a general point: although wind systems in corresponding latitudinal belts around the world may give rise to analogous migrations, these are always heavily influenced and constrained by the geography of the region concerned.

Movements within temperate regions are likely to be in a variety of directions because of the variability of winds. For many species this variability is unimportant, because the sought-after resource is not confined to a particular geographical direction. For example, the adults of the Large Pine Weevil *Hylobius abietis* move moderate distances (several tens of kilometres) in order to locate recently killed conifers on which they oviposit (Solbreck, 1985). Such trees may occur in any direction within the insects' migration range. Another example is provided by host-alternating aphids, where the autumn migrants (gynoparae and males) migrate from the secondary hosts to recolonise the relatively dispersed primary host (often a woody shrub). This is well illustrated by autumn flights of the Hop Aphid *Phorodon humuli* from two foci of hops in Britain (Taylor, Woiwod & Taylor, 1979). Although the median flight distance was only 15–20 km, a few individuals reached the far corners of the island (400–600 km), but there was no suggestion that this species (at least in its autumn forms) shows any preference to fly on winds from a particular direction. The very high reproductive rate of some aphids enables them to bear the costs of a strategy involving the production of vast numbers of airborne individuals dispersing in all directions, presumably with a high mortality rate (Cammell, Tatchell & Woiwod, 1989).

In all seasons, there is evidence that long-distance migrants can be taken into areas at higher or lower latitudes that are unfavourable for survival or reproduction. For example, brief spells of warm poleward-blowing winds in the Mediterranean and northern temperate zones take a variety of Lepidoptera beyond areas suitable for breeding. Such 'overshooting' is recognisable in the first spring arrivals in northern Europe, but it is more common there in autumn, when subsequent changes in wind direction are often accompanied by temperature falls which inhibit flight, thus stranding the migrants. The frequency and significance of such 'Pied Piper' movements has been much debated (see e.g. Walker, 1980; Dingle, 1982; McNeil, 1987), but the proportion of the population that becomes stranded in this way remains unknown. It has been suggested (e.g. by Johnson, 1987) that when population losses occur they may be an unavoidable consequence of the need for movement between ephemeral habitats in breeding areas at lower latitudes.

Movements in particular compass directions (which cannot be easily accomplished by simple downwind migration) may have apparent advantages for some species. As mentioned above, some large day-flying insects, particularly butterflies, seem to exert a degree of control over the direction of movement by flying within their boundary layer. However, many large migrants are nocturnal and these generally fly high in the air (Drake & Farrow, 1988). Although large night-flying insects often show a degree of common orientation off the wind direction (Riley & Reynolds, 1986), by far the most effective way that windborne migrants could achieve displacements in 'preferred' compass directions would be to fly selectively on appropriate winds. However, there is as yet little evidence for this (Schaefer, 1976; Taylor & Reling, 1986), and the indications from our review are that, apart from the effects of temperature on flight thresholds, long-distance windborne movements are generally made irrespective of wind direction. The evidence from case studies shows: (*a*) that downwind movement can occur in wind directions markedly different from the seasonal normal; (*b*) that a succession of arrivals by the same species show a variety of accompanying wind directions, sometimes only days apart; and (*c*) that numerous arrivals are associated with winds that have brought insects into areas where breeding is highly unlikely (e.g. poleward of sources in autumn).

Evidence from radar studies, which might be expected to shed light on this matter, has generally been ambiguous. Marked night-to-night variation does occur between the number of insects flying on winds from different directions, and this could indicate a tendency to take off, or to prolong flight, in winds with certain characteristics (e.g. of temperature or humidity) (J.R. Riley and D.R. Reynolds, unpublished data; see also Schaefer, 1976; Farrow, 1990). However, the effect could be due merely to variations in the sources of flight-ready individuals in different directions from the radar, and the data are inadequate to determine which of these two explanations is correct.

The simplest hypothesis accounting for long-distance windborne migration is that flight behaviour is not affected by wind direction as such, although variation of air temperature with direction may influence take-off and flight duration. This interpretative approach reduces emphasis on seasonality of 'return' migrations by 'choice' (e.g. equatorward in autumn), and increases emphasis on the continual windborne mixing of populations throughout the year. We therefore propose this hypothesis with a view to encouraging others to find evidence to the contrary, and so narrow the range of taxa to which it might apply. The degree of control of displacement direction in long-range migrants has been a long-standing source of controversy (Johnson, 1969; Baker, 1978), and it remains one of the most interesting unsolved problems in insect migration today.

## References

Anon. (1980). *The Times Atlas of the World – Comprehensive Edition.* London: Times Books/Bartholomew.

Aubert, J., Aubert, J.-J. & Pury, P. (1973). Les sphingides, bombyces et noctuides du col de Bretolet (Val d'Illiez, Alpes valaisannes). *Bulletin de la Murithienne*, **90**, 75–112.

Aubert, J., Goeldlin, P. & Lyon, J.-P. (1969). Essais de marquage et de reprise d'insectes migrateurs en automne 1968. *Mitteilungen Schweizerschen Entomologischen Gesellschaft*, **42**, 140–66. In French, English summary.

Baker, R.H.A., Guillet, P., Sékétéli, A., Poudiougo, P., Boakye, D., Wilson, M.D. & Bissan, Y. (1990). Progress in controlling the reinvasion of windborne vectors into the western area of the Onchocerciasis Control Programme in West Africa. *Philosophical Transactions of the Royal Society of London B*, **328**, 731–50.

Baker, R.R. (1968). A possible method of evolution of the migratory habit in butterflies. *Philosophical Transactions of the Royal Society of London B*, **253**, 309–41.

Baker, R.R. (1978). *The Evolutionary Ecology of Animal Migration*. London: Hodder & Stoughton.

Bowden, J. & Johnson, C.G. (1976). Migrating and other terrestrial insects at sea. In *Marine Insects*, ed. L. Cheng, pp. 97–117. Amsterdam: North-Holland.

Bretherton, R.F. (1982). Lepidoptera immigration to the British Isles 1969 to 1977. *Proceedings and Transactions of the British Entomological & Natural History Society*, **15**, 98–110.

Bretherton, R.F. & Chalmers-Hunt, J.M. (1989). The immigration of Lepidoptera to the British Isles in 1988. *Entomologist's Record and Journal of Variation*, **101**, 153–7, 225–30.

Bretherton, R.F. & Chalmers-Hunt, J.M. (1990). The immigration of Lepidoptera into the British Isles in 1989. *Entomologist's Record and Journal of Variation*, **102**, 153–9, 215–24.

Brown, E.S., Betts, E. & Rainey, R.C. (1969). Seasonal changes in distribution of the African armyworm, *Spodoptera exempta* (Wlk.) (Lepidoptera: Noctuidae), with special reference to eastern Africa. *Bulletin of Entomological Research*, **58**, 661–728.

Cammell, M.E., Tatchell, G.M. & Woiwod, I.P. (1989). Spatial pattern of abundance of the black bean aphid, *Aphis fabae*, in Britain. *Journal of Applied Ecology*, **26**, 463–72.

Cayrol, R.A. (1972). Famille des Noctuidae. In *Entomologie Appliquée à l'Agriculture. Tome II. Lépidoptères*, vol. 2. Zygaenoidea, Pyraloidea, Noctuidea, ed. A.S. Balachowsky, pp. 1255–520. Paris: Masson.

Dingle, H. (1982). Function of migration in the seasonal synchronisation of insects. *Entomologia Experimentalis et Applicata*, **31**, 36–48.

Dingle, H. (1985). Migration. In *Comprehensive Insect Physiology, Biochemistry and Pharmacology*, vol. 9. *Behaviour*, ed. G.A. Kerkut & L.I. Gilbert, pp. 375–415. Oxford: Pergamon.

Drake, V.A. & Farrow, R.A. (1988). The influence of atmospheric structure and motions on insect migration. *Annual Review of Entomology*, **33**, 183–210.

Duranton, J.F., Launois, M., Launois-Luong, M.H. & Lecoq, M. (1979). Biologie et écologie de *Catantops haemorrhoidalis* en Afrique de l'Ouest (Orthopt. Acrididae). *Annales de la Société Entomologique de France (N. S.)*, **15**, 319–43. In French, English summary.

Duviard, D. (1977). Migration of *Dysdercus* spp. (Hemiptera: Pyrrhocoridae) related to movements of the inter-tropical convergence zone in West Africa. *Bulletin of Entomological Research*, **67**, 185–204.

24 *Long-range migration: Africa and Europe*

Edwards, J.S. (1987). Arthropods of alpine aeolian ecosystems. *Annual Review of Entomology*, **32**, 163–79.
Farrow, R.A. (1990). Flight and migration in acridoids. In *Biology of Grasshoppers*, ed. R.F. Chapman & A. Joern, pp. 227–314. New York: John Wiley & Sons.
Fishpool, L.D.C. & Popov, G.B. (1984). The grasshopper faunas of the savannas of Mali, Niger, Benin and Togo. *Bulletin de l'Institut Fondamental d'Afrique Noire*, **43**, sér. A, 3–4, 275–410.
Garms, R., Cheke, R.A. & Sachs, R. (1991). A temporary focus of savanna species of the *Simulium damnosum* complex in the forest zone of Liberia. *Tropical Medicine & Parasitology*, **42**, 157–318.
Garms, R. & Walsh, J.F. (1987). The migration and dispersal of black flies: *Simulium damnosum s. l.*, the main vector of human onchocerciasis. In *Black Flies: Ecology, Population Management and Annotated World List*, ed. K.C. Kim & R. W. Merritt, pp. 201–14. University Park: Pennsylvania State University.
Gatter, D. & Gatter, W. (1976). Schmetterlingswanderungen durch die Sahara. *Atalanta*, **8**, 241–6. In German, English summary.
Gerbier, N.E. (1965). *Analysis of the Relationship between Meteorology and Movements of Gregarious Desert Locust Swarms in the Western Invasion Area*. Report UNSF/DL/RFS/4. Rome: FAO.
Ghauri, M.S.K. (1983). A case of long-distance dispersal of a leafhopper. In *1st International Workshop on Leafhoppers & Planthoppers of Economic Importance*, ed. W.J. Knight, N.C. Pant, T.S. Robertson & M.R. Wilson, pp. 249–55. London: Commonwealth Institute of Entomology.
Hachler, M. (1989). Le ver gris, *Agrotis ipsilon* Hufn. (Lep.: Noctuidae), ravageur en grand culture, en culture maraîchère et en culture de petits fruits. II. Piégeage des adultes au moyen du piège lumineux. *Revue Suisse d'Agriculture*, **21**, 159–68. In French, English summary.
Haggis, M.J. (1984). *Distribution, Frequency of Attack and Seasonal Incidence of the African Armyworm* Spodoptera exempta *(Walk.) (Lep.: Noctuidae), with Particular Reference to Africa and Southwestern Arabia*. Tropical Development and Research Institute Report L69. London: Tropical Development and Research Institute.
Heathcote, G.D. (1984). Did aphids come to Suffolk from the Continent in 1982? *Transactions of the Suffolk Naturalists' Society*, **20**, 45–51.
Hurst, G.W. (1969a). Insect migrations to the British Isles. *Quarterly Journal of the Royal Meteorological Society*, **95**, 435–9.
Hurst, G.W. (1969b). Meteorological aspects of insect migrations. *Endeavour*, **28**, 77–81.
Johnson, C.G. (1969). *Migration & Dispersal of Insects by Flight*. London: Methuen.
Johnson, C.G. & Bowden, J. (1973). Problems related to the transoceanic transport of insects, especially between the Amazon and Congo areas. In *Tropical Forest Ecosystems in Africa and South America: a Comparative Review*, ed. B.J. Meggars, E.S. Ayensu & D. Duckworth, pp. 207–22. Washington: Smithsonian Institution Press.
Johnson, S.J. (1987). Migration and the life history strategy of the fall armyworm, *Spodoptera frugiperda* in the western hemisphere. In *Recent Advances in Research on Tropical Entomology*, ed. M.F.B. Chaudhury, pp. 543–9. *Insect Science and its Application*, vol. 8, Special Issue. Nairobi: ICIPE Science Press.

Kettlewell, H.B.D. & Heard, M.J. (1961). Accidental radioactive labelling of a migratory moth. *Nature*, **189**, 676–7.

Kutsenin, B.A. (1980). [The cabbage moth on the Kola Peninsula] *Zashchita Rastenii*, 1980 no. 51. In Russian; English summary in *Review of Applied Entomology A*, **69**, 6021.

Labonne, G. (1978). Note sur la présence de fortes populations de *Sitobion avenae* sur un sommet des Pyrénées centrales (Hom. Aphididae). *L'Entomologiste*, **34**, 143–5.

Launois, M. (1979). An ecological model for the study of the grasshopper *Oedaleus senegalensis* in West Africa. *Philosophical Transactions of the Royal Society of London B*, **287**, 345–55.

Lecoq, M. (1978). Biologie et dynamique d'un peuplement acridien de zone soudanienne en Afrique de l'Ouest (Orthoptera, Acrididae). *Annales de la Société Entomologique de France (N. S.)*, **14**, 603–81. In French, English summary.

Liston, A.D. & Leslie, A.D. (1982). Insects from high-altitude summer snow in Austria 1981. *Entomologische Gesellschaft (Basel)*, **32**, 42–7.

Lokki, J., Malmstrom, K.K. & Suomalainen, E. (1978). Migration of *Vanessa cardui* and *Plutella xylostella* (Lep.) to Spitzbergen in the summer of 1978. *Notulae Entomologicae*, **58**, 121–3.

Magor, J.I. & Rosenberg, L.J. (1980). Studies of winds and weather during migrations of *Simulium damnosum* Theobald (Diptera: Simuliidae), the vector of onchocerciasis in West Africa. *Bulletin of Entomological Research*, **70**, 693–716.

Malicky, H. (1981). Ein phänologischer vergleich von wanderfalten zwischen mittel- und süd-europa. *Acta Entomologica Jugoslavica*, **17**, 55–63. In German, English abstract.

Markkula, M. (1979). Pests of cultivated plants in Finland in 1978. *Annales Agriculturae Fenniae*, **18**, 92–5.

McNeil, J.N. (1987). The true armyworm, *Pseudaletia unipuncta*: a victim of the Pied Piper or a seasonal migrant? In *Recent Advances in Research on Tropical Entomology*, ed. M.F.B. Chaudhury, pp. 591–7. *Insect Science and its Application*, vol. 8, Special Issue. Nairobi: ICIPE Science Press.

Mel'nichenko, A.N. (1936). [Regularities of mass flying of the adults of *Loxostege sticticalis* L. and the problem of the prognosis of their flight migrations.] *Bulletin of Plant Protection, Leningrad*, Ser. 1, no. 17, 1–56. In Russian, English summary.

Mikkola, K. (1970). The interpretation of long-range migrations of *Spodoptera exigua* Hb. (Lep.: Noctuidae). *Journal of Animal Ecology*, **39**, 593–8.

Mikkola, K. (1986). Direction of insect migrations in relation to the wind. In *Insect Flight: Dispersal and Migration*, ed. W. Danthanarayana, pp 152–71. Heidelburg: Springer-Verlag.

Mikkola, K. & Salmensuu, P. (1965). Migration of *Laphygma exigua* (Lep.: Noctuidae) in northwestern Europe in 1964. *Annales Zoologici Fennici*, **2**, 124–39.

Nasr, E.A., Tucker, M.R. & Campion, D.G. (1984). Distribution of moths of the Egyptian cotton leafworm, *Spodoptera littoralis* (Boisduval) (Lepidoptera: Noctuidae), in the Nile Delta interpreted from catches in a pheromone trap network in relation to meteorological factors. *Bulletin of Entomological Research*, **74**, 487–94.

Odiyo, P.O. (1975). *Seasonal distribution and migrations of* Agrotis ipsilon *(Hufnagel) (Lep.: Noctuidae)*, Tropical Pest Bulletin 4. London: Centre for Overseas Pest Research.

Pedgley, D.E. (ed.) (1981). *Desert Locust Forecasting Manual*, vols. 1 and 2. London: Centre for Overseas Pest Research.

Pedgley, D.E. (1982). *Windborne Pests and Diseases: Meteorology of Airborne Organisms*. Chichester: Ellis Horwood.

Pedgley, D.E. (1985). Windborne migration of *Heliothis armigera* (Hübner) (Lepidoptera: Noctuidae) to the British Isles. *Entomologist's Gazette*, **36**, 15–20.

Pedgley, D.E. (1986). Windborne migration in the Middle East by the moth *Heliothis armigera* (Lepidoptera: Noctuidae). *Ecological Entomology*, **11**, 467–70.

Pedgley, D.E., Page, W.W., Mushi, A., Odiyo, P., Amisi, J., Dewhurst, C.F., Dunstan, W.R., Fishpool, L.D.C., Harvey, A.W., Megenasa, T. & Rose, D.J.W. (1989). Onset and spread of an African armyworm upsurge. *Ecological Entomology*, **14**, 311–33.

Pedgley, D.E., Reynolds, D.R., Riley J.R. & Tucker, M.R. (1982). Flying insects reveal small-scale wind systems. *Weather*, **37**, 295–306.

Pedgley, D.E. & Yathom, S. (1993). Windborne moth migration over the Middle East. *Ecological Entomology*, **18**, 67–72.

Poitout, S., Cayrol, R., Causse, R. & Anglade, P. (1974). Déroulement du programme d'études sur les migrations de Lépidoptères Noctuidae réalisé en montagne et principaux résultats acquis. *Annales de Zoologie Écologie Animale*, **6**, 585–7. In French, English summary.

Ragge, D.R. (1972). An unusual case of mass migration by flight in *Gryllus bimaculatus* Degeer (Orthoptera Gryllidae). *Bulletin de l'Institut Fondamental d'Afrique Noire*, **34**, sér. A, 869–78.

Rainey, R.C. (1963). *Meteorology and the Migration of Desert Locusts: Applications of Synoptic Meteorology in Locust Control*. WMO Technical Note 54. Geneva: World Meteorological Organization.

Rainey, R.C. (1976). Flight behaviour and features of the atmospheric environment. In *Insect Flight*, ed. R.C. Rainey, pp. 75–112. Symposia of the Royal Entomological Society of London, no. 7. Oxford: Blackwell Scientific Publications.

Reynolds, D.R. & Riley, J.R. (1988). A migration of grasshoppers, particularly *Diabolocatantops axillaris* (Thunberg) (Orthoptera: Acrididae), in the West African Sahel. *Bulletin of Entomological Research*, **78**, 251–71.

Rezbanyai, L. (1980). Wanderfalter in der Schweiz 1978. Fangergebrisse aus sieben Lichtfallen sowie Weitere Meldungen. *Atalanta*, **11**, 81–119.

Riley, J.R. & Reynolds, D.R. (1979). Radar-based studies of the migratory flight of grasshoppers in the middle Niger area of Mali. *Proceedings of the Royal Society of London B*, **204**, 67–82.

Riley, J.R. & Reynolds, D.R. (1983). A long-range migration of grasshoppers observed in the Sahelian zone of Mali by two radars. *Journal of Animal Ecology*, **52**, 167–83.

Riley, J.R. & Reynolds, D.R. (1986). Orientation at night by high-flying insects. In *Insect Flight: Dispersal and Migration*, ed. W. Danthanarayana, pp. 71–87. Heidelberg: Springer-Verlag.

Riley, J.R. & Reynolds D.R. (1990). Nocturnal grasshopper migration in West Africa: transport and concentration by the wind, and the implications for air-to-air control. *Philosophical Transactions of the Royal Society of London B*, **328**, 655–72.

Riley, J.R., Reynolds, D.R. & Farmery, M.J. (1983). Observations of the flight behaviour of the armyworm moth, *Spodoptera exempta*, at an emergence site using radar and infra-red optical techniques. *Ecological Entomology*, **8**, 395–418.

Ritchie, J.M. (1978). Melanism in *Oedaleus senegalensis* and other oedipodines (Orthoptera, Acrididae). *Journal of Natural History*, **12**, 153–62.

Ritchie, J.M. (1981). A taxonomic revision of the genus *Oedaleus* Fieber (Orthoptera: Acrididae). *Bulletin of the British Museum (Natural History), Entomology Series*, **42**, 83–183.

Ritchie, M. & Pedgley, D. (1989). Desert locusts cross the Atlantic. *Antenna*, **13**, 10–12.

Rose, D.J.W., Page, W.W., Dewhurst, C.F., Riley, J.R., Reynolds, D.R., Pedgley, D.E. & Tucker, M.R. (1985). Downwind migration of the African armyworm moth, *Spodoptera exempta*, studied by mark-and-capture and by radar. *Ecological Entomology*, **10**, 299–313.

Schaefer, G.W. (1976). Radar observations of insect flight. In *Insect Flight*, ed. R.C. Rainey, pp. 157–97. Symposia of the Royal Entomological Society of London, no. 7. Oxford: Blackwell Scientific Publications.

Schmutterer, H. (1969). *Pests of Crops in Northeast and Central Africa*. Stuttgart: Gustav Fischer Verlag.

Sellers, R.F., Gibbs, E.P.J., Herniman, K.A.J., Pedgley, D.E. & Tucker, M.R. (1979). Possible origin of the bluetongue epidemic in Cyprus, August 1977. *Journal of Hygiene (Cambridge)*, **83**, 547–55.

Sellers, R.F. & Pedgley, D.E. (1985). Possible windborne spread to Western Turkey of bluetongue virus in 1977 and of Akabane virus in 1979. *Journal of Hygiene (Cambridge)*, **95**, 149–58.

Sellers, R.F., Pedgley, D.E. & Tucker, M.R. (1977). Possible spread of African horse sickness on the wind. *Journal of Hygiene (Cambridge)*, **79**, 279–98.

Sellers, R.F., Pedgley, D.E. & Tucker, M.R. (1978). Possible windborne spread of bluetongue to Portugal, June–July 1956. *Journal of Hygiene (Cambridge)*, **81**, 189–96.

Sellers, R.F., Pedgley, D.E. & Tucker, M.R. (1982). Rift Valley fever, Egypt 1977: disease spread by windborne insect vectors? *The Veterinary Record*, **110**, 73–7.

Shaw, B. (1965). Depressions and associated desert locust swarm movements in the Middle East. In *Meteorology and the Desert Locust*. WMO Technical Note 69, pp. 194–8. Geneva: World Meteorological Organization.

Shaw, M.W. (1962). The diamondback moth migration of 1958. *Weather*, **17**, 221–34.

Solbreck, C. (1985). Insect migration strategies and population dynamics. In *Migration: Mechanisms and Adaptive Significance*, ed. M.A. Rankin, pp. 641–62. Contributions in Marine Science, vol. 27 (Suppl.). Port Aransas: Marine Science Institute, The University of Texas at Austin.

Southwood, T.R.E. (1962). Migration of terrestrial arthropods in relation to habitat. *Biological Reviews*, **37**, 171–214.

Steedman, A. (ed.) (1990). *Locust Handbook*, 3rd edn. Chatham: Natural Resources Institute.

Sutton, S.L. (1966). South Caspian insect fauna, 1961. II. Migration, status and distribution of certain insect species in northern Persia. *Transactions of the Royal Entomological Society of London*, **118**, 51–72.

Tatchell, G.M. (1991). Monitoring and forecasting aphid problems. In *Aphid-Plant Interactions: Populations to Molecules*. Oklahoma Agricultural Experiment Station Miscellaneous Publication 132, ed. D.C. Peters, J.A. Webster & C.S. Chlouber, pp. 215–31. Stillwater: Oklahoma State University.

Taylor, L.R. (1974). Insect migration, flight periodicity and the boundary layer. *Journal of Animal Ecology*, **43**, 225–38.

Taylor, L.R. (1985). An international standard for the synoptic monitoring and dynamic mapping of migrant pest aphid populations. In *The Movement and Dispersal of Agriculturally Important Biotic Agents*, ed. D.R. MacKenzie, C.S. Barfield, G.G. Kennedy, R.D. Berger & D.J. Taranto, pp. 337–80. Baton Rouge: Claitor's.

Taylor, L.R. (1986). Synoptic dynamics, migration and the Rothamsted Insect Survey. *Journal of Animal Ecology*, **55**, 1–38.

Taylor, L.R., Woiwod, I.P. & Taylor, R.A.J. (1979). The migratory ambit of the hop aphid and its significance in aphid population dynamics. *Journal of Animal Ecology*, **48**, 955–72.

Taylor, R.A.J. & Reling, D. (1986). Preferred wind direction of long-distance leafhopper (*Empoasca fabae*) migrants and its relevance to the return migration of small insects. *Journal of Animal Ecology*, **55**, 1103–14.

Tucker, M.R. (1984). Possible sources of outbreaks of the armyworm, *Spodoptera exempta* (Walker) (Lepidoptera: Noctuidae), in East Africa at the beginning of the season. *Bulletin of Entomological Research*, **74**, 599–607.

Tucker, M.R. (1994). Inter- and intra-seasonal variation in outbreak distribution of the armyworm, *Spodoptera exempta* (Walker) (Lepidoptera: Noctuidae), in Eastern Africa. *Bulletin of Entomological Research*, **84**, 275–87.

Tucker, M.R. & Pedgley, D.E. (1983). Rainfall and outbreaks of the African armyworm, *Spodoptera exempta* (Walker) (Lepidoptera: Noctuidae). *Bulletin of Entomological Research*, **73**, 195–9.

Uvarov, B.P. (1977). *Grasshoppers and Locusts: a Handbook of General Acridology*, vol. II. London: Centre for Overseas Pest Research.

Walker, T.J. (1980). Migrating Lepidoptera: are butterflies better than moths? *Florida Entomologist*, **63**, 79–98.

Walker, T.J. (1985). Butterfly migration in the boundary layer. In *Migration: Mechanisms and Adaptive Significance*, ed. M.A. Rankin, pp. 704–23. Contributions in Marine Science, vol. 27 (Suppl.). Port Aransas: Marine Science Institute, The University of Texas at Austin.

Walker, T.J. (1991). Butterfly migration from and to peninsular Florida. *Ecological Entomology*, **16**, 241–52

Walker, T.J. & Riordan, A.J. (1981). Butterfly migration: are synoptic wind systems important? *Ecological Entomology*, **6**, 433–40.

Waloff, Z. & Pedgley, D.E. (1986). Comparative biogeography and biology of the South American locust, *Schistocerca cancellata* (Seville), and the South African desert locust, *S. gregaria flaviventris* (Burmeister) (Orthoptera: Acrididae): a review. *Bulletin of Entomological Research*, **76**, 1–20.

Wiktelius, S. (1977). The importance of southerly winds and other weather data on the incidence of sugar beet yellowing viruses in southern Sweden. *Swedish Journal of Agricultural Research*, **7**, 89–95.

Wiktelius, S. (1981). Wind dispersal of insects. *Grana*, **20**, 205–7.

Wiktelius, S. (1984). Long-range migration of aphids into Sweden. *International Journal of Biometerology*, **28**, 185–200.

Williams, C.B. (1930). *The Migration of Butterflies*. Edinburgh: Oliver & Boyd.

Williams, C.B. (1958). *Insect Migration*. London: Collins.

Wiltshire, E.P. (1946). Studies in the geography of Lepidoptera. III. Some Middle East migrants, their phenology and ecology. *Transactions of the Royal Entomological Society of London*, **96**, 163–82.

Zaazou, M.H., Fahmy, H.S.M., Kamel, A.A.M. & El-Hemaesy, A.H. (1973). Annual movement and host plants of *Agrotis ipsilon* (Hufn.) in Egypt. *Bulletin, Société Entomologique d'Egypte*, **57**, 175–80.

# 2

# Insect migration in North America: synoptic-scale transport in a highly seasonal environment[1]

## S. J. JOHNSON

### Introduction

Many North American insects, including several that are important pests, undertake long-distance migrations (Table 2.1). Knowledge of the role of migration in the population processes of these species, and of the relation of migration to climate and atmospheric phenomena, has developed in four broad phases. At first it was simply recognised that serious pest problems in the northern latitudes of the USA were sometimes initiated by populations that appeared suddenly in association with southerly winds. Then in the 1920s and 1930s, airplane sampling of the upper air became possible (Felt, 1928, 1937; Glick, 1939) and it was found that insects, spiders, and mites were commonly to be found at heights up to 1.5 km and occasionally (<1% of specimens obtained) at greater altitudes, even up to ~4.6 km. Over 700 species were caught in these airplane samples, but certainly not all were long-distance migrants. Seasonality, light intensity, temperature, wind speed and direction, and some other meteorological variables were found to be important factors affecting the distribution of insects in the upper air. In the third phase of migration research, during the 1940s, 1950s and 1960s, the association between the sudden appearance of a species in an area and the passage of synoptic-scale (200–2000 km) weather systems was documented in a series of studies. Finally, a conference held in Raleigh, North Carolina in 1979 (Rabb & Kennedy, 1979) stimulated the current era of research on insect migration in North America, in which modern technologies, such as ground-based (Wolf, Westbrook & Sparks, 1986; Beerwinkle *et al.* 1991) and airborne (Wolf *et al.*, 1990) radar, trajectory analysis (Wolf *et al.*, 1987; Smelser *et al.*, 1991; Carlson *et al.*, 1991), aerial collection

[1] Approved for publication by the director of Louisiana State University Agricultural Experiment Station as manuscript no. 94-178010.

devices (Beerwinkle *et al.*, 1989; Hollinger *et al.*, 1991), mark–release–recapture studies (Showers *et al.*, 1989*a*, 1993) and modelling (Hendrie *et al.*, 1985) have greatly increased our knowledge of the relation of long-distance migration to synoptic-scale weather phenomena. This renewed effort has identified many new migrants (Table 2.1) and closed the gap in knowledge about insect migration between North America and other parts of the world.

The climate of the mid-latitudes of North America is highly seasonal, with severe winters. Each spring there is a gradual poleward advance of the warmer weather required for plant growth, insect survival and breeding. Southerly airflows capable of transporting migrants northwards occur frequently, and the direction of insect migration at this season is predominantly towards the north. Both establishment on resources at higher latitudes as soon as they become available, and escape from unfavourable conditions in the southern source regions, could be advantageous for migrant individuals. The large number of species showing such springtime movements suggests that there is strong selective pressure for this behaviour.

Conversely, in autumn there is a retreat of habitat availability towards the equator. Winds suitable for transport southwards also occur. However, sudden appearances (or increases in species density) have not been evident in autumn, perhaps because of the difficulty of detecting immigrants from oversummering populations; it has therefore not been possible to confirm that migrations to lower latitudes occur, or to correlate them with particular types of airflow. Since evidence for a return migration to lower latitudes has been lacking (at least until very recently), species that migrate northwards in spring have been termed 'Pied Piper' migrants (Rabb & Stinner, 1979). It has been suggested that the descendants of these migrants must perish during the harsh mid-latitude winters, but this appears to create an evolutionary dilemma as the genes for migratory behaviour would then be lost from the population (Walker, 1980; Dingle, 1982). However, it has been pointed out that these genes could be maintained, and Pied Piper migrants could continue to be produced, if selection pressure for migration exists within some part of the species' lower-latitude range (Oku and Kobayashi, 1978; Johnson, 1987).

This chapter reviews the features of the climate and weather associated with long-distance insect migration in North America in spring, summer and autumn. The review focuses primarily on the geographic area between the Rocky and Appalachian Mountains (Fig. 2.1*a*). Here the topography and weather combine to create the greatest opportunity for

Table 2.1. *Long-distance migratory pests in North America*

| Order, family, genus and species | Common name | Reference[a] |
|---|---|---|
| **Orthoptera** | | |
| Acrididae | | |
| *Melanoplus sanguinipes* (F.) | Migratory grasshopper | Parker, Newton & Shotwell (1955) |
| **Homoptera** | | |
| Cicadellidae | | |
| *Circulifer tenellus* Baker | Beet leafhopper | Dorst & Davis (1937) |
| *Empoasca fabae* Harris | Potato leafhopper | Medler (1957) |
| *Macrosteles fascifrons* Stål | Aster leafhopper | Chiykowski & Chapman (1965) |
| Aphididae | | |
| *Macrosiphum avenae* (F.) | English grain aphid | Wallin, Peters & Johnson (1967) |
| *Rhopalosiphum maidis* (Fitch) | Corn leaf aphid | Medler (1962) |
| *Rhopalosiphum padi* (L.) | Oat bird-cherry aphid | Medler (1962) |
| *Schizaphis graminum* (Rondani) | Greenbug | Medler & Smith (1960) |
| **Coleoptera** | | |
| Chrysomelidae | | |
| *Diabrotica undecimpunctata howardi* Barber | Spotted cucumber beetle | Smith & Allen (1932) |
| *Diabrotica virgifera virgifera* LeConte | Western corn rootworm | Grant & Seevers (1989) |
| **Lepidoptera** | | |
| Gelechiidae | | |
| *Pectinophora gossypiella* (Saunders) | Pink bollworm | Glick (1967) |
| Plutellidae | | |
| *Plutella xylostella* (L.) | Diamondback moth | Smith & Sears (1982) |
| Pyralidae | | |
| *Diaphania nitidalis* (Stoll) | Pickleworm | Reid & Cuthbert (1956) |
| *Homoeosoma electellum* (Hulst) | Sunflower moth | Arthur & Bauer (1981) |
| Lasiocampidae | | |
| *Malacosoma disstria* Hübner | Forest tent caterpillar | Brown (1965) |
| Noctuidae | | |
| *Agrotis ipsilon* (Hufnagel) | Black cutworm | von Kaster & Showers (1982) |

Table 2.1. *(cont.)*

| Order, family, genus and species | Common name | Reference[a] |
|---|---|---|
| *Alabama argillacea* (Hübner) | Cotton leafworm | Riley (1885) |
| *Anticarsia gemmatalis* (Hübner) | Velvetbean caterpillar | Watson (1916) |
| *Euxoa auxiliaris* (Grote) | Army cutworm | Pruess (1967) |
| *Helicoverpa zea* (Boddie) | Cotton bollworm | Hartstack *et al.* (1982) |
| *Heliothis virescens* (F.) | Tobacco budworm | Raulston (1979) |
| *Peridroma saucia* (Hübner) | Variegated cutworm | Buntin, Pedigo & Showers (1990) |
| *Plathypena scabra* (F.) | Green cloverworm | Wolf *et al.* (1987) |
| *Pseudaletia unipuncta* (Haw.) | True armyworm | Forbes (1954) |
| *Pseudoplusia includens* (Walker) | Soybean looper | Mitchel *et al.* (1975) |
| *Spodoptera exigua* (Hübner) | Beet armyworm | Mitchel (1979) |
| *Spodoptera frugiperda* (J.E. Smith) | Fall armyworm | Luginbill (1928) |
| *Trichoplusia ni* (Hübner) | Cabbage looper | Lingren, Henneberry & Sparks (1979) |

[a] First publication containing some evidence or indication of long-distance migration by these species on synoptic-scale winds.

synoptic-scale insect migrations in North America. The insects considered are mainly homopterans and moths. Historical as well as recent studies are considered, but the review does not deal with migratory flight within the insects' flight boundary layer (*sensu* Taylor, 1958) or movements associated with local or mesoscale weather systems, even though there are outstanding examples of these in North America (see Johnson, 1969; Pedgley, 1982).

## North America as an environment for insect migration
### Geography

The North American continent spans nearly sixty degrees of latitude and has some distinctive geographical features (see Fig. 2.1*a* for names). The West Coast is backed by the Rocky and Sierra Nevada Mountains, with some peaks rising to over 4300 m. These ranges lie across the path of depressions in the mid-latitude westerlies and prevent the extension of maritime influences inland. The East Coast is backed by the Appalachian

Mountains with some peaks rising to over 1800 m. This range also lies across the depression path, but the extension of maritime influences inland from the East Coast is limited more by the prevalence of westerly airflows at this latitude than by the Appalachians. In west-central North America there is a sloping region of valleys and plains stretching east from the base of the Rockies for ~600 km and extending from Texas in the south to Alberta, Canada in the north: the Great Plains. To the east of these plains, and extending to the Appalachians, lies the Mississippi River Drainage Basin (MRDB). There are no significant obstructions to air movement within either of these last two regions, and air masses from the Arctic or the Gulf of Mexico can sweep southwards or northwards for a thousand kilometres or more; this perhaps is one of the ultimate reasons why insect migration is so prevalent here.

### Climate and host availability

The mid-latitudes of North America, and especially the MRDB, have a highly seasonal climate with an extreme range of temperature variation between summer and winter. Winter temperatures, as indicated by January isotherms (Fig. 2.2*a*), clearly reveal the moderating influences of the Pacific Ocean and the barrier to those influences formed by the Rocky Mountains. The near absence of a moderating influence on the East Coast is obvious. The widest latitudinal range in mean January temperatures occurs in the MRDB, with values below −12 °C in southern Canada but near 16 °C along the Gulf of Mexico. The harsh winters of the higher mid-latitudes restrict the northern distribution for nearly all migrant species. The limits for overwintering of ten selected species are shown in Fig. 2.2*a*. There is considerable variation between species in their ability to tolerate low winter temperatures, but none occurs north of a subzero isotherm. The two species known to have an adult diapause, *P. scabra*[1] (L.D. Newsom, unpublished data) and *E. fabae* (Taylor, 1993), have two of the most northern winter distributions, while the four species with the lowest latitudinal overwintering limit, *A. gemmatalis*, *S. frugiperda*, *D. nitidalis* and *A. argillacea*, apparently have neither a diapause nor a diapause-like condition. Summer temperatures in the MRDB, as indicated by average July isotherms (Fig. 2.2*b*), are very high. The northern limit of distribution in summer extends into Canada for most of the same

---

[1] See Table 2.1 for generic and common names.

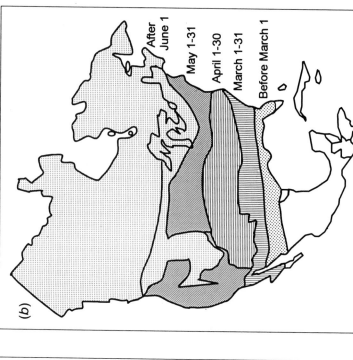

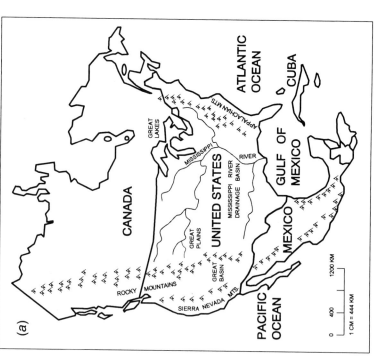

Fig. 2.1. (a) Map of part of North America with major geographical features. (b) Average dates of last killing frost in North America. The white areas indicate regions where frost does not occur, except perhaps at high elevations. (After Visher (1966).)

ten selected migrant species (Fig. 2.2*b*). It is apparent that the winter and summer distributions of these ten species reflect the highly seasonal climate of North America, and especially of the MRDB.

In spring there is an obvious poleward advance of vegetation growth, and hence of host availability; this coincides approximately with the advance of the last killing frost (Fig. 2.1*b*). In the MRDB, recolonisation in spring will be most successful if timed to coincide with the renewed availability of host plants. If migrations occur too early, the immigrants will find no hosts, and may themselves be killed by a late cold spell; late immigrants on the other hand may find all hosts occupied, and natural enemies well established. Further south, in areas where overwintering is possible, immigrants could get a jump on resident populations by colonising the host plants before the locally produced insects have emerged.

In autumn there is a reverse equatorward retreat of vegetation, and hence host availability, which coincides with the advance of the first killing frost. Migrants either perish in the northern areas of their expanded summer distribution or make return migrations to lower latitudes. There they can pass the winter in diapause, or they can continue developing but at a slowed pace, or, if host plants are available, they can breed.

### Wind systems

Long-distance insect migrants are transported on the winds. Atmospheric circulation processes are important determinants of the favourability of winds for transport. Their importance in North America was noted early on by Wellington (1954). Air masses, fronts, and depressions are primary features of these circulation processes. They occur within (and are confined to) the troposphere, the zone extending upwards from the surface for 8–16 km.

An air mass is a large body of air that has more or less uniform temperature and moisture content and extends horizontally for hundreds of kilometres. (See Barry & Chorley (1992) for more information on the atmospheric phenomena described in this and the next section.) Air masses are known by their region of origin, which is classified on the basis of two primary factors: latitude, resulting in *arctic*, *polar*, or *tropical* air; and the type of surface, *maritime* or *continental*. The principal North American air-mass types, and the approximate locations of their source regions, are shown in Fig. 2.3 (from which arctic and transitional types have been omitted). Cold air masses from polar regions can be either

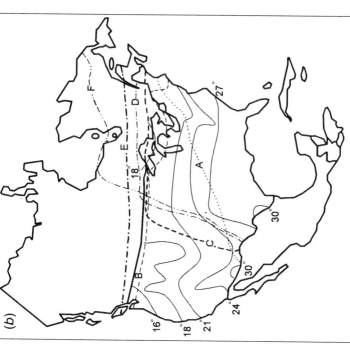

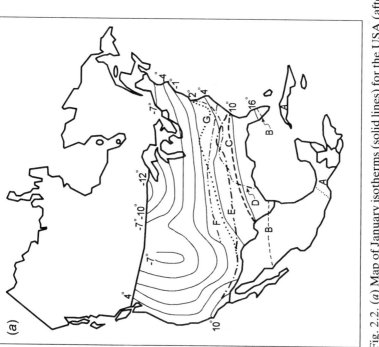

Fig. 2.2. (a) Map of January isotherms (solid lines) and northern overwintering limits (broken lines) for the USA (after Visher, 1966) for A *A. argillacea* (Pyenson, 1940); B *A. gemmatalis* (Buschman *et al.*, 1977; Gregory *et al.*, 1990); D. *nitidalis* (Reid & Cuthbert, 1956) and *S. frugiperda* (Luginbill, 1928); C *M. fascifrons* (Chiykowski & Chapman, 1965); E *A. ipsilon* (Smelser *et al.*, 1991); F *R. maidis* and *S. graminum* (Kieckhefer *et al.*, 1974); G *P. scabra* (Wolf *et al.*, 1987). (b) Map of July isotherms (solid lines) for the USA (after Visher, 1966) and northern distribution limits (broken lines) for *A. A. gemmatalis* (Ford *et al.*, 1975); B *D. nitidalis* (Reid & Cuthbert, 1956); C *A. argillacea* (Pyenson, 1940); D *E. fabae* (Delong, 1938); E *S. frugiperda* (Johnson, 1987); *M. fascifrons* (Nielson, 1968), *A. ipsilon* (Pires *et al.*, 1976), *R. maidis* (Smith & Parron, 1979) and *S. graminum* & P.

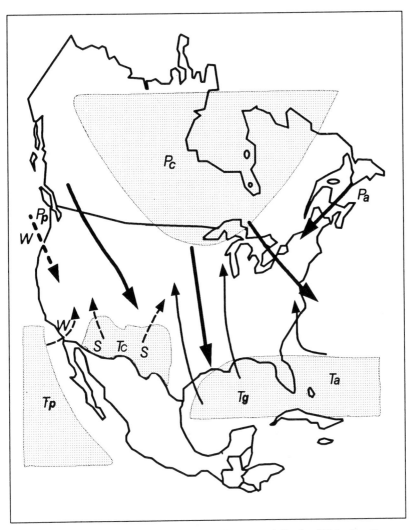

Fig. 2.3. Air masses affecting North America, especially the USA. Continuous lines indicate flows that occur throughout the year; seasonal flows are shown dashed and are labelled *S* (summer) or *W* (winter); source areas are labelled *Pa* (Polar Atlantic), *Pc* (Polar continental), *Pp* (Polar Pacific), *Ta* (Tropical Atlantic), *Tg* (Tropical Gulf), *Tc* (Tropical continental) and *Tp* (Tropical Pacific). (After Visher (1966).)

marine or continental. Polar-continental air is dry and cold; the temperature and humidity of polar-Pacific air depends on its previous trajectory, but often after crossing the Rocky Mountains it is almost as dry as air of polar-continental origin. Warm air masses come from tropical regions

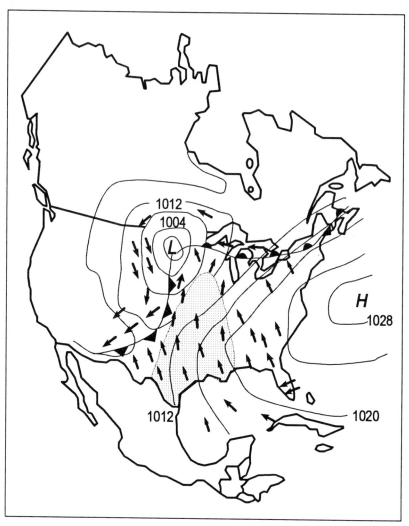

Fig. 2.4. Typical PTSWS (see text) weather situation for North America in spring or early summer, showing the anti-clockwise wind flow around a mid-latitude depression. The shaded area shows the region where low-level jets occur.

and include tropical-maritime air from the Pacific, the Gulf of Mexico and the Atlantic, and tropical-continental air from the arid southwestern part of the continent. Tropical-Gulf air is warm and moist even when modified considerably over time and after moving large distances inland. Tropical-continental air, which is produced mainly during the summer months, is usually hot and dry but has often become humid by the time it reaches Canada. In the MRDB, poleward migrations in spring occur in tropical-Gulf air, while in summer they occur on both tropical-Gulf and tropical-continental airflows; equatorward migrations in autumn occur on polar-continental and polar-Pacific flows.

Weather fronts are the boundaries between different air masses, and many of the day-to-day changes in mid-latitude weather are associated with their formation and movements. There are discontinuities in temperature, wind direction, humidity and other physical phenomena at a front. Fronts are of two types: warm, where warm air replaces cold; and cold, where cold air replaces warm. The frontal surface of a warm front has a very gentle slope, while that for a cold front is steep. At the passage of a warm front the wind veers, the temperature rises, the pressure stops falling, and rain becomes intermittent or ceases. The passage of a cold front is marked by a short period of more intense bad weather; the wind also veers sharply, the temperature falls, and the pressure begins to rise.

A mid-latitude depression (also termed an extratropical cyclone) is an area of relatively low pressure around which the air circulates in an anticlockwise direction (in the northern hemisphere; see Fig. 2.4). Depressions form at the interface between contrasting air masses and they move steadily eastwards at average speeds of $32$ km h$^{-1}$ in summer and $48$ km h$^{-1}$ in winter, under the influence of a westerly airflow at higher altitudes; however, there are spells when depressions move very little and wind patterns persist for several days. The air-mass interface takes the form of a leading warm front and a following cold front. Depressions link up into families of three or four, forming what is known as a frontal wave on the polar front. Wind direction changes continually as a depression passes over and there is usually rain or a storm.

Depression tracks shift from season to season and form natural groupings associated with particular source regions (Fig. 2.5a); however, they all track northeastwards. In summer the polar front shifts some $10°$ poleward, and the mid-latitudes come under the influence of a persistent subtropical high-pressure cell in the Atlantic Ocean (the Bermuda High). The combination of these influences causes the depression tracks to shift polewards to pass over the northern USA and the Great Lakes. There is also a summer track with an origin in the desert southwest.

An anticyclone is an area of high pressure with clockwise circulation (in the northern hemisphere). Winds also change continually as an anticyclone passes over, but the weather is usually settled. Anticyclones

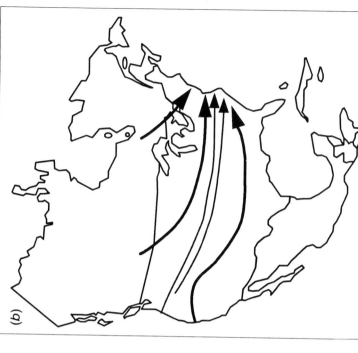

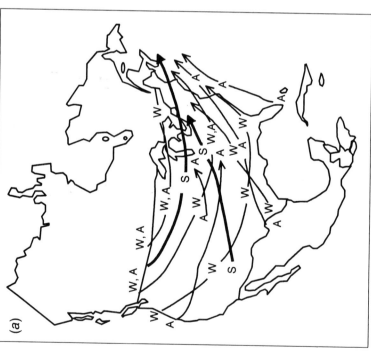

Fig. 2.5. (a) Approximate tracks of mid-latitude depressions (low-pressure areas) by seasons (W winter, S summer, A autumn or spring). (b) Approximate tracks of anticyclones (high-pressure areas). Line thickness indicates relative frequency. (After Visher (1966).)

tend to move southeastwards over the MRDB carrying polar-Pacific and polar-continental air southwards (Fig. 2.5*b*).

### The planetary boundary layer

The direct effects of the Earth's surface on the atmosphere are most apparent in the lowest 1–2 km of the troposphere, a zone known as the Planetary Boundary Layer (PBL) (Drake & Farrow, 1988). The PBL exists in two forms: *convective*, in which there is vertical mixing driven by thermal circulatory air motions; and *stable*, in which mixing is suppressed and the atmosphere becomes stratified. The PBL generally becomes stable around sunset over land areas. After sunset a nocturnal temperature inversion (i.e. a zone in which the temperature increases with height) can form, especially under high-pressure conditions when skies are clear and surface winds are light. The nocturnal inversion is shallow at first but gradually becomes deeper. In both the stable and convective PBL, surface wind speeds in the lowest few tens of metres increase approximately logarithmically with height. Wind speed continues to increase with altitude, and in extratropical regions of the northern hemisphere there is a gradual clockwise change in direction, the Ekman spiral. However, at the top of the PBL there are no frictional effects and the wind is usually approximately geostrophic, i.e. it has the speed expected from the local gradient of atmospheric pressure and blows almost parallel to the isobars.

Night-time long-distance migrations undertaken by nocturnal insects such as moths as well as planthoppers and other Homoptera usually commence at dusk. Also, microinsects such as aphids that probably start their migrations in the daytime on convective updrafts sometimes continue their flights into the night. Long-distance migrations above the surface layer are essentially downwind, because the wind speeds are usually much greater than the insects' flight speed. Because of the Ekman-spiral phenomenon, the speed and direction of migration in middle latitudes change gradually with height up to altitudes of about 500 m (Drake & Farrow, 1988). Radar studies suggest that insect migrants typically become concentrated into layers at heights of 200–600 m, usually near the top of the surface temperature inversion (Wolf *et al.*, 1986, 1990; Beerwinkle *et al.*, 1991). These layers may not form until about 23.00 LST, well after migratory flight was initiated ~30 min after sunset. Insect numbers at all altitudes often decrease rapidly at dawn, indicating that the insects land in response to daylight.

In the MRDB the night-time wind profile frequently exhibits a wind-speed maximum at an altitude of a few hundred metres, a phenomenon termed the low-level jet (LLJ). Air in a LLJ moves at supergeostrophic speeds, and this adds an extra 20–30% to the speed of migration and therefore to the distance covered in a given flying time (Drake & Farrow, 1988). Maximum windspeeds often exceed 21 m s$^{-1}$ (74 km h$^{-1}$). LLJs are particularly frequent in the MRDB because of the influence of the Rocky Mountains to the west. They are often associated with southerly airflows, and are typically located near the region of the highest night-time air temperatures (the top of the surface inversion) at altitudes between 200 and 1000 m (Wolf *et al.*, 1990; Carlson *et al.*, 1991).

Radar observations have shown that insect migrations commonly occur in LLJs (Drake & Farrow, 1988; Wolf *et al.*, 1990). Insects show some crosswind orientation in the LLJ, but the strength of the wind is so much greater than their flight speed that their trajectory is still close to downwind (Wolf *et al.*, 1990).

### Spring and summer insect migrations on synoptic-scale winds

Despite the harsh winters of the Great Plains and all but the southern fringe of the MRDB, many migrant species are distributed well into Canada by summer's end. The poleward displacements that lead to the annual recolonisation of this region occur on southerly winds associated with a specific synoptic weather pattern. A description of this weather type and its variations, and case histories of the associated insect migrations, follow.

### *Weather system and transporting airflows*

There is one primary synoptic weather pattern that draws warm southerly tropical-Gulf and tropical-continental air into the MRDB in spring and summer. It consists of a high-pressure area located on or just east of the Atlantic Coast of the USA and a mid-latitude depression with a trailing cold front approaching from the west (Fig. 2.4). This configuration will be termed a poleward-transporting synoptic weather system (PTSWS). On the western side of the high-pressure system there is a large-scale flow of tropical air towards the northwest, north and northeast; the flow has a cross-isobar component, outwards from the anticyclone and inwards towards the depression. The southerly airflow is termed either Gulf return (GR) or, near the approaching front where lifting and convergence is evident, frontal Gulf return (FGR) (Muller & Willis, 1983). GR and FGR airflows occur 50–60% of the time at the representative southern cities of Shreveport, Louisiana and Brownsville, Texas during March (Muller & Tucker, 1986). The speed of the airflow in the warm sector of the depression is typically relatively high. LLJs often form in these systems in spring (Fig. 2.4): they consist of a stream of super-geostrophic winds only 50–100 km wide.

The depressions tend to approach along particular tracks, of which there are three in spring and two in summer (Fig. 2.5a). The track followed, as well as the location of the frontal boundary, will affect a migrant's trajectory and deposition site. Also, slowly moving depressions may allow for greater poleward displacement than average- or fast-moving depressions. The part of the pattern in which there is potential for poleward transport normally has a northeast/southwest orientation; it lies parallel to, and to the southeast of, the surface cold front and coincides approximately with the region where a LLJ, if present, will be located (Fig. 2.4).

In late spring and during summer this weather type still occurs, and there is still potential for poleward transport, but the distances that can be covered in a night are much smaller, because the southerly winds (and the atmospheric pressure gradients that produce them) are much weaker. In summer the Bermuda high usually becomes a dominant weather feature. It creates southerly and southeasterly flows from the Gulf into the MRDB, providing additional atmospheric transport opportunities for migrant insects. Also, there are periods when the western extension of the Bermuda high is displaced southwards over the Gulf and the airflow becomes southwesterly.

### Homoptera

*E. fabae* overwinters in the southern USA (Fig. 2.2a) and its sudden appearance in the eastern and north-central USA each year is generally associated with occurrences of the PTSWS weather pattern (Huff, 1963; Pienkowski & Medler, 1964; Carlson *et al.*, 1991). In the two early studies, the southerly winds averaged 9 m s$^{-1}$ (33 km h$^{-1}$) between the surface and 1600 m, and temperatures were generally above normal and above the species' flight threshold of 15 °C; the average duration of the weather pattern was 3.2 days. Backtracking of an airflow that brought *E. fabae* into Michigan on 1 June 1989 demonstrated that the wind direction shifted from southerly to southwesterly as the cold front

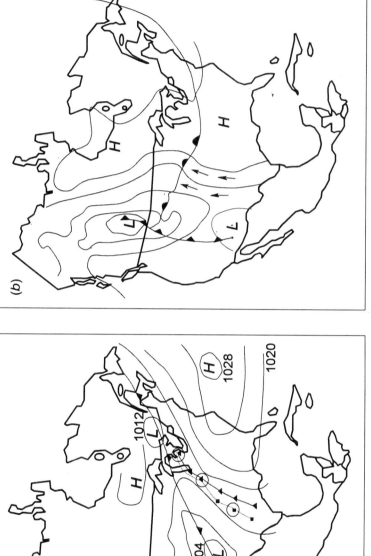

Fig. 2.6. (a) Backtracks of *E. fabae* migration into Michigan on 31 May 1989 at 900 hPa, with synoptic weather map for 30 May 1989 superimposed. Trajectories are for continuous flight (▲–▲) and flight only at night (■–■). The markers on the trajectories denote 6-h positions; circled markers denote overnight location. (After Carlson *et al.* (1991).) (b) Trajectories for *M. avenae* and *S. graminum* migration into upper MRDB (4–5 days of airflow), with synoptic weather map for 18 May 1964 superimposed. (After Kieckhefer *et al.*

advanced eastwards (Fig. 2.6*a*) (Carlson *et al.*, 1991). The southwesterly flow persisted for four days, and apparently carried the leafhoppers from probable source areas in eastern Texas, eastern Oklahoma, Arkansas, or Louisiana. Wind speeds were 16–26 m s$^{-1}$, and by the evening of 29 May the temperature at the 850-hPa level (~1600 m above sea level (ASL)) over southern Michigan was above 15 °C. The migration may have occurred on a nocturnal LLJ that was present at ~600–1000 m. If flight was continuous, transport time from Texas to Michigan would have been 36 h, but if flight was only at night, then three consecutive nights would have been required.

   *M. fascifrons* is a vector of the mycoplasma-like organism that causes the plant disease Aster Yellows, and is a pest of many susceptible crops and ornamentals. This insect has a long history of sudden appearances in areas of Canada and the north-central USA that are far removed from its overwintering and breeding grounds in the central and south-central states. First arrivals at Winnipeg, Manitoba, between 1954 and 1964 were backtracked to their likely source area. Most first arrivals were in May and they were always associated with the warm southerly airflow of a PTSWS in which the depression was located in the Great Basin (i.e. just west of the Rocky Mountains). Back-trajectories over 2–3 nights extended for 400–900 km at the surface (with probable sources in Kansas, Nebraska and South Dakota) or 660–1500 km at 1600 m ASL (with probable source areas in Mississippi, Texas, Oklahoma and Kansas) (Nichiporick, 1965). In a separate study, field surveys during the springs of 1953–58 demonstrated the northward movement of large numbers of these leafhoppers into Wisconsin from probable source areas in Missouri, Arkansas, Texas, Oklahoma, Kansas, Nebraska and South Dakota (Chiykowski & Chapman, 1965).

   *C. tenellus* has been suspected of undertaking long-distance migration for many years, because of the sudden appearance of adults. Except for areas in New Mexico and western Texas, most of its permanent breeding grounds in North America are west of the Continental Divide. Most influxes of this leafhopper into Grand Valley, Colorado between 1930 and 1938 were associated with southerly winds that blow when a low pressure area moves eastwards from the west coast through Nevada towards Colorado (Cook, 1967). Backtracks usually suggested an origin in southern Arizona.

   Studies with aphids such as *S. graminum*, *M. avenae*, *R. maidis* and *R. padi* indicate that they too utilise southerly winds and LLJs. Aphid migrations into South Dakota, Iowa, Illinois and other northern states in

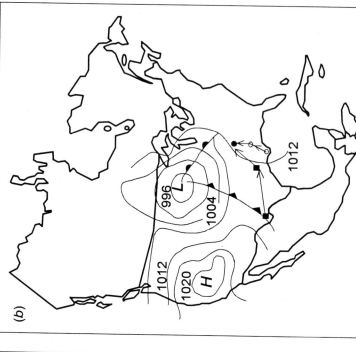

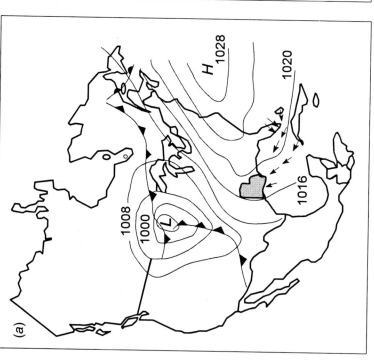

Fig. 2.7. (a) Trajectory of *A. gemmatalis* migration into Louisiana (three days of airflow), with synoptic weather map for 2 June 1990 superimposed. (After Johnson *et al.* (1991).) (b) Backtrack of *H. zea* migration into Arkansas on 31 March 1981 at two altitudes, 800 hPa (■–■) and the surface (O—O), during two 12-h night-time periods. (After Hartstack *et al.* (1982).)

the MRDB are associated with the same PTSWS weather pattern already identified for leafhopper migrations (Fig. 2.6*b*; Wallin, Peters & Johnson, 1967; Berry & Taylor, 1968; Kieckhefer, Lytle & Spuhler, 1974; Irwin & Thresh, 1988). Indications are that aphids initiate flight during the day on convective currents, but will continue flight during the night if on a favourable southerly airflow (Berry & Taylor, 1968). Migratory flights can be in early April and May, leading to infestation of seedling wheat and other cereal grains. Later flights also occur: *R. maidis* appeared suddenly at Sault Ste Marie, Ontario on 6 August 1973 during a warm southerly airflow in a PTSWS (Rose, Silversides & Lindquist, 1975): the synoptic situation was similar to that during immigrations of the same species earlier in the season. Backtracking from synoptic charts suggested that take-off occurred late the previous afternoon, and that a distance of 400 km was covered in 9 h from a source region around Green Bay, Wisconsin.

### *Moths*

There are many North American moth species that make long-distance poleward migrations in spring and summer (Table 2.1) (Johnson & Mason, 1985). A few case histories will demonstrate the similarities to the examples given for homopterans and also show the effects on migration trajectories of variations in depression track and frontal orientation.

*A. gemmatalis* is a tropical moth that cannot overwinter north of southern Florida and southern Texas (Fig. 2.2*a*; Gregory *et al.*, 1990). Its first appearance in coastal Louisiana in the springs of 1987–90 occurred during PTSWS episodes such as that illustrated in Fig. 2.7*a* (Johnson *et al.*, 1991). Although there is transport opportunity further northward, the migration appears to terminate along the Gulf Coast where suitable legume host plants grow wild (Johnson *et al.*, 1991). The long flight time over the Gulf from source regions in Cuba or southern Florida, estimated at 1.5 to 2.5 days, could explain why the migrants apparently land on reaching the coast. Transport from these source regions does not occur if the depression moves along the southern depression track, because the usual southeasterly flow is then turned sharply into Florida. Also, if the Bermuda High is displaced southwards over the Gulf of Mexico, the local airflow becomes southwesterly and there is no possibility of landfall on the Gulf coast.

*H. zea* has been suspected of long-distance migration for many years: extremely high densities in certain areas often could not be explained by

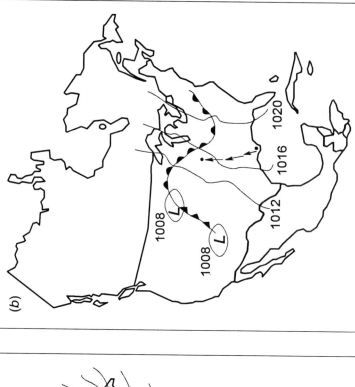

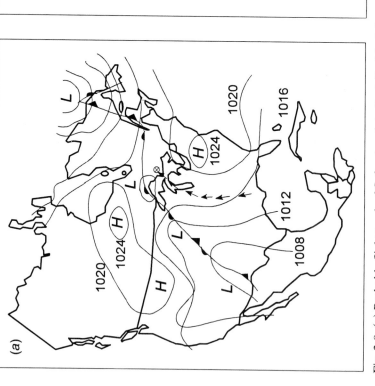

Fig. 2.8. (a) Probable flight track of *S. frugiperda* to Sault Ste Marie between 3 and 4 September 1973, with synoptic weather map for 2 September 1973 superimposed. (After Rose *et al.* (1975).) (b) Backtrack of marked *A. ipsilon* from recapture site in Iowa to release point in Crowley, Louisiana during three nights (9–12 June 1984), with synoptic weather map for 12 June 1984 superimposed. (After Showers *et al.* (1989a,b).)

local population increase; and adults were often present prior to emergence of overwintering populations (Raulston *et al.*, 1986). In the spring of 1981, large numbers of *H. zea* moths appeared in Portland, Arkansas 33 days ahead of local diapause emergence. Their appearance coincided with a southerly airflow that formed part of a PTSWS. Backtracking at 850 hPa indicated that these moths had flown from south Texas, which they may have reached after an earlier windborne migration from Mexico (Fig. 2.7*b*) (Hartstack *et al.*, 1982); however, if the backtracks were calculated with near-surface winds, the source area was southern Louisiana (Fig. 2.7*b*). A mixed population of *H. zea* and *S. frugiperda* moths that took off in the early evening of 20 June 1989 from 200 000 ha of infested maize in the lower Rio Grande Valley, following peak emergence of the two species 2–3 days earlier, was tracked by airborne radar for almost 8 h (Wolf *et al.*, 1990). The moths moved over 400 km downwind (towards the north), flying at heights of 200–700 m and speeds of 12–25 m s$^{-1}$; there was a LLJ wind maximum at the migration altitude at least part of the time.

*S. frugiperda* is a well-known migrant (Luginbill, 1928; Johnson, 1987) which has occasional outbreak years. The last major outbreak in the southeastern USA was in 1977. An intrusion of abnormally cold air penetrated as far south as Homestead, Florida in January of that year, and probably killed all populations north of this latitude. However, warm weather followed, and allowed colonisation of the southeastern USA and the buildup of populations to epidemic levels. Backtracking from Tifton, Georgia suggested transport on southeasterly winds from southern Florida, Cuba and the Bahama Islands (Westbrook & Sparks, 1986). By autumn, *S. frugiperda* is established throughout the MRDB, and migration can carry the adults as far north as Canada (Fig. 2.2*b*). Large numbers of moths appeared suddenly around Sault Ste Marie, Ontario on 3 September 1973, and began depositing egg masses (Rose *et al.*, 1975). The appearance was associated with a PTSWS and another east/west front just north of the destination area. Transport apparently occurred on a strong southerly flow, with a LLJ at an altitude of 1000–1500 m, that persisted for more than 36 h. Backtracking suggested that the migrants originated in Mississippi, where populations were at a high level (Fig. 2.8*a*).

*A. ipsilon* is an important pest of corn seedlings in the midwestern USA in spring. Increased captures in pheromone traps in Iowa during spring were associated with strong southerly winds blowing at speeds of 13–27 km h$^{-1}$ (von Kaster & Showers, 1982; Domino *et al.*, 1983). Transport

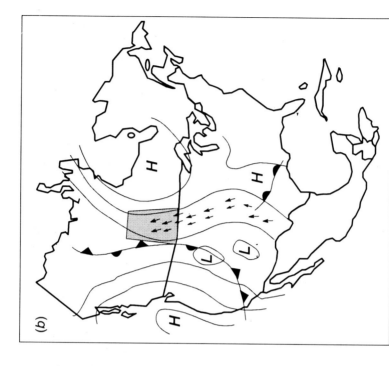

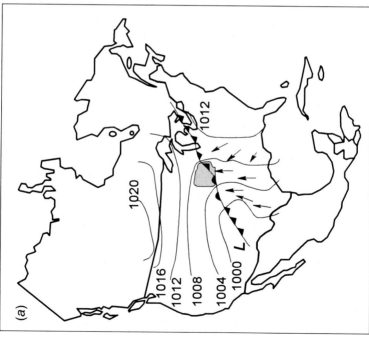

Fig. 2.9. (a) Trajectories for *P. scabra* migration into Iowa (five days of airflow), with synoptic weather map for 9 May 1979 superimposed. (After Wolf *et al.* (1987).) (b) Trajectories for *H. electellum* migration into Saskatchewan (four days of airflow), with synoptic weather map for 2 July 1976 superimposed. (After Arthur & Bauer (1981).)

from suspected source areas of southern Missouri, Arkansas, Mississippi, Louisiana and Texas occurred on southerly PTSWS airflows. Moths were suspected of flying several hundred kilometres per night and apparently needed only a few nights to complete the journey. These movements have now been confirmed with a mark–release–recapture experiment (Fig. 2.8*b*; Showers *et al.*, 1989*a,b*).

The first appearances of *P. scabra* in central Iowa during the springs of 1976–85 coincided with southerly PTSWS airflows (Fig. 2.9*a*; Wolf *et al.*, 1987). The number of potential migration days between 1 April and 30 June each year ranged from 17 to 43, and the major difference between outbreak and non-outbreak years was that there were more such days during the former than the latter.

Sudden appearances of *H. electellum* in Saskatchewan in summer (early July) during 1976–79 were associated with southerly winds and a form of the PTSWS in which the whole system is located further to the west and the frontal boundary is on a north/south axis (Fig. 2.9*b*) (Arthur & Bauer, 1981). Slow movement of the front allows for a long poleward trajectory on winds of 20–30 km h$^{-1}$ over 2–3 days.

*A. argillacea* is historically one of the most well-known migrants in North America (Riley, 1885; Parencia & Rainwater, 1964). However, it is apparently no longer a regular migrant into the MRDB, since it has not been collected in Louisiana or other southern states since 1982, possibly because of the elimination of cotton growing in the probable former source regions in northern Mexico[1].

### Autumn insect migrations on synoptic-scale winds

By autumn, several migrant species have colonised regions far to the north of their overwintering range (Fig. 2.2*a,b*). Do these populations perish, or do they make southward return migrations before winter? Equatorward airflows suitable for carrying migrants south occur during autumn. Although many entomologists have suspected that these species return to southern overwintering areas, the supporting data have been absent or circumstantial. The synoptic weather pattern that provides southward transport opportunities in autumn is now described, and the few instances of autumnal long-distance migration

---

[1] Another North American insect that appears to have ceased migrating is the grass-hopper *M. sanguinipes*. This is probably due to insecticide use: suppression is so effective that numbers never build up to the levels where migratory behaviour starts to be exhibited (Farrow, 1990).

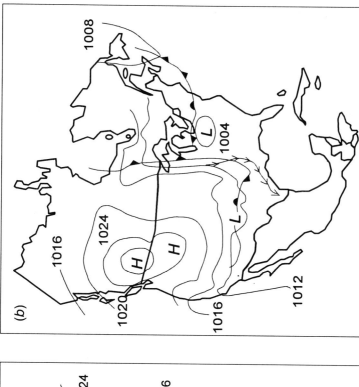

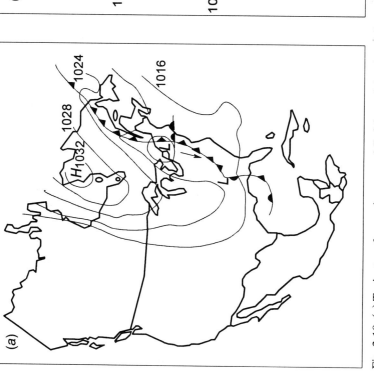

Fig. 2.10. (a) Trajectory for an airmass and *E. fabae* originating at State College, Pennsylvania immediately after a cold front passed on 6 September 1981. (After Taylor & Reling (1986).) (b) Backtrack of marked *A. ipsilon* from recapture site in Brownsville, Texas around 25 September 1987 to release point in Ankeny, Iowa on 17 September 1987, with synoptic weather map for 18 September 1987 superimposed. (After Showers *et al.* (1993).)

above the flight boundary layer reported from North America are reviewed.

### Weather systems and transporting airflows

Northerly airflows over the MRDB in autumn are produced by a synoptic weather pattern consisting of an eastward-moving depression with a high-pressure ridge to its west towards the Rocky Mountains (Fig. 2.10*a*,*b*). This pattern will be termed the equatorward-transporting synoptic weather system (ETSWS). The air-mass source region, the position of the high-pressure ridge and the location of the cold front can vary considerably, and Muller & Willis (1983) recognised three forms of this general type. The first form, Pacific high, has a very deep surface depression; the air mass originates in the Pacific and the high-pressure ridge is over the Rocky Mountains. Pacific high air is mild and dry, with winds from the west to northwest. The second form, continental high, has northerly to northeasterly winds; the air-mass source region is Canada or Alaska and the high-pressure ridge is east of the Rocky Mountains, over either Colorado or Canada. The third form, frontal overrunning, also has northeasterly winds, but is characterised by cloud and rain along the Gulf coast as the cold front becomes stationary over the Gulf of Mexico.

The opportunity for southward migration greatly increases immediately after the passage of the cold front, because northerly near-surface winds average 18–35 km h$^{-1}$ (Muller, 1985). Downwind displacement could exceed 300 km in one night. Except during ETSWS episodes, northerly winds in the northern MRDB are weak (5–10 km h$^{-1}$) in autumn. As the season progresses, the frequency of mid-latitude depressions increases to about one every 7–10 days, and the depression tracks shift southwards along with the polar front and the Bermuda High. As the weather cools, air temperatures become potentially as important determinants of transport potential as frontal frequency. Opportunity for southward migration may persist for only a few hours in the falling temperatures behind the cold front. There are probably very few opportunities for southward migration out of the northern MRDB after October.

### Homoptera

There is good evidence that *E. fabae* makes a southward migration to overwintering areas. In Pennsylvania, the largest aerial collections of

*E. fabae* in August and September of 1981 and 1982 were associated with northerly winds immediately after the passage of a cold front, even though westerly winds prevail during these months (Fig. 2.10*a*; Taylor & Reling, 1986). Even if the migration was only a few hundred kilometres, the next generation would have had time to catch another front and be transported southwards to overwintering areas. The story for *E. fabae* has recently been completed with two new findings (Taylor, 1993). First, this species overwinters much further north than previously believed: as well as southern Louisiana and Mississippi, it occurs in other south-eastern states and a narrow Atlantic coastal band up to Maryland and southern New Jersey (Fig. 2.2*a*). Second, in autumn, late-instar nymphs moult into an adult reproductive diapause. The autumn adults thus enter diapause and make a return migration to overwintering grounds that are close enough to be reached by a series of short-duration flights on northerly winds behind successive cold fronts.

### *Moths*

Sudden increases of insects in an area in association with northerly winds following the passage of a cold front have been reported sporadically. Insect densities as high as $1200/10^6 m^3$ have been observed with radar in east central Texas in September and October (Beerwinkle *et al.*, 1991). These migrants, mostly moths, were most numerous at 250 m and were moving southwards on northerly airflows. Large numbers of different kinds of insects, including many migrant moth species, were collected on oil platforms in the Gulf of Mexico up to 160 km off the Louisiana coast during September 1973 (Sparks, Jackson & Allen, 1975; Sparks *et al.*, 1986). The catches were associated with northerly airflows that became established following the passage of cold fronts. *H. zea* suddenly increased in density at Brownsville, Texas in mid-September 1984 after two cold fronts passed over (Pair *et al.*, 1987); northerly winds could have transported these moths from more northern areas of the MRDB. *P. unipuncta* adults have also been observed to increase in density in September at sites along the east coast of the USA in association with northerly winds at near-surface (100–300 m ASL) and 850-hPa (1500 m ASL) levels (McNeil, 1987). In the northern latitudes of the MRDB and southern Canada, at distances up to 1000 km north of the trap sites, *P. unipuncta* becomes sexually inactive (the females delay calling and males are not responsive to sex pheromone) in late summer (Turgeon & McNeil, 1983); this appears to be a pre-migratory condition that is entered prior to a flight to the south.

A physiological and behavioural pre-migratory condition much like that of *P. unipuncta* is also exhibited by *A. ipsilon*. In the northern latitudes of the MRDB, the males of this species are not attracted to sex-pheromone lures in late summer and autumn, and the females remain unmated (von Kaster & Showers, 1982). There is now direct evidence from a mark–release–recapture experiment that *A. ipsilon* makes a return migration on northerly near-surface (100–300 m) winds after the passage of multiple cold fronts (Showers *et al.*, 1993). In September and October 1986, two marked moths were recaptured at distances of 400 and 460 km downwind from the release point at Ankeny, Iowa. A Pacific high airflow produced 28 km h$^{-1}$ northwesterly winds at the release point, but at the latitude of central Missouri the moths ran into a southerly airflow and with no following cold fronts they were apparently stranded. However, in 1987 southward migration from Ankeny to Brownsville, Texas occurred between 17 and 25 September on 19–30 km h$^{-1}$ northerly and northeasterly near-surface airflows following the passage of two succeeding cold fronts (continental high) and two frontal overrunning conditions (Fig. 2.10*b*). Migration was southwards even though there were several opportunities for northward movements on southerly pre-frontal winds (Showers *et al.*, 1993).

The results of the *A. ipsilon* mark–release–recapture experiment demonstrate fundamental differences between spring and autumn migrations. Spring poleward migrations are usually on fast-moving southerly LLJ winds at altitudes of 200–1000 m and take 2–3 nights. Autumn equatorward migrations occur nearer the surface (100–300 m altitude) on slower northerly airflows, and are limited by how rapidly temperatures fall behind the front. They also require the passage of successive fronts to enable the migrants to reach overwintering areas.

Although there are many examples of spring migrations poleward, these are the only documented examples of autumn equatorward migrations in North America. However, this does not mean that other species do not make such flights. For example, large autumn populations of adults of the migrant moth species *P. includens*, *A. gemmatalis* (and occasionally also *A. argillacea* (Johnson *et al.*, 1985)), and *S. frugiperda* suddenly and dramatically disappear from Baton Rouge, Louisiana immediately after the passage of the first cold front in September or October (S.J. Johnson, unpublished data). It seems possible that these moths are transported southwards, probably across the Gulf of Mexico to southern Florida, Cuba, or Mexico.

Not all autumn migrations are southward. Warm southerly pre-frontal airflows associated with eastward-moving depressions occur in autumn as well as spring. There are some documented cases of wrong-way poleward migrations at this season. For example, *S. frugiperda* migrated into Sault Ste Marie on 3 September 1973 (Rose *et al.*, 1975), and *A. argillacea* has appeared in New York, where it does not even have host plants, in September and October (Grossbeck, 1911; Chapman & Lienk, 1981).

## Discussion

### *Seasonal migration and natural selection*

A life-history strategy that incorporates migration allows species to escape from unfavourable conditions in both space and time. Long-distance migrants exploiting the summer habitats of the northern MRDB must recolonise the region each year, because they cannot survive the harsh winters there. The overwintering ranges that these species leave behind apparently become unfavourable (or at least less favourable) than the northern habitats that they fly to. For all species studied, the recolonising poleward migration is achieved by transport on the strong southerly winds and LLJs found in the warm sector of a PTSWS depression. Use of this type of southerly airflow for transport is so widespread that it seems reasonable to suppose it has been adaptively selected.

Recolonisation is likely to be most successful if arrival is timed to follow the last killing frost and to coincide with the start of the spring-time flush of host-plant growth. These phenomena occur in a predictable poleward sequence each year. The frequency of different types of synoptic weather and the associated airflows also varies with the season. Features of the environment that change in a predictable way can serve as cues that forewarn of seasonal changes. In temperate North America, contrasting temperature and photoperiod changes are well-known determinants of seasonality which these migrants may use (see Chapter 13, this volume).

Except in the case of migrants like *S. frugiperda* that may be adapted to seasonally arid areas of Central America (Johnson, 1987), these north-ward migrations in spring present a potential evolutionary dilemma, the 'Pied Piper' phenomenon (Walker, 1980). If there is no return southward migration by the progeny of the spring migrants, each successive genera-tion in the southern source region would be expected to have fewer migratory genes. However, the migration trait appears to be maintained very well, suggesting that it is part of a highly successful seasonally synchronised life-history strategy (Dingle, 1982). The dilemma has apparently been resolved for a few species with the recent demonstration that they do have a southward return migration in autumn. The principal question remaining is how universal this phenomenon is. There are many examples of return southward migrations by butterfly species (Urquhart & Urquhart, 1978, 1979; Walker, 1978). Are butterflies, which normally migrate within their flight boundary layer, often in a fixed direction, really better migrants than moths, as suggested by Walker (1980)? Or are their migrations just more apparent, since they are colourful and fly by day at low altitudes?

If long-distance migrants do make return southward migrations, then they can be expected to undergo physiological and behavioural changes to enable them to deal with the unfavourable conditions that they will experience prior to moving to overwintering habitats (Tauber, Tauber & Masaki, 1986). Such changes have now been documented for *E. fabae*, *P. unipuncta* and *A. ipsilon*, and may be present in other species (see e.g. Mason, Johnson & Woodring, 1989). Identification of an autumn premigratory condition in springtime colonist species would provide strong circumstantial evidence that they do make a return migration. Another indication of an adaptation for southward migration in autumn would be a preference for flying on northerly winds at this season. *E. fabae* and *A. ipsilon* seem to exhibit just this. 'Wrong-way' (northward) flights are also known to occur in autumn, but they may be unimportant to the overall success of a species' migratory strategy so long as they are infrequent.

### *Implications for pest management*

Understanding long-distance movement is vital to improving pest-management systems. Strategies for the development of pest-resistant crops, management of pesticide use to avoid the development of resistance, and systems for pest and for crop-loss forecasting would all benefit from an improved knowledge of insect migration. The introduction of genetically engineered organisms places added emphasis on the importance of migration, because a migrant pest could greatly expand the spatial scale of the organism's impact in a short time.

We know the identity of many long-distance migrants but we rarely have the capability to forecast density, destinations and arrival dates with sufficient accuracy to be useful for initiating a pest-management action. We can forecast trajectories and arrival dates for a few species that move on PTSWS airflows in spring (Carlson *et al.*, 1991; Smelser *et al.*, 1991). In summer, however, forecasting is more difficult, because temperatures are high enough for transport to be possible on almost any type of airflow.

Even though the problems and obstacles are great, we need to push ahead with additional mark–release–recapture experiments, ground-based and airborne radar observations and meteorological studies, so that we can gain a better understanding of the entire migration process. A strategy has been developed for studying the long-distance aerial movements of insects in North America through the creation of the Alliance For Aerobiology Research (Isard & Irwin, 1993). This organisation of scientists from many disciplines aims to provide the concerted effort required to advance our understanding of the movement process and to improve our predictive capabilities.

## Acknowledgements

I would like to thank Alistair Drake, Jim Fuxa, Abner Hammond, Elson Shields, Bill Showers and Paul Taylor for reviewing earlier drafts of this manuscript and improving it with their comments. Also, a special thanks to Xikui Wei for putting the figures in their final form.

## References

Arthur, A.P. & Bauer, D.J. (1981). Evidence of the northerly dispersal of the sunflower moth by warm winds. *Environmental Entomology*, **10**, 528–33.

Barry, R.G. & Chorley, R.J. (1992). *Atmosphere, Weather and Climate*, 6th edn. London: Routledge.

Beerwinkle, K.R., Lopez, J.D., Bouse, L.F. & Brusse, J.C. (1989). Large helicopter-towed net for sampling airborne arthropods. *Transactions of the American Society of Agricultural Engineers*, **32**, 1847–52.

Beerwinkle, K.R, Lopez, J.D. Jr, Witz, J.A., Schleider, P.G. & Eyster, R.S. (1991). Seasonal radar observations of upper-air nocturnal insect activity and related meteorology in east-central Texas. In *Preprints, Tenth Conference on Biometeorology and Aerobiology, 10–13 September 1991, Salt Lake City, Utah*, pp. 107–13. Boston: American Meteorological Society.

Berry, R.E. & Taylor, L.R. (1968). High altitude migration of aphids in maritime and continental climates. *Journal of Animal Ecology*, **37**, 713–22.

Brown, C.E. (1965). Mass transport of forest tent caterpillar moths, *Malacosoma disstria* Hübner, by a cold front. *The Canadian Entomologist*, **97**, 1073–5.

Buntin, G.D., Pedigo, L.P. & Showers, W.B. (1990). Temporal occurrence of the variegated cutworm (Lepidoptera: Noctuidae) adults in Iowa with evidence for migration. *Environmental Entomology*, **19**, 603–8.

Buschman, L.L., Whitcomb, W.H., Neal, T.M. & Mays, D.L. (1977). Winter survival and hosts of the velvetbean caterpillar in Florida. *Florida Entomologist*, **60**, 267–73.

Carlson, J.D., Whalon, M.E., Landis, D.A. & Gage, S.H. (1991). Evidence for long-range transport of potato leafhopper into Michigan. In *Preprints, Tenth Conference on Biometeorology and Aerobiology, 10–13 September 1991, Salt Lake City, Utah*, pp. 123–8. Boston: American Meteorological Society.

Chapman, P.J. & Lienk, S.E. (1981). *Flight Periods of Adults of Cutworms, Armyworms, Loopers, and Others (Family Noctuidae) Injurious to Vegetable and Field Crops.* Search: Agriculture (New York) no. 14. Geneva, New York: New York State Agricultural Experiment Station.

Chiykowski, L.N. & Chapman, R.K. (1965). *Migration of the Six-spotted Leafhopper* Macrosteles fascifrons *(Stål). Part 2. Migration of the Six-spotted Leafhopper in Central North America.* Research Bulletin of the University of Wisconsin Agricultural Experiment Station no. 261, pp. 21–45.

Cook, W.C. (1967). *Life History, Host Plants, and Migrations of the Beet Leafhopper in the Western United States.* Technical Bulletin of the U.S. Department of Agriculture no. 1365. Washington: U.S. Government Printing Office.

Delong, D.M. (1938). *Biological Studies of the Leafhopper,* Empoasca fabae *as a Bean Pest.* Technical Bulletin of the U.S. Department of Agriculture no. 618. Washington: U.S. Government Printing Office.

Dingle, H. (1982). Function of migration in the seasonal synchronization of insects. *Entomologia Experimentalis et Applicata*, **31**, 36–48.

Domino, R.P. Showers, W.B., Taylor, S.E. & Shaw, R.H. (1983). Spring weather pattern associated with suspected black cutworm moth (Lepidoptera: Noctuidae) introduction to Iowa. *Environmental Entomology*, **12**, 1863–71.

Dorst, H.E. & Davis, E.W. (1937). Tracing long-distance movement of beet leafhopper in the desert. *Journal of Economic Entomology*, **30**, 948–54.

Drake, V.A. & Farrow, R.A. (1988). The influence of atmospheric structure and motions on insect migration. *Annual Review of Entomology*, **33**, 183–210.

Farrow, R.A. (1990). Flight and migration in acridoids. In *Biology of Grasshoppers*, ed. R.F. Chapman & A. Joern, pp. 227–314. New York: John Wiley & Sons.

Felt, E.P. (1928). *Dispersal of Insects by Aircurrents.* New York State Museum Bulletin no. 274, pp. 59–129.

Felt, E.P. (1937). Dissemination of insects by air currents. *Journal of Economic Entomology*, **30**, 458–61.

Forbes, W.T.M. (1954). *Lepidoptera of New York and Neighboring States. Noctuidae. Part 3.* Cornell University Agricultural Experiment Station Memoir no. 329. Ithaca: New York State College of Agriculture.

Ford, B.J., Strayer, J.R., Reid, J. & Godfrey, G.L. (1975). *The Literature of Arthropods Associated with Soybeans IV. A Bibliography of the Velvetbean Caterpillar,* Anticarsia gemmatalis Hübner *(Lepidoptera: Noctuidae).* Biological Notes Illinois Natural History Survey no. 92. Urbana: State of Illinois.

Glick, P.A. (1939). *The Distribution of Insects, Spiders, and Mites in the Air.* Technical Bulletin of the U.S. Department of Agriculture no. 673. Washington: U.S. Government Printing Office.

Glick, P.A. (1967). *Aerial Dispersal of the Pink Bollworm in the United States and Mexico.* Production Research Report of the U.S. Department of Agriculture no. 96. Washington: U.S. Government Printing Office.

Grant, R.H. & Seevers, K.P. (1989). Local and long-range movement of adult western corn rootworm (Coleoptera: Chrysomelidae) as evidenced by washup along southern Lake Michigan shores. *Environmental Entomology*, **18**, 266–72.

Gregory, B.M. Jr, Johnson, S.J., Levins, A.W., Hammond, A.M. & Delgado-Salinas, A. (1990). A midlatitude survival model of *Anticarsia gemmatalis* (Lepidoptera: Noctuidae). *Environmental Entomology*, **19**, 1017–23.

Grossbeck, J.A. (1911). Migration of *Alabama argillacea* Hübner. *Journal of the New York Entomological Society*, **19**, 259–61.

Hartstack, A.W., Lopez, J.D., Muller, R.A., Sterling, W.L., King, E.G., Witz, J.A. & Eversull, A.C. (1982). Evidence of long range migration of *Heliothis zea* (Boddie) into Texas and Arkansas. *The Southwestern Entomologist*, **7**, 188–201.

Hendrie, L.K., Irwin, M.E., Liquido, N.J., Ruesink, W.G., Mueller, E.A., Voegtlin, D.J., Achtemeier, G.L., Steiner, W.M. & Scott, R.W. (1985). Conceptual approach to modeling aphid migration. In *The Movement and Dispersal of Agriculturally Important Biotic Agents*, ed. D.R. MacKenzie, C.S. Barfield, G.G. Kennedy, R.D. Berger & D.J. Taranto, pp. 541–82. Baton Rouge: Claitor's.

Hollinger, S.E., Sivier, K.R., Irwin, M.E. & Isard, S.A. (1991). A helicopter-mounted isokinetic aerial insect sampler. *Journal of Economic Entomology*, **84**, 476–83.

Huff, F.A. (1963). Relation between leafhopper influxes and synoptic weather conditions. *Journal of Applied Meteorology*, **2**, 39–43.

Irwin, M.E. & Thresh, J.M. (1988). Long-range aerial dispersal of cereal aphids as virus vectors in North America. *Philosophical Transactions of the Royal Society of London B*, **321**, 421–46.

Isard, S.A. & Irwin, M.E. (1993). A strategy for studying the long-distance aerial movements of insects. *Journal of Agricultural Entomology*, **10**, 283–97.

Johnson, C.G. (1969). *The Migration and Dispersal of Insects by Flight.* London: Methuen.

Johnson, S.J. (1987). Migration and the life history strategy of the fall armyworm, *Spodoptera frugiperda* in the western hemisphere. In *Recent Advances in Research on Tropical Entomology*, ed. M.F.B. Chaudhury, pp. 543–9. *Insect Science and its Application*, vol. 8, Special Issue. Nairobi: ICIPE Science Press.

Johnson, S.J., Foil, L.D., Hammond, A.M., Sparks, T.C. & Church, G.E. (1985). Effects of environmental factors on phase variation in larval cotton leafworms, *Alabama argillacea* (Lepidoptera: Noctuidae). *Annals of the Entomological Society of America*, **78**, 35–40.

Johnson, S.J., Gregory, B.M. Jr, Hsu, S.A. & Hammond, A.M. (1991). Migration paradigm for the velvetbean caterpillar. In *Preprints, Tenth Conference on Biometeorology and Aerobiology, 10–13 September 1991, Salt Lake City, Utah*, pp. 88–92. Boston: American Meteorological Society.

Johnson, S.J. & Mason, L.J. (1985). The Noctuidae: a case history. In *The Movement and Dispersal of Agriculturally Important Biotic Agents*, ed. D.R. MacKenzie, C.S. Barfield, G.G. Kennedy, R.D. Berger & D.J. Taranto, pp. 421–33. Baton Rouge: Claitor's.

Kieckhefer, R.W., Lytle, W.F. & Spuhler, W. (1974). Spring movement of cereal aphids into South Dakota. *Environmental Entomology*, **3**, 347–50.

Lingren, P.D., Henneberry, T.J. & Sparks, A.N. (1979). Current knowledge and research on movement of the cabbage looper and related looper species. In *Movement of Highly Mobile Insects: Concepts and Methodology in Research*, ed. R.L. Rabb & G.G. Kennedy, pp. 394–405. Raleigh, North Carolina: University Graphics.

Luginbill, P. (1928). *The Fall Armyworm*. Technical Bulletin of the U.S. Department of Agriculture no. 34. Washington: U.S. Government Printing Office.

Mason, L.J., Johnson, S.J. & Woodring, J.P. (1989). Seasonal and ontogenetic examination of the reproductive biology of *Pseudoplusia includens*. *Environmental Entomology*, **18**, 980–5.

McNeil, J.N. (1987). The true armyworm, *Pseudaletia unipuncta*: a victim of the Pied Piper or a seasonal migrant? In *Recent Advances in Research on Tropical Entomology*, ed. M.F.B. Chaudhury, pp. 591–7. *Insect Science and its Application*, vol. 8, Special Issue. Nairobi: ICIPE Science Press.

Medler, J.T. (1957). Migration of the potato leafhopper – a report on a cooperative study. *Journal of Economic Entomology*, **50**, 493–7.

Medler, J.T. (1962). Long-range displacement of Homoptera in the central United States. *Transactions of the XI International Congress of Entomology, Vienna, 17–25 August 1960*, vol. 3, pp. 30–5.

Medler, J.T. & Smith, P.W. (1960). Greenbug dispersal and distribution of barley yellow dwarf virus in Wisconsin. *Journal of Economic Entomology*, **53**, 473–4.

Mitchell, E.R. (1979). Migration by *Spodoptera exigua* and *S. frugiperda*, North American style. In *Movement of Highly Mobile Insects: Concepts and Methodology in Research*, ed. R.L. Rabb & G.G. Kennedy, pp. 386–93. Raleigh, North Carolina: University Graphics.

Mitchell, E.R., Chalfant, R.B., Green, G.L. & Creighton, C.S. (1975). Soybean looper: populations in Florida, Georgia, and South Carolina, as determined with pheromone baited B.L. traps. *Journal of Economic Entomology*, **68**, 747–50.

Muller, R.A. (1985). The potential for the atmospheric transport of moths from the perspective of synoptic climatology. In *The Movement and Dispersal of Agriculturally Important Biotic Agents*, ed. D.R. MacKenzie, C.S. Barfield, G.G. Kennedy, R.D. Berger & D.J. Taranto, pp. 179–202. Baton Rouge: Claitor's.

Muller, R.A. & Tucker, N. (1986). Climatic opportunities for the long-range migration of moths. In *Long-Range Migration of Moths of Agronomic Importance to the United States and Canada: Specific Examples of Occurrence and Synoptic Weather Patterns Conducive to Migration. ARS-43*, ed. A.N. Sparks, pp. 61–83. Washington, DC: United States Department of Agriculture, Agricultural Research Service.

Muller, R.A. & Willis, J.E. (1983). *New Orleans Weather: A Climatology by Means of Synoptic Weather Types*. Miscellaneous Publication 83–1. Baton Rouge: Louisiana State University School of Geoscience.

Nichiporick, W. (1965). The aerial migration of the six-spotted leafhopper and the spread of the virus disease aster yellows. *International Journal of Biometeorology*, **9**, 219–27.

Nielson, M.W. (1968). *The Leafhopper Vectors of Phytopathogenic Viruses (Homoptera, Cicadellidae) Taxonomy, Biology, and Virus Transmission*.

Technical Bulletin of the U.S. Department of Agriculture no. 1382. Washington: U.S. Government Printing Office.

Oku, T. & Kobayashi, T. (1978). Migratory behaviors and life-cycle of noctuid moths (Insecta, Lepidoptera), with notes on the recent status of migrant species in northern Japan. *Bulletin of the Tohoku National Agricultural Experiment Station* no. 58, 97–209. In Japanese, English summary.

Pair, S.D., Raulston, J.R., Rummel, D.R., Westbrook, J.K., Wolf, W.W., Sparks, A.N. & Schuster, M.F. (1987). Development and production of corn earworm and fall armyworm in the Texas High Plains: evidence for reverse fall migration. *The Southwestern Entomologist*, **12**, 89–99.

Parencia, C.R. & Rainwater, C.F. (1964). First finding of cotton leafworm larvae in the United States 1922 to 1963. *Journal of Economic Entomology*, **57**, 432.

Parker, J.R., Newton, R.C. & Shotwell, R.L. (1955). *Observations on the Mass Flight and Other Activities of the Migratory Grasshopper.* Technical Bulletin of the U.S. Department of Agriculture no. 1109. Washington: U.S. Government Printing Office.

Pedgley, D.E. (1982). *Windborne Pests and Diseases: Meteorology of Airborne Organisms.* Chichester: Ellis Horwood.

Pedigo, L.P. (1994). Green Cloverworm. In *Handbook of Soybean Insect Pests*, ed. L.E. Higley & D.J. Boethel, pp. 61–2. Lanham, Maryland: Entomological Society of America.

Pienkowski, R.L. & Medler, J.T. (1964). Synoptic weather conditions associated with long-range movement of the potato leafhopper, *Empoasca fabae*, into Wisconsin. *Annals of the Entomological Society of America*, **57**, 588–91.

Pruess, K.P. (1967). Migration of the army cutworm, *Chorizagrotis auxiliaris* (Lepidoptera: Noctuidae). I. Evidence for a migration. *Annals of the Entomological Society of America*, **60**, 910–20.

Pyenson, L. (1940). The cotton leafworm in the western hemisphere. *Journal of Economic Entomology*, **33**, 830–3.

Rabb, R.L. & Kennedy, G.G. (ed.) (1979). *Movement of Highly Mobile Insects: Concepts and Methodology in Research.* Raleigh, North Carolina: University Graphics.

Rabb, R.L. & Stinner, R.E. (1979). The role of insect dispersal and migration in population processes. In *Radar, Insect Population Ecology, and Pest Management.* NASA Conference Publication 2070, ed. C.R. Vaughn, W.W. Wolf & W. Klassen, pp. 3–16. Wallops Island, Virginia: NASA Wallops Flight Centre.

Raulston, J.R. (1979). *Heliothis virescens* migration. In *Movement of Highly Mobile Insects: Concepts and Methodology in Research*, ed. R.L. Rabb & G.G. Kennedy, pp. 412–19. Raleigh, North Carolina: University Graphics.

Raulston, J.R., Pair, S.D., Pedraza Martinez, F.A., Westbrook, J., Sparks, A.N. & Sanchez Valdez, V.M. (1986). Ecological studies indicating the migration of *Heliothis zea, Spodoptera frugiperda*, and *Heliothis virescens* from northeastern Mexico and Texas. In *Insect Flight: Dispersal and Migration*, ed. W. Danthanarayana, pp. 204–20. Berlin: Springer-Verlag.

Reid, W.J. Jr & Cuthbert, F.P. Jr (1956). Biology studies of the pickleworm. *Journal of Economic Entomology*, **49**, 870–3.

Riley, C.V. (1885). *Fourth Report of the United States Entomological Commission, being a Revised Edition of Bulletin no. 3, and the Final*

*Report on the Cotton Worm together with a Chapter on the Boll Worm.*
Washington: Government Printing Office.

Rings, R.W., Musick, G.J., Keaster, A.J. & Luckman, W.H. (1976).
Geographical distribution and economic importance of the black, glassy,
and bronzed cutworms. *U.S. Department of Agriculture Cooperative Plant
Pest Report*, **1**, 173–9.

Rose, A.H., Silversides, R.H. & Lindquist, O.H. (1975). Migration flight by
an aphid, *Rhopalosiphum maidis* (Hemiptera: Aphididae), and a noctuid,
*Spodoptera frugiperda* (Lepidoptera: Noctuidae). *The Canadian
Entomologist*, **107**, 567–76.

Showers, W.B., Keaster, A.J., Raulston, J.R., Hendrix, W.H. III, Derrick,
M.E., McCorcle, M.D., Robinson, J.F., Way, M.O., Wallendorf, M.J. &
Goodenough, J.L. (1993). Mechanism of southward migration of a
noctuid moth [*Agrotis ipsilon* (Hufnagel)]: a complete migrant. *Ecology*,
**74**, 2303–14.

Showers, W.B., Smelser, R.B., Keaster, A.J., Whitford, F., Robinson, J.F.,
Lopez, J.D. & Taylor, S.E. (1989a). Recapture of marked black cutworm
(Lepidoptera: Noctuidae) males after long-range transport.
*Environmental Entomology*, **18**, 447–58.

Showers, W.B., Whitford, F., Smelser, R.B., Keaster, A.J., Robinson, J.F.,
Lopez, J.D. & Taylor, S.E. (1989b). Direct evidence for meteorologically
driven long-range dispersal of an economically important moth. *Ecology*,
**70**, 987–92.

Smelser, R.B., Showers, W.B., Shaw, R.H. & Taylor, S.E. (1991).
Atmospheric trajectory analysis to project long-range migration of black
cutworm (Lepidoptera: Noctuidae) adults. *Journal of Economic
Entomology*, **84**, 879–85.

Smith, C.E. & Allen, N. (1932). The migratory habit of the spotted cucumber
beetle. *Journal of Economic Entomology*, **25**, 53–7.

Smith, C.F. & Parron, C.S. (1978). *An Annotated list of Aphididae
(Homoptera) of North America.* Technical Bulletin of the North Carolina
Agricultural Experiment Station no. 255. Raleigh: North Carolina
Agricultural Experiment Station.

Smith, D.B. & Sears, M.K. (1982). Evidence for dispersal of diamondback
moth, *Plutella xylostella* (Lepidoptera: Plutellidae), into southern
Ontario. *Proceedings of the Entomological Society of Ontario*, **113**, 21–7.

Sparks, A.N., Jackson, R.D. & Allen, C.L. (1975). Corn earworms: capture
of adults in light traps on unmanned oil platforms in the Gulf of Mexico.
*Journal of Economic Entomology*, **68**, 431–2.

Sparks, A.N., Jackson, R.D., Carpenter, J.E. & Muller, R.A. (1986). Insects
captured in light traps in the Gulf of Mexico. *Annals of the Entomological
Society of America*, **79**, 132–9.

Tauber, M.J., Tauber, C.A. & Masaki, S. (1986). *Seasonal Adaptations of
Insects.* Oxford: Oxford University Press.

Taylor, L.R. (1958). Aphid dispersal and diurnal periodicity. *Proceedings of
the Linnean Society of London*, **169**, 67–73.

Taylor, P.S. (1993). *Phenology of Empoasca fabae (Harris) (Homoptera:
Cicadellidae), and Development of Springtime Migrant Source
Populations.* PhD thesis, Cornell University, Ithaca, New York.

Taylor, R.A.J. & Reling, D. (1986). Preferred wind direction of long-distance
leafhopper (*Empoasca fabae*) migrants and its relevance to the return
migration of small insects. *Journal of Animal Ecology*, **55**, 1103–14.

Turgeon, J.J. & McNeil, J.N. (1983). Modification in the calling behaviour of *Pseudaletia unipuncta* (Haw.) (Lepidoptera: Noctuidae), induced by temperature conditions during pupal and adult development. *The Canadian Entomologist*, **115**, 1015–22.

Urquhart, F.A. & Urquhart, N.R. (1978). Autumnal migration routes of the eastern population of the monarch butterfly (*Danaus p. plexippus* L.; Danaidae; Lepidoptera) in North America to the overwintering site in the Neo-volcanic Plateau of Mexico. *Canadian Journal of Zoology*, **56**, 1759–64.

Urquhart, F.A. & Urquhart, N.R. (1979). Vernal migration of the monarch butterfly (*Danaus p. plexippus*, Lepidoptera: Danaidae) in North America from the overwintering site in the Neo-volcanic Plateau of Mexico. *The Canadian Entomologist*, **11**, 15–18.

Visher, S.S. (1966). *Climatic Atlas of the United States*. Cambridge: Harvard University Press.

von Kaster, L. & Showers, W.B. (1982). Evidence of spring immigration and autumn reproductive diapause of the adult black cutworm in Iowa. *Environmental Entomology*, **11**, 306–12.

Walker, T.J. (1978). Migration and re-migration of butterflies through north peninsular Florida: quantification with malaise traps. *Journal of the Lepidopterists' Society*, **32**, 178–90.

Walker, T.J. (1980). Migrating Lepidoptera: are butterflies better than moths? *Florida Entomologist*, **63**, 79–98.

Wallin, J.R., Peters, D. & Johnson, L.C. (1967). Low-level jet winds, early cereal aphid and barley yellow dwarf detection in Iowa. *Plant Disease Reporter*, **51**, 527–30.

Watson, J.R. (1916). Life history of the velvetbean caterpillar (*Anticarsia gemmatalis* Hübner). *Journal of Economic Entomology*, **9**, 521–8.

Wellington, W.G. (1954). Atmospheric circulation processes and insect ecology. *The Canadian Entomologist*, **86**, 312–33.

Westbrook, J.K. & Sparks, A.N. (1986). The role of atmospheric transport in the economic fall armyworm (Lepidoptera: Noctuidae) infestations in the southeastern United States in 1977. *Florida Entomologist*, **69**, 492–502.

Wolf, R.A., Pedigo, L.P., Shaw, R.H. & Newsom, L.D. (1987). Migration/transport of the green cloverworm, *Plathypena scabra*, (F.) (Lepidoptera: Noctuidae), into Iowa as determined by synoptic scale weather patterns. *Environmental Entomology*, **16**, 1169–74.

Wolf, W.W., Westbrook, J.K., Raulston, J., Pair, S.D. & Hobbs, S.E. (1990). Recent airborne radar observations of migrant pests in the United States. *Philosophical Transactions of the Royal Society of London B*, **328**, 619–30.

Wolf, W.W., Westbrook, J.K. & Sparks, A.N. (1986). Relationship between radar entomological measurements and atmospheric structure in south Texas during March and April 1982. In *Long-Range Migration of Moths of Agronomic Importance to the United States and Canada: Specific Examples of Occurrence and Synoptic Weather Patterns Conducive to Migration. ARS-43*, ed. A.N. Sparks, pp. 84–97. Washington, DC: United States Department of Agriculture, Agricultural Research Service.

# 3

# Migration of the Brown Planthopper *Nilaparvata lugens* and the White-backed Planthopper *Sogatella furcifera* in East Asia: the role of weather and climate

## R. KISIMOTO AND K. SOGAWA

### Introduction

Of over one hundred species of planthoppers (Homoptera: Delphacidae) occurring in East Asia, only three, the Brown Planthopper *Nilaparvata lugens* (BPH), the White-backed Planthopper *Sogatella furcifera* (WBPH), and the Small Brown Planthopper *Laodelphax striatellus*, both reproduce successfully on rice and migrate long distances (Kisimoto, 1981). Two of these species, BPH and WBPH, are serious pests of rice throughout the region. Outbreaks of both species have been recorded in Japan since the mid-1700s (Suenaga & Nakatsuka, 1958), but in the tropics, where BPH is the more serious pest, outbreaks did not occur until the mid-1960s (Kisimoto, 1984).

The significance of migration in initiating BPH and WBPH outbreaks was first tentatively recognised by Murata & Hirano (1929) who proposed that infestations in the Japanese mainland (i.e. the four main islands of Japan: Hokkaido, Honshu, Kyushu and Shikoku (Fig. 3.1)) are initiated each year by immigrations from the south during the *Bai-u* (*Mei-yu* in the Chinese literature) rainy season (June–July). This hypothesis was revived following the observation of a mass of WBPH and a few BPH swarming around a weather ship at location 'Tango' (29° N, 135° E) in the Pacific Ocean (Fig. 3.1) in July 1967 (Asahina & Turuoka, 1968). At about the same time, catches of planthoppers in pole-mounted tow-nets set at Chikugo (33°12′ N, 130°30′ E), on the western coast of Kyushu, were found to be correlated with warm and moist southwesterly or west-south-westerly winds that occurred in the warm sectors of northeastward-moving depressions (Kisimoto, 1976). Systematic trapping on ships plying the East China Sea (Kisimoto, 1971, 1981; Mochida, 1974), and further observations at 'Tango' (Asahina & Turuoka, 1970; Itakura, 1973), soon established the migrant status of both species in the central part of East Asia.

Over the last two decades, planthopper migration has been studied in several regions of East Asia, with a variety of techniques (Kisimoto 1987, 1991). The most important of these movements are now sufficiently well understood that forecasts of planthopper invasions are provided on an operational basis in both China (Chapter 17, this volume) and Japan (Chapter 18, this volume). In this chapter, our present understanding of planthopper migration, and its relation to weather and climate, is described for the whole of the East Asian region.

### Planthoppers as pests of rice in East Asia

Both as nymphs and adults, BPH and WBPH damage rice by sucking the plant sap, causing 'hopperburn' (browning of the leaves or, when damage

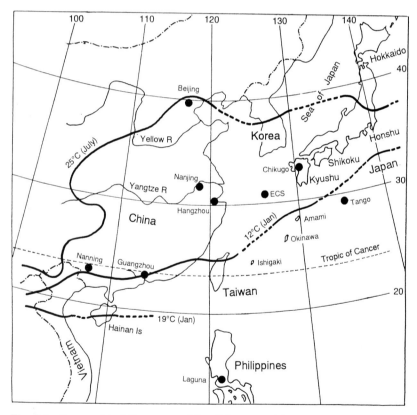

Fig. 3.1. Geographical features and localities named in the text, and monthly average temperatures. ECS and Tango are localities where ocean weather ships keep station.

is severe, withering of the whole plant). In addition, WBPH females damage young rice plants by laying their eggs inside the tissue of the leaf sheath, while in the tropics (but only sporadically in temperate East Asia) BPH transmits the viruses of Rice Grassy-Stunt and Rice Ragged-Stunt diseases.

The development of BPH outbreaks in the tropics followed the introduction of modern rice varieties: severe damage was first recorded in 1964 in paddy-fields of the International Rice Research Institute, Laguna, Philippines, where new high-yielding varieties were being developed. Devastating hopperburns on an unprecedented scale soon followed through much of tropical Asia wherever modern varieties had been introduced. Changes in rice-cropping practices such as the introduction of staggered rice growing with dense transplanting, heavy nitrogen application, and intensive application of insecticides, have also influenced the occurrence of both species.

In temperate East Asia, rice is grown once a year: it is transplanted mostly in May–June and harvested in September–October. In the subtropics, there are two crops each year and these may be grown either consecutively or 'staggered' (i.e. the second crop is transplanted before the first is harvested): first-crop rice is transplanted in March–April and second-crop in July–August. In the tropics, rice is grown at any season depending on the water supply (rainfall or irrigation).

Rice crops are vulnerable to planthopper invasion during the first 1–2 months after transplanting. Young rice plants are apparently selected preferentially by landing planthoppers, and the immigrant populations have time to multiply to damage-causing densities before the plants mature and become tolerant to feeding. The economic significance of a planthopper immigration therefore depends on the growth stage of the rice crops in the invaded region. With WBPH, immigrants are often sufficiently numerous to cause damage themselves, or within 1–2 generations. BPH migrates in smaller numbers and damage occurs after 2–3 generations.

## Planthopper migratory flight

### *Flight capacity*

BPH and WBPH have both long-winged (macropterous) and short-winged (brachypterous) forms, although in WBPH brachyptery is confined to the females. Brachypterous females cannot fly but start to lay their eggs sooner than macropters and therefore multiply more rapidly. Most WBPH adults are long-winged and move actively. In BPH, there is considerable geographical variation in the proportion of macropters and brachypters in both males and females (Nagata & Masuda, 1980; Iwanaga, Nakasuji & Tojo, 1987; Kisimoto & Rosenberg, 1994). Some populations in the Philippines, Indonesia and southern Vietnam produce a very high proportion of brachypters, whereas many populations in Japan, Taiwan, and the coastal region of China produce the long-winged form in high or medium proportions. In Japan, most female and a proportion of male BPH are short-winged in the first and second generation, but long-winged forms appear in the third generation as the rice plant starts to senesce. In WBPH, populations from the Philippines produce a higher proportion of brachypters than populations from Thailand, and Thai populations in turn produce them in higher proportions than populations from Japan.

Tethered-flight experiments in the laboratory have shown that macropterous planthoppers maintain active wing beating for long periods. Female BPH that fly for more than 30 min continue on average for 12 h (maximum 23 h), and males for 9 h; for WBPH, the equivalent figures are 11 h (23 h maximum) and 7.5 h, respectively (Ohkubo, 1981). The values for BPH are similar to the 10–14 h estimated from measured fuel-consumption rates (Padgham, 1983*a*). The pattern of post-flight fuel synthesis suggests that they are unlikely to make more than one major flight (Padgham, 1983*b*). (However, the possibility that planthoppers alight temporarily during migrations over land areas has not been adequately investigated; at sea, they have been observed to take off again after landing on the surface (Asahina & Turuoka, 1968).) Flight duration is reduced when humidity is low (Ohkubo, 1981), and may be limited by water as much as by fuel (Hirao, 1979). Little is known about how the migratory capacity of planthoppers varies between different regions and seasons.

### Migratory flight and atmospheric conditions

All three stages of a planthopper migration – takeoff, horizontal displacement, and landing – are affected by the state of the atmosphere at the time of the flight. In particular, because of their low airspeeds[1], planthoppers

---

[1] Casual laboratory and field observations suggest horizontal airspeeds do not exceed ~0.5 m s$^{-1}$ (R. Kisimoto, unpublished data).

drift downwind even when there is only a light breeze. They are therefore completely dependent on transport by the wind for making long-distance movements.

## Takeoff

Young adult planthoppers of the long-winged form take off actively after a teneral period of 1–2 days. The insects move to the tip of the rice plant and, if conditions are calm, fly vertically upwards (Ohkubo & Kisimoto, 1971). Nearby insects often launch themselves into the air simultaneously. Takeoff occurs at both dusk and dawn during summer. In the cooler weather of autumn, the dawn takeoff is much reduced; by late autumn it is completely suppressed, and the dusk takeoff occurs earlier (i.e. at a higher light intensity) than in summer (Ohkubo & Kisimoto, 1971; Riley, Reynolds & Farrow, 1987; Riley *et al.*, 1991). The dusk takeoff is also reduced in windy conditions: it is completely inhibited when the windspeed (measured at 10 m) exceeds $\sim$3 m s$^{-1}$.

## Horizontal displacement

Because it occurs at altitudes up to 2.5 km, the main displacement stage of planthopper migration can be studied directly only from aircraft (by sampling with interception traps) or by radar. However, major displacements can sometimes be inferred from catches in a trap network, and mark–recapture experiments have provided information about the total displacement achieved.

Trapping from an aircraft flying over southern China (Guangzhou and Nanning districts and Hainan Island) between late April and early May showed that both BPH and WBPH were migrating at altitudes of 0.5–2 km when the wind was from the southwest, but that only WBPH were present in southeasterlies (Zhu *et al.*, 1982). In central China ($\sim$27–30° N and $\sim$115–117° E), BPH and WBPH were found to be migrating at altitudes of 0.3–2.5 km in mid- to late July and at 0.1–1.5 km between late September and late October (Dung, 1981). In July, both species were caught when the wind was from the southwest, south, southeast, or east, with the biggest catches being made in southwesterlies that had windspeeds of 8–14 m s$^{-1}$ at $\sim$1.5 km. There were no catches in northwesterly or northerly winds, presumably because areas to the north of the sampling site were free of planthoppers at that season. In October, planthoppers were caught when the wind was from almost any direction, but the largest catches were made when northeasterlies with speeds of 8–12 m s$^{-1}$ at 1.5 km were blowing. Therefore in autumn the main source of migrants appears to be towards the northeast. The density of migrants was greatest at 1.5–2 km in July and 0.5–1.0 km in October. Densities of both species were typically in the range $10^{-4}$–$10^{-3}$ m$^{-3}$, with WBPH outnumbering BPH in July but BPH being more numerous in autumn. The highest density encountered, $4.8 \times 10^{-3}$ m$^{-3}$ for BPH, was recorded during a flight across a cold front at a height of 1.5 km. However, samples taken in rougher weather tended to be dominated by WBPH. The samples also frequently contained *L. striatellus*, *Sogatella vibix* (a planthopper that feeds on weeds growing near rice paddies), and many other small insects as well as moths, flies, etc. Both WBPH and BPH have been caught in July at altitudes between 0.1 and 2 km over the East China Sea to the west of Kyushu (Research Council, Ministry of Agriculture, Forestry and Fisheries, Japan, 1989).

Radar observations (accompanied by sampling with a kite-borne net) near Nanjing, China, during late September showed that migrating BPH frequently form a dense layer concentration above the top of the surface temperature inversion, at altitudes of 0.4–1.0 km. The layer appears 1–2 h after dusk and persists for several hours (Riley *et al.*, 1991). The temperature at the top of the layer (the flight ceiling) was ~16 °C, which corresponds closely to the laboratory measurement of 16.5 °C for the active-flight threshold of this species (Ohkubo, 1981). If flights lasted all night (~12 h), the 7 m s$^{-1}$ winds that were observed regularly during this period would have resulted in displacements of ~300 km overnight. Similar layers also formed after a mass takeoff at dawn but lasted only 1–2 h. In contrast, at Laguna, Philippines during the dry season, the radar showed that the small insects flying (mainly rice hemipterans, including BPH and WBPH) were confined to altitudes below ~400 m during the dusk and dawn flights, both of which lasted only ~30 min (Riley *et al.*, 1987).

A mass immigration of BPH and WBPH into the Japanese mainland in late June 1969 provided evidence of movements of ~500 km in a night in winds associated with a depression that was moving rapidly northeast-wards (see Fig. 3.4*b*, below) (Kisimoto, 1976). Immigrants were first caught in the pole-mounted net traps at Chikugo (130°30′ E) in the late afternoon of 25 June, when the windspeed was ~30 km h$^{-1}$. Overnight catches in light traps showed that they had spread throughout the western half of the Japanese mainland, as far as 135° E, by the following dawn. A day later they were also to be found in the eastern part of the mainland, to ~140° E. Following a similar immigration early the next month, densities in rice paddies (estimated by sweep-net sampling) decreased linearly from west to east over a distance of 1000–1500 km (Kisimoto, 1979). BPH densities were lower than those of WBPH, and they decreased more rapidly with distance towards the east, which suggests that WBPH is more migratory than BPH. Early in the migration season, if the temperature is in the range 17–20 °C, only WBPH arrive (Kisimoto, 1976).

In southern China (between 22–26° N and 110–114° E), four BPH and one WBPH marked by spraying field populations with a dye solution before takeoff were captured in light traps between ~300 and 720 km northeast of the marking site during trials in June and July (Nanjing Agricultural College Department of Plant Protection. . . , 1981). A similar trial in October resulted in a single BPH being recaptured after a flight of ~180 km to the southwest.

## Landing

Little is known about what causes migrating planthoppers to land. Landing appears to be associated with rainfall during the *Bai-u* season, and often occurs at cold fronts as these move eastwards (Kisimoto, 1971, 1976; Jiang *et al.*, 1981, 1982). Turbulence at the front may suppress active flight and force the migrants to land: laboratory studies show that tethered BPH stop wing-beating when exposed to airflows faster than 5.5 m s$^{-1}$. The often inclement weather associated with landing contrasts with the near-calm conditions in which takeoff occurs.

During the *Bai-u* rains, BPH, WBPH and other planthoppers have been found drowned at various locations in the East China Sea (Mochida, 1974). Large numbers of immigrants sometimes arrive in non-host habitats (e.g. lawns and vegetable plots) and on remote islands where no rice grows. These observations suggest that migrations may end almost randomly if the conditions for flight become unfavourable. A similar outcome may result if the airflow transporting the migrants slows: swarms have been observed around ships during periods of light winds (Asahina & Turuoka, 1968), and trap catches at sea are increased when windspeeds at the height of migration fall (Kisimoto, 1991).

Topography also influences where planthoppers land: in Japan, immigration densities are higher on the eastern (lee) sides of hills and at the upper ends of windward-facing valleys (Noda & Kiritani, 1989).

## The climate of East Asia and its relation to planthopper migration

The winter climate of East Asia is dominated by the Siberian continental anticyclone. By January, dry and cold northwesterly to northeasterly airflows spread out over the whole region, extending as far as Korea, Japan and Taiwan. In Korea, the Japanese Mainland, and central and northern China, neither BPH nor WBPH can survive the low temperatures and the absence of rice plants during winter. The 12 °C January average isotherm, which marks the northern limit of overwintering for BPH (Cheng *et al.*, 1979; Chapter 17, this volume), lies to the south of Kyushu and passes through the southern coastal region of the Chinese mainland at about the latitude of the Tropic of Cancer (Fig. 3.1). WBPH can overwinter slightly further north, to ~26° N (National Coordinated Research Group for White-backed Planthoppers, 1981). Overwintering on ratoon or volunteer rice plants is physiologically possible in the subtropics, at least during less severe winters. It occurs in northern Vietnam (Ich, 1991), southern China (Zhao-ching District Crop Pest Prognostic Station. . . , 1977; Cheng *et al.*, 1979), Taiwan (Cheng & Lu, 1990), and Okinawa (Tsurumachi & Yasuda, 1989), but only in the first of these does it appear to lead to significant populations later in the season.

BPH and WBPH have their permanent breeding areas, where they can reproduce even in winter and populations are present in fluctuating numbers throughout the year, in the tropics. For BPH, the northern limit follows the 19 °C January isotherm, which passes through northern Vietnam and Hainan Island, and to the north of the Philippines (Fig. 3.1).

In March and April, the continental anticyclone gradually weakens and incursions of warm and moist tropical air from the southeast replace the cold northeasterlies. A front that often lies from the south coast of China to near Japan forms at the boundary between these airflows. When a depression moves eastwards along this front, the southerlies surge northwards, raising temperatures by over 20 °C and bringing rainfall that is known as the 'spring rain'. Sporadic northward movements of BPH and WBPH out of the tropics occur on these airflows (see below).

By late May or early June, the continental polar westerlies have retreated northwards sufficiently for the mid-latitude westerlies to shift their route from the southern to the northern side of the Tibetan Plateau (Fig. 3.2). This retreat allows an influx of warm and humid tropical air (the southwest monsoon) to extend into East Asia from the Bay of Bengal (Kurashima, 1990). This becomes wedged between the continental air from the north and the moist tropical air from the south, transforming the boundary between them into the '*Bai-u* front', along which rainfall is heavier and more sustained. A core of strong southwesterly winds, known as the '*Bai-u* low-level jet', frequently forms 200–300 km south of

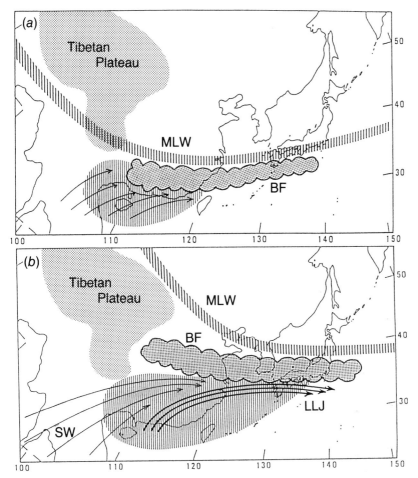

Fig. 3.2. Airflows during the *Bai-u* season. (*a*) Late May; (*b*) early June. MLW, mid-latitude westerlies (at ~3 km altitude); BF, *Bai-u* front; LLJ, *Bai-u* low-level jet; vertically hatched area, region where migration occurs. After Sogawa & Watanabe (1992).

the front (Matsumoto, Ninomiya & Yoshizumi, 1971). The *Bai-u* thus provides both the rain needed for growing paddy rice and the airflows that may transport BPH and WBPH northwards or northeastwards into temperate East Asia.

From mid-May to early June, the *Bai-u* front extends from southern China through Okinawa, passing south of the Japanese mainland (Fig. 3.3, $F_1$). As the season progresses, it advances northwards discontinuously, usually becoming stationary over the Yangtze River region of China and the southern part of the Japanese mainland from late June to mid-July (Fig. 3.3, $F_2$). The northward movement then resumes until late July, when the front becomes stationary over the Yellow River region of northeastern China (45–50° N) for about a month (Fig. 3.3, $F_3$) (Gao & Xiu, 1975). In the central part of the East China Sea, where planthopper migration has been studied with ship-borne traps (Kisimoto, 1981), the temperature at sea level during the migration period rises from 22–23 °C just south of the *Bai-u* front and 19–20 °C just to its north in late June to

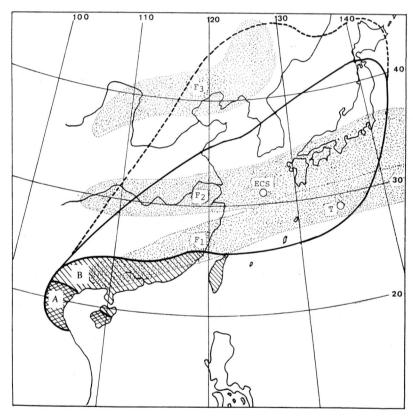

Fig. 3.3. Source and destination regions of BPH and WBPH migrations in East Asia, and the belts of persistent *Bai-u*-season rains. A, suspected source area of the migrants; B, region in which overwintering occurs sporadically; solid line, boundary of region into which BPH and WBPH regularly migrate (the main rice-growing area of temperate East Asia); dashed line, boundary of region infested mainly by WBPH but sporadically also by BPH (region of less intensive rice production); $F_1$, region where *Bai-u* front stagnates from mid-May to early June, bringing sustained heavy rainfall; $F_2$, same for late June to early July; $F_3$, same for mid-July to mid-August. Climate data modified from Gao & Xiu (1975).

26–27 °C and 22–24 °C, respectively in July. Migration was not detected in this study when northerly winds were blowing; this can be attributed to an absence of BPH and WBPH in potential source regions to the north of the front at this season, rather than to the lower temperatures of these airflows.

In summer (mid-July to August), the weather of East Asia is dominated by the North Pacific subtropical anticyclone. Hot moist southwesterly to southeasterly flows of maritime air prevail over western Japan, Korea and most parts of China during this period. During July, average temperatures of 25 °C or more occur throughout the Yellow River region (35–40° N), in southern Korea, and in the central and southern parts of the Japanese mainland. These high temperatures favour fast development of the planthoppers. North of the 25 °C isotherm, only WBPH is of much economic significance, as the temperature accumulation (~500 day-degrees above 10 °C per generation) required for BPH to attain damage-causing densities is unlikely to be achieved (Kisimoto, 1981). The summer weather may be disturbed temporarily by tropical depressions or cyclones (typhoons) originating at low latitudes or by depressions that develop sporadically around the Sea of Japan. Movements of planthoppers in summer and early autumn sometimes occur on the winds associated with these systems.

In the autumn, the Siberian continental anticyclone extends its influence southwards from ~40° N in early to mid-September to 20–23° N in mid-October. There is no influx of the southwesterly monsoon during this *shu-rin* (autumn rainy season) period, and only moderate rainfall occurs at the front marking the southern boundary of the Siberian air. These rains move southwards over the eastern coastal regions of China and the Japanese mainland. Winds with a northerly component behind this front may provide a transport system allowing planthopper populations to retreat southwards at this season.

## The annual cycle of planthopper migration in East Asia

### *The initial northward movement out of the tropics in spring*

BPH and WBPH first appear north of the permanent breeding area in the tropics in March and April, when small numbers sporadically reach the region extending from Hongkong through Taiwan to Okinawa following northward incursions of warm air from the south (Fig. 3.4*a*) (Kisimoto & Dyck, 1976; Liu, 1984; Tsurumachi & Yasuda, 1989). Small numbers of

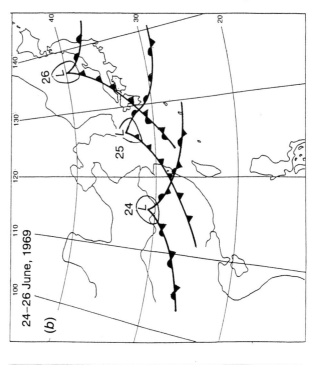

(b) 24–26 June, 1969

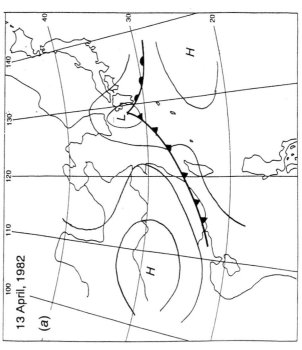

(a) 13 April, 1982

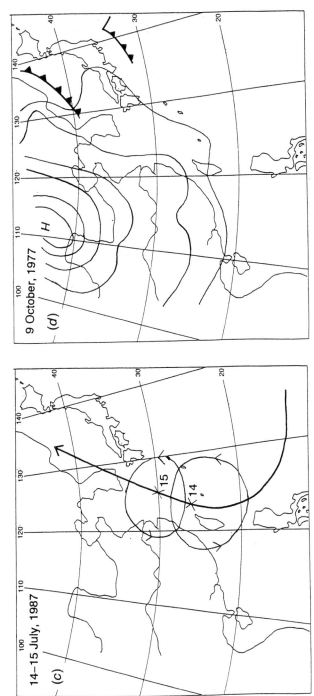

Fig. 3.4. Examples of synoptic weather conditions during migrations of rice planthoppers. (*a*) Spring movement (of sporadic occurrence) into Taiwan (from Liu (1984)); (*b*) *Bai-u*-season mass migration into western and central Japan during northeastward passage of a temperate depression (after Kisimoto (1976)); (*c*) movement into Taiwan on winds circulating around a northward-moving typhoon (crosses mark location of typhoon centre on days when many BPH and WBPH were trapped) (after Cheng & Lu (1990)); (*d*) autumn southward movement on northeasterlies to the southeast of the continental anticyclone (after Jiang *et al.* (1982)).

planthoppers have appeared in Okinawa as early as March (Tsurumachi & Yasuda, 1989).

The first major northward migrations occur from late April to mid-May and carry planthoppers into the area of the Chinese mainland between 20 and 23° N (Zhao-ching District Crop Pest Prognostic Station. . . , 1977; Cheng *et al.*, 1979). These migrants appear to originate south of 19° N, i.e. in southern Hainan Island or northern Indo-China. They are carried on a southwesterly airflow (Fig. 3.5*a,b*) to the southeast of the *Bai-u* front (which is just starting to form at this season). Small numbers of these migrants reach Taiwan and Ishigaki (Tsurumachi & Yasuda, 1989; Cheng & Lu, 1990) and (though only sporadically) southern Kyushu (Hara *et al.*, 1971). There may be further similar migrations between mid-May and early June, and these may extend a little further north (Cheng *et al.*, 1979; Chapter 18, this volume).

The most likely source of these spring migrants appears to be winter-spring rice (transplanted in December–January and harvested in May–June) grown in northern Vietnam (Zhu *et al.*, 1982; Sogawa & Watanabe, 1992). Evidence supporting this is provided by the change in biotype of BPH immigrants into Japan that occurred in 1990 (Sogawa, 1992). This followed a similar change in northern Vietnam (Ich, 1991) and southern China (Yu, Wu & Tao, 1991; Zhang, Tan & Pan, 1991) over the years 1987–1991.

## *Subsequent northward movements within the subtropical and temperate zones during the* Bai-u *season*

The planthopper migrations occurring during the *Bai-u* season (June-July) at latitudes between 25 and 35° N are the most prominent in terms of intensity, distance and area covered of the entire year. The migrations take place exclusively on southwesterly or west-southwesterly airflows. The source region of these migrants is the southernmost part of the Chinese mainland where long-winged BPH and WBPH, the progeny of both the initial immigrants from the south and local overwintering populations, increase in number from late May to July on early transplanted rice (Zhao-ching District Crop Pest Prognostic Station. . . , 1977; National Coordinated Research Group for White-backed Planthoppers, 1981). Sudden disappearances of the adults, often simultaneously at several localities, are believed to be due to emigration flights (Cheng *et al.*, 1979). The planthoppers apparently mostly move northwards, but some may land nearby and infest middle- to late-maturing rice.

The region invaded by these migrants extends gradually northwards as the southwesterly airflow advances. By early June the mainland from 24–26° N is infested, and by mid- to late July the area around the Yangtze river (30–33° N) (Cheng *et al.*, 1979). The southwesterlies also carry BPH and WBPH over the Formosa Strait and East China Sea to Taiwan, Ishigaki and Okinawa in early to late June (Tsurumachi & Yasuda, 1989; Cheng & Lu, 1990; Watanabe *et al.*, 1991), to the Japanese mainland in late June to mid-July (Kisimoto 1971, 1976; Watanabe & Seino, 1991; Chapter 18, this volume), and into the Korean peninsula during July (Uhm *et al.*, 1988). By late July and early August, the region between 26 and 33° N has become a source region from which a second or later generation of WBPH (but usually not the less migratory BPH) have migrated into northeastern China (Liaoning, Jilin and Heilongjiang provinces as far as 46° N; Tsuchiyama, 1940; Ma, 1993) and the central part of Hokkaido. These last movements carry WBPH to the northern limit of rice growing in East Asia.

In Taiwan and Okinawa the immigrant populations cause little damage, because the rice plants are already well developed. However, in the temperate areas of China, Korea and Japan, the middle- to late-transplanting rice is very vulnerable, and even at the limit of rice growing damage is sporadically severe.

The synoptic weather patterns in which the long-distance migrations take place (Kisimoto, 1976; Jiang *et al.*, 1981; Cheng & Lu, 1990) can be classified into two broad types: 'frontal' and 'subtropical anticyclone'. In the former, which is the more frequent of the two, the migrants are transported on the southwesterly to west-southwesterly airflow in the warm sector of a northeastwards-moving temperate depression (Fig. 3.4*b*). In the latter, transport occurs on the southwesterly airflow in the northwestern quadrant of the Pacific subtropical anticyclone; as this system moves slowly, the associated southwesterlies persist for several days. This type of transport is most likely to occur at the end of the *Bai-u* season as the influence of the anticyclone begins to extend over the whole region; however, it accounts for only ~10% of *Bai-u*-season migrations into Japan (Kisimoto, 1976), and in some years is completely absent.

Seino *et al.* (1987) proposed that planthopper migration occurred in the low-level jet wind that blows along the *Bai-u* front. The presence of west-southwesterly winds with speeds over 20 knots (~10 m s$^{-1}$) at the 850-hPa standard pressure level (~1.5 km altitude) to the south of the *Bai-u* front has proved a useful criterion for predicting atmospheric transport of planthoppers to Japan (Watanabe *et al.*, 1991; Watanabe & Seino, 1991;

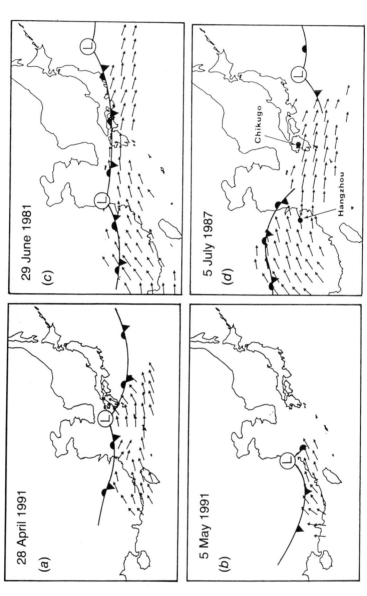

Fig. 3.5. Windfields during major planthopper migrations. Only winds exceeding 20 knot at the 850-hPa pressure level (~1.5 km) are shown. (*a,b*) Occasions when migrations into the spring rice crop of southern China occurred; (*c,d*) occasions when both the Yangtze delta and Japan were invaded. In the last two, there is a well-developed *Bai-u* low-level jet.

Chapter 18, this volume). When a jet of this type developed, mass immigrations of planthoppers occurred simultaneously in Hangzhou on the Chinese mainland, the weather station 'ECS', and Chikugo, Kyushu (Sogawa & Watanabe, 1991; Fig. 3.5*c,d*).

### Movements during summer and early autumn, after the **Bai-u** *season*

During late July and August, after the *Bai-u* front has advanced far to the north (Fig. 3.3) and WBPH has spread northeastwards to the limit of rice growing, movements of both species continue to occur as successive generations mature. Northeastward migrations take place on weather systems that are similar to those of the *Bai-u* season but smaller in scale (Sogawa & Watanabe, 1990). For example, on 4 August 1987, large numbers of WBPH were caught at Chikugo, Kyushu in a pole-mounted tow-net during a period of strong southwesterly winds associated with the passage of a northeastward-moving temperate depression. On the next day, large numbers of this species were trapped at light near Sapporo, Hokkaido, suggesting a migration of over 1000 km to the northeast. On the other hand, when the take-off occurs in calm weather the distances covered may be insignificant. Catches in light traps then cease 1–2 h after sunset, suggesting that the planthoppers fly only for this time. Post-*Bai-u* movements are usually not economically significant in temperate regions as the single rice crop is already well advanced by the time they occur.

In the subtropics, where post-*Bai-u* movements are often associated with the passage of a tropical depression or typhoon, the immigrants can cause damage to second-crop rice which has just been transplanted. Because of the nature of typhoons, such movements are sporadic, and of limited range and duration. However, Taiwan and Ishigaki are sometimes invaded in this way by planthoppers carried from the Chinese mainland on northwesterly winds (Fig. 3.4*c*). Such invasions account for ~35% of BPH and WBPH immigrations into Taiwan during July and August (Cheng & Lu, 1990). Similar typhoon-associated immigrations into Ishigaki and neighbouring islands occur approximately once per month between August and October (Tsurumachi & Yasuda, 1989); an immigration in October 1985 led to severe damage after one generation.

### Southward movements during autumn, and the return to the overwintering areas

From September onwards, the general direction of planthopper migration is predominantly towards the south. These movements occur to the north of the *shu-rin* front, the migrants being carried on the northeasterly airflow that forms on the southeastern side of the Siberian continental anticyclone (Jiang *et al.*, 1982; Fig. 3.4*d*). On the Chinese mainland, BPH moves southwards from ~35° N to around the Tropic of Cancer in three stages between late August and late October (Cheng *et al.*, 1979). Radar observations near Nanjing in late September have shown that large numbers of BPH were flying 300–400 km overnight into areas where late rice crops would have been vulnerable to infestation (Riley *et al.*, 1991). Later generations could then have continued the southwestward migration to the permanent breeding areas in the tropics. During September and October, planthoppers and other small insects have been carried over 500 km to 'Tango', probably from western Japan, on these northwesterly airflows (Asahina & Turuoka, 1970; Itakura, 1973). Temperatures to the north of the front were 22.5–27 °C, which is well above the threshold for flight.

Local movements and migrations in other directions also occur at this season. BPH and WBPH have sometimes been found in large numbers on non-host plants at elevations up to 2.7 km in the mountains of Kyushu, Honshu and Hokkaido during September and October (Section of Plant Protection, Ministry of Agriculture and Forestry, 1965). These flights, which seem most likely to have originated in nearby lowland areas where rice is grown extensively, often occurred during the eastward passage of a depression along the *shu-rin* front. Immigrations of BPH and WBPH into Kyushu that occur frequently at this season are more likely to have been carried from the Chinese mainland on westerly or southwesterly winds on the southern side of the front (Wada *et al.*, 1987).

### Movements within the permanent breeding area

Planthopper movements within the tropics are mainly local in scale. Traps operated on ships plying the territorial waters of the Philippine archipelago in both wet and dry seasons caught only small numbers of BPH and WBPH (Saxena & Justo, 1984), and the circumstances of the catches were not indicative of long-distance migration. Radar observations in the Philippines in March and April showed that the brief dusk flight would have taken the migrants no further than ~30 km, and the dawn flight even shorter distances (Riley *et al.*, 1987). Light trapping and counts of the number of insects on host plants in southern Vietnam (Ich, 1991) and central Thailand (Tsurumachi, 1986) have shown that BPH and WBPH fly throughout the year; however, sudden increases in catch size, which in temperate regions are characteristic of major immigrations from distant source areas, have not been detected.

Due to the influence of the continental anticylone, northeasterly winds prevail in this area in November and December. These may allow the southwestward autumn movements of BPH that have been observed in China to continue into the subtropics, and perhaps even the tropics, and thus to complete the return migration. However, no evidence to support this has so far been presented.

## Discussion

The range of BPH and WBPH in East Asia extends from the equator to 45° N, and spans a series of climatic zones from the permanently hot tropics to highly seasonal climates with brief growing seasons and long harsh winters. Neither species can overwinter in the mid-latitude parts of the range, which are recolonised each year over 2–3 generations by windborne immigrants from further south. Both species exhibit varying degrees of migratory activity, and produce flightless brachypters in varying proportions: in general (and unsurprisingly), populations at higher latitudes are more migratory. The question arises as to whether the often spectacular migrations of these two species are adaptive, or whether they are examples of 'Pied Piper' flights (Rabb & Stinner, 1979), with the offspring produced in the temperate zone unable to return south and doomed to be extinguished by the winter.

The northward and northeastward movements of these planthoppers in spring and summer appear to be of at least short-term benefit to the migrants. The warm and humid *Bai-u* air provides a favourable environment for sustained flight, at least during the initial stages of the migration; during landing, weather conditions are often so rough that it is difficult to believe that planthoppers are advantaged by being caught in them. The *Bai-u* heralds the coming growing season of the species' host plants, and the persistent and rapid northward transport it provides allows the migrants to exploit habitats that are located well beyond the range of their powered flight. With rice grown so extensively in temperate East Asia, there is scope for enormous population growth by a succession of generations as the season advances northwards.

However, for these movements to be adaptive, the offspring of at least some of the migrants flying north must survive in the long as well as the short term, and they can do this only by returning south before the onset of the harsh temperate-zone winter. If they fail to do so, then the annual recolonisation of the species' higher-latitude ranges can be accounted for only as a side effect of adaptive migrations within the tropics and subtropics. For this scenario to be plausible, the posited migrations at low latitudes must confer significant advantages to balance the ultimately fatal effects of chance migrations into the temperate zone; if they do not, the genes for migration will be consistently selected against and eventually eliminated from the population. Such side effects perhaps provide the most plausible explanation for the annual reinvasion of offshore islands such as Ishigaki and Okinawa (and perhaps also the Japanese mainland and peripheral regions of the species' distribution on continental East Asia), as the probability of success for migration to and from such areas must be very low.

In parts of the temperate range, southward movements do occur in autumn, at least for BPH. The northerly winds required for a return migration prevail at this season, but being cool and dry they are not as favourable for sustained flight as the southwesterlies of spring; they also tend to induce the decline of host vegetation rather than its growth. Nevertheless, in at least some situations, large numbers of planthoppers are carried several hundred kilometres southwards. At lower latitudes, where more than one rice crop is grown each year, these immigrants from the north may arrive just when the late rice is at its susceptible stage, so that a further generation is possible. As the season advances the northerlies extend further south, and it can be speculated (but has not yet been shown) that they provide transport right back to the permanent breeding area in the tropics. Whether any such return migrants would be able to compete against resident populations that may already have occupied all suitable habitats is unclear.

The northward movements of BPH and WBPH in spring and summer, and (if they are confirmed) the return movements southward in autumn, appear to represent a remarkable adaptation to the geography, climate, weather and agricultural practices of East Asia. The region perhaps provides especially favourable conditions for such a migration system to be maintained: winds suitable for transport prevail (or at least occur frequently) during both the outward and the return sectors of the loop, and the principal host plant, rice, is grown extensively over the whole latitudinal range. The capacity of the migrant populations for sustained flight, together with the relatively short time between generations, allows these two species to exploit the available opportunities to the full, and ensures their status as severe and regular pests.

In the tropics, where staggered rice crops provide an almost perennial habitat, both BPH and WBPH produce a high proportion of brachypters. The long-winged forms have the capacity to fly for many hours (Baker *et al.*, 1980; Padgham, Perfect & Cook, 1987) but, at least in the Philippines, do not appear to use it (Riley *et al.*, 1987). These traits may represent adaptations to an environment in which only short-distance flights from senescing to newly transplanted rice crops are essential for survival; in such a situation, the relative risks and advantages for each generation of remaining in place or relocating are apparently altered, and a degree of brachyptery, with its higher multiplication rate, confers a significant selective advantage.

Geographical variations in biotype, insecticide resistance, transmission of plant viruses and the proportion of brachypters, suggest that mixing between tropical and migratory populations of the two species may be slight. The migrants that colonise the temperate region each year are believed to originate mainly from northern Indo-China, and possibly also from the southern part of Hainan Island. Whether the offspring of these migrants eventually return to these likely source regions, and reinitiate the cycle the following year, is perhaps the major outstanding question of planthopper entomology today.

## Acknowledgements

We wish to thank Alistair Drake and David Pedgley for their helpful comments on earlier drafts. R. Kisimoto acknowledges support from the Ministry of Education, Science and Culture, Japan (Grants-in-Aid for Scientific Research no. 63480042 and no. 02660047).

## References

Asahina, S. & Turuoka, Y. (1968). Records of the insects visited a weather ship located at the Ocean Weather Station 'Tango' on the Pacific, II. *Kontyu*, **36**, 190–202. In Japanese, English summary.

Asahina, S. & Turuoka, Y. (1970). Records of the insects visited a weather-ship located at the Ocean Weather Station 'Tango' on the Pacific, V. Insects captured during 1968. *Kontyu*, **38**, 318–30. In Japanese, English summary.

Baker, P.S., Cooter, R.J., Chang, P.-M. & Hashim, H.B. (1980). The flight capabilities of laboratory and tropical field populations of the brown planthopper, *Nilaparvata lugens* (Stål) (Hemiptera: Delphacidae). *Bulletin of Entomological Research*, **70**, 589–600.

Cheng, C.-H. & Lu, J.-L. (1990). Detection of the trans-oceanic immigration of rice planthoppers, *Nilaparvata lugens* Stål and *Sogatella furcifera* Horváth to the southwestern Taiwan and their relative weather conditions. *Chinese Journal of Entomology*, **10**, 301–24. In Chinese, English summary.

Cheng, S.-N., Chen, J.-C., Si, H., Yan, L.-M., Chu, T.-L., Wu, C.-T., Chien, J.-K. & Yan, C.-S. (1979). Studies on the migrations of brown planthopper, *Nilaparvata lugens* Stål. *Acta Entomologica Sinica*, **22**, 1–21. In Chinese, English summary.

Dung, W.-S. (1981). A general survey on seasonal migrations of *Nilaparvata lugens* (Stål) and *Sogatella furcifera* (Horváth) (Homoptera: Delphacidae) by means of airplane collections. *Acta Phytophylacica Sinica*, **8**, 73–82. In Chinese, English summary.

Gao, Y.-X. & Xiu, S.-Y. (1975). Advancement and retreat of the East Asian monsoon and the beginning and end of the rainy season. In *Rain and Climate of China*, ed. T. Yoshino, pp. 111–21. Tokyo: Taimeido. In Japanese.

Hara, K., Horikiri, M., Fukamachi, S., Imamura, M., Nagashimada, Y., Muranaga, H., Ikeda, K., Waki, K., Oshikawa, M. & Makino, S. (1971). Surveys of mass flight of white-backed planthopper and brown planthopper by net traps. *Proceedings of the Association for Plant Protection of Kyushu*, **17**, 110–12. In Japanese.

Hirao, J. (1979). Tolerance for starvation by three rice planthoppers (Hemiptera: Delphacidae). *Applied Entomology and Zoology*, **14**, 121–2.

Ich, B.V. (1991). The occurrence and migration of brown plant hopper in Vietnam. In *Migration and Dispersal of Agricultural Insects*, pp. 183–204. Tsukuba, Japan: National Institute of Agro-Environmental Sciences.

Itakura, H. (1973). [Relation between planthopper migration and meteorological conditions at the Ocean Weather Station 'Tango' during 1973.] *Shokubutsu Boeki [Plant Protection (Japan)]*, **27**, 489–92. In Japanese.

Iwanaga, K., Nakasuji, F. & Tojo, S. (1987). Wing polymorphism in Japanese and foreign strains of the brown planthopper, *Nilaparvata lugens*. *Entomologia Experimentalis et Applicata*, **43**, 3–10.

Jiang, G.-H., Tan, H.-Q., Shen, W.-Z., Cheng, X.-N. & Chen, R.-C. (1981). The relation between long-distance northward migration of the brown planthopper (*Nilaparvata lugens* Stål) and synoptic weather conditions. *Acta Entomologica Sinica*, **24**, 251–61. In Chinese, English summary.

Jiang, G.-H., Tan, H.-Q., Shen, W.-Z., Cheng, X.-N. & Chen, R.-C. (1982). The relation between long-distance southward migration of the brown planthopper (*Nilaparvata lugens* Stål) and synoptic weather conditions. *Acta Entomologica Sinica*, **25**, 147–55. In Chinese, English summary.

Kisimoto, R. (1971). Long-distance migration of planthoppers, *Sogatella furcifera* and *Nilaparvata lugens*. *Symposium on Rice Insects*. Tropical Agriculture Research Center, Ministry of Agriculture and Forestry, Japan, pp. 201–16.

Kisimoto, R. (1976). Synoptic weather conditions inducing long-distance immigration of planthoppers, *Sogatella furcifera* Horváth and *Nilaparvata lugens* Stål. *Ecological Entomology*, **1**, 95–109.

Kisimoto, R. (1979). Brown planthopper migration. In *Brown Planthopper: Threat to Rice Production in Asia*, pp. 113–24. Manila: International Rice Research Institute.

Kisimoto, R. (1981). Development, behaviour, population dynamics and control of the brown planthopper, *Nilaparvata lugens* Stål. *Review of Plant Protection Research*, **14**, 26–58.

Kisimoto, R. (1984). Insect pests of the rice plant in Asia. *Protection Ecology*, **7**, 83–104.

Kisimoto, R. (1987). Ecology of planthopper migration. In *Proceedings of the 2nd International Workshop on Leafhoppers and Planthoppers of Economic Importance, Provo, Utah, 28 July–1 August 1986*, ed. M.R. Wilson & L.R. Nault, pp. 41–54. London: C.A.B. International Institute of Entomology.

Kisimoto, R. (1991). Long-distance migration of rice insects. In *Rice Insects: Management Strategies*, ed. E.A. Heinrichs & T.A. Miller, pp. 167–95. New York: Springer-Verlag.

Kisimoto, R. & Dyck, V.A. (1976). Climate and rice insects. In *Climate and Rice*, pp. 367–91. Los Banos: The International Rice Research Institute.

Kisimoto, R. & Rosenberg, L.J. (1994). Long-distance migration in Delphacid planthoppers. In *Planthoppers: Their Ecology, Genetics, and Management*, ed. R.F. Denno & J. Perfect, pp. 302–22. New York: Chapman & Hall.

Kurashima, A. (1990). *Climate in Japan*. Tokyo: Kokinshoin. In Japanese.

Liu, C.-H. (1984). Study on the long-distance migration of the brown planthopper in Taiwan. *Chinese Journal of Entomology*, **4**, 49–54.

Ma, C.-S. (1993). Occurrence of rice planthoppers in northeastern China and their northward migration. In *International Workshop: Establishment, Spread, and Management of the Rice Water Weevil and Migratory Rice Insect Pests in East Asia*, ed. K. Hirai, pp. 307–22. Tsukuba, Japan: National Agriculture Research Center.

Matsumoto, S., Ninomiya, K. & Yoshizumi, S. (1971). Characteristic features of 'Baiu' front associated with heavy rainfall. *Journal of the Meteorological Society of Japan*, **49**, 267–81.

Mochida, O. (1974). Long-distance movement of *Sogatella furcifera* and *Nilaparvata lugens* (Homoptera: Delphacidae) across the East China Sea. *Rice Entomology Newsletter*, **1**, 18–22.

Murata, T. & Hirano, I. (1929). [On the rice planthoppers]. *Byochugai-zasshi* [*Journal of Plant Diseases and Insects, Japan*], **16**, 518–28, 597–611. In Japanese.

Nagata, T. & Masuda, T. (1980). Insecticide susceptibility and wing-form ratio of the brown planthopper, *Nilaparvata lugens* (Stål) (Hemiptera: Delphacidae) and the white-backed planthopper, *Sogatella furcifera* (Horváth) (Hemiptera: Delphacidae) of southeast Asia. *Applied Entomology and Zoology*, **15**, 10–19.

Nanjing Agricultural College Department of Plant Protection, Guangdong Academy of Agriculture Institute of Plant Protection, Chenzhou Prefecture (Hunan Province) Institute of Agriculture & Guilin Prefecture (Guangxi Autonomous Region) Bureau of Agriculture (1981). Test of the releasing and recapturing of marked planthoppers, *Nilaparvata lugens* and *Sogatella furcifera*. *Acta Ecologica Sinica*, **1**, 49–53. In Chinese, English summary.

National Coordinated Research Group for White-backed Planthoppers (1981). Studies on the migration of white-backed planthoppers (*Sogatella furcifera* Horváth). *Scientia Agricultura Sinica*, **1981**, (5), 25–31. In Chinese, English summary.

Noda, T. & Kiritani, K. (1989). Landing places of migratory planthoppers, *Nilaparvata lugens* (Stål) and *Sogatella furcifera* (Horváth) (Homoptera: Delphacidae) in Japan. *Applied Entomology and Zoology*, **24**, 59–65.

Ohkubo, N. (1981). Behavioural and ecological studies on the migratory flight of rice planthoppers. PhD thesis, Kyoto University. In Japanese.

Ohkubo, N. & Kisimoto, R. (1971). Diurnal periodicity of flight behaviour of the brown planthopper, *Nilaparvata lugens* Stål, in the 4th and 5th emergence periods. *Japanese Journal of Applied Entomology and Zoology*, **15**, 8–16. In Japanese, English abstract.

Padgham, D.E. (1983*a*). Flight fuels in the brown planthopper *Nilaparvata lugens*. *Journal of Insect Physiology*, **29**, 95–9.

Padgham, D.E. (1983*b*). The influence of the host-plant on the development of the adult brown planthopper, *Nilaparvata lugens* (Stål) (Hemiptera: Delphacidae), and its significance in migration. *Bulletin of Entomological Research*, **73**, 117–28.

Padgham, D.E., Perfect, T.J. & Cook, A.G. (1987). Flight behaviour in the brown planthopper, *Nilaparvata lugens* (Stål) (Homoptera: Delphacidae). *Insect Science and its Application*, **8**, 71–5.

Rabb, R.L. & Stinner, R.E. (1979). The role of insect dispersal and migration in population processes. In *Radar, Insect Population Ecology, and Pest Management*. NASA Conference Publication 2070, ed. C.R. Vaughn, W.W. Wolf & W. Klassen, pp. 3–16. Wallops Island, Virginia: NASA Wallops Flight Center.

Research Council, Ministry of Agriculture, Fisheries and Forestry, Japan (1989). [*Development of Techniques Predicting Long-distance Migration of Migratory Pest Insects.* Research Report] 217. Tokyo: Ministry of Agriculture, Fisheries and Forestry. In Japanese.

Riley, J.R., Cheng, X.-N., Zhang, X.-X., Reynolds, D.R., Xu, G.-M., Smith, A.D., Cheng, J.-Y., Bao, A.-D. & Zhai, B.-P. (1991). The long-distance migration of *Nilaparvata lugens* (Stål) (Delphacidae) in China: radar observations of mass return flight in the autumn. *Ecological Entomology*, **16**, 471–89.

Riley, J.R., Reynolds, D.R. & Farrow, R.A. (1987). The migration of *Nilaparvata lugens* (Stål) (Delphacidae) and other Hemiptera associated with rice during the dry season in the Philippines: a study using radar, visual observations, aerial netting and ground trapping. *Bulletin of Entomological Research*, **77**, 145–69.

Saxena, R.C. & Justo, H.D. Jr (1984). Trapping airborne insects aboard inter-island ships in the Philippine archipelago, with emphasis on the brown planthopper (BPH). *International Rice Research Newsletter*, **9** (5), 16.

Section of Plant Protection, Ministry of Agriculture and Forestry, Japan (1965). [*Physiological and Ecological Investigations in Relation to Forecasting of the Brown Planthopper and the White-backed Planthopper.* Special Report on the Pests Forecasting Programme] 20. Tokyo: Ministry of Agriculture and Forestry. In Japanese.

Seino, H., Shiotsuki, Y., Oya, S. & Hirai, H. (1987). Prediction of long-distance migration of rice planthoppers to northern Kyushu considering low-level jet stream. *Journal of Agricultural Meteorology*, **43**, 203–8.

Sogawa, K. (1992). Rice brown planthopper (BPH) immigrants in Japan change biotype. *International Rice Research Newsletter*, **17** (2), 26.

Sogawa, K. & Watanabe, T. (1990). Migration of the rice planthoppers in the post-baiu season. *Proceedings of the Association for Plant Protection of Kyushu*, **36**, 90–4. In Japanese.

Sogawa, K. & Watanabe, T. (1991). Comparison of immigration of the rice planthoppers in Hangzhou, Zhejiang, China and Chikugo, Fukuoka,

Japan during the Baiu season. *Proceedings of the Association for Plant Protection of Kyushu*, **37**, 91–4. In Japanese.

Sogawa, K. & Watanabe, T. (1992). Redistribution of rice planthoppers and its synoptic monitoring in East Asia. *Technical Bulletin of the Food and Fertilizer Technology Center (Taipei)*, **131**, 1–9.

Suenaga, H. & Nakatsuka, K. (1958). [*Review on the Forecasting of Rice Planthoppers and Leafhoppers*. Special Report on the Pests Forecasting Programme] 1. Tokyo: Ministry of Agriculture and Forestry. In Japanese.

Tsuchiyama, T. (1940). [Occurrence of several agricultural insect pests in North Manchuria.] *Manshu Nogakkaishi* [*Journal of Manchurian Agriculture*], **2**, 524–30. In Japanese.

Tsurumachi, M. (1986). Population growth pattern of the brown planthopper in Thailand. *Tropical Agriculture Research Series (Tsukuba)*, **19**, 209–19.

Tsurumachi, M. & Yasuda, K. (1989). [Long-distance immigration and infestation of the rice planthoppers and the rice leaffolder in Okinawa.] *Research Reports on Tropical Agriculture (Tsukuba)*, **66**, 50–4. In Japanese.

Uhm, K.-B., Park, J.-S., Lee, Y.-I., Choi, K.-M., Lee, M.-H. & Lee, J.-O. (1988). Relationship between some weather conditions and immigration of the brown planthopper, *Nilaparvata lugens* Stål. *Korean Journal of Applied Entomology*, **27**, 200–10. In Korean, English abstract.

Wada, T., Seino, H., Ogawa, Y. & Nakasuga, T. (1987). Evidence of autumn overseas migration in the rice planthoppers, *Nilaparvata lugens* and *Sogatella furcifera*: analysis of light trap catches and associated weather patterns. *Ecological Entomology*, **12**, 321–30.

Watanabe, T. & Seino, H. (1991). Correlation between the immigration area of rice planthoppers and the low-level jet stream in Japan. *Applied Entomology and Zoology*, **26**, 457–62.

Watanabe, T., Sogawa, K., Hirai, Y., Tsurumachi, M., Fukamachi, S. & Ogawa, Y. (1991). Correlation between migratory flight of rice planthoppers and the low-level jet stream in Kyushu, southwestern Japan. *Applied Entomology and Zoology*, **26**, 215–22.

Yu, X., Wu, G. & Tao, T. (1991). Virulence of brown planthopper (BPH) populations collected in China. *International Rice Research Newsletter*, **16** (3), 26.

Zhang, Y., Tan, Y. & Pan, Y. (1991). A population of brown planthopper (BPH) biotypes 1 and 2 mixture in Guangdong, China. *International Rice Research Newsletter*, **16** (5), 22–3.

Zhao-ching District Crop Pest Prognostic Station & Yun-fu County (Kwangtung Province) Crop Pest Prognostic Station (1977). On the occurrence and control of the brown planthopper *Nilaparvata lugens* Stål in rice fields. *Acta Entomologica Sinica*, **20**, 279–88. In Chinese, English summary.

Zhu, S.-X., Wu, C.-Z., Du, J.-Y., Huang, X.-G., Shua, Y.-H. & Hong, F.-L. (1982). [A summary report on the long-distance migration of the brown planthopper]. *Guangdong Agricultural Sciences*, **4**, 22–4. In Chinese.

# 4

# Migration of the Oriental Armyworm *Mythimna separata* in East Asia in relation to weather and climate. I. Northeastern China

R.-L. CHEN, Y.-J. SUN, S.-Y. WANG,
B.-P. ZHAI AND X.-Z. BAO

## Introduction

The Oriental Armyworm *Mythimna separata* is a severe pest of cereal crops, especially wheat, maize, millet and rice, throughout eastern China. Historical records suggest that the species has been a pest on millet and wheat for thousands of years (Zou, 1956). In the northeastern provinces of Liaoning, Jilin and Heilongjiang (Fig. 4.1), where the climate is highly seasonal, with severe winters and a growing season lasting only from April to September, an initial generation of *M. separata* larvae regularly causes severe damage to spring wheat, millet and maize in July, and a second generation damages millet and maize, and sometimes also rice, in August. The initial infestation in the northeast arises from egg-lays by moths that appear every year, often in large numbers, in late May and early June (Chen, 1962). An intensive and comprehensive programme of research on this key pest of Chinese agriculture during the late 1950s and early 1960s prompted the proposal, in 1959, that these moths were immigrants from the south (Cai, 1990). The validity of this hypothesis was soon established and it was shown that the migrations were achieved by transport on the wind. In this chapter, we first review the earlier work on *M. separata* migration, and then summarise the results of a more recent research programme in which the migration has been observed directly with entomological radar.

## Previous research

The migratory status of *M. separata* was demonstrated by a variety of techniques. Extensive searches of possible overwintering sites in northeastern China failed to find the species in any stage (Li *et al.*, 1965), and cage and laboratory experiments showed conclusively that no stage could

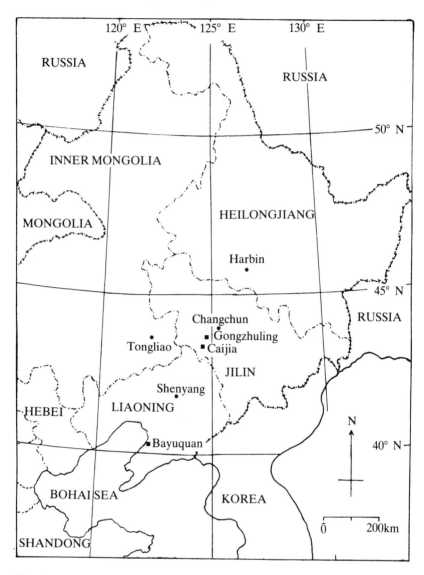

Fig. 4.1. Map of northeastern China, showing locations named in the text and (■) radar observation sites.

survive the sustained low winter temperatures (average minima around -20 °C during December–February) typical of the northeast (Chen *et al.*, 1965). Along with several other species, *M. separata* was caught on ships crossing the Bohai Sea during May to August (Hsia, Tsai & Ten, 1963).

The capacity for long-distance migration into the northeast was conclusively demonstrated in 1963 when large numbers of *M. separata* moths were marked at Linyi (Shandong province) and Xuzhou (Jiangsu province) and six were recaptured in Liaoning and Jilin provinces at distances of 600–1200 km to the northeast (Li, Wong & Woo, 1964).

An analysis of the seasonal occurrence of *M. separata* throughout China showed an annual northward advance over three generations from winter to early summer and a southward retreat over two generations in summer and autumn (Fig. 4.2) (Chen, 1962; Chen *et al.*, 1965). In the southern provinces of Guangdong, Guangxi and Fujian (21–27° N), larvae develop on early ripening winter wheat from January to March and migrate northwards in March and April. These moths initiate a second generation in Zhejiang, Jiangsu, Anhui, Hubei, Henan and Shandong provinces (30–38° N) which develops on middle-ripening winter wheat from April to May. The moths from this region invade the northeast (and Shanxi and Hebei provinces and eastern Inner Mongolia, all at 39–49° N) and initiate the third generation there in late May and early June. In all three of these generations, the eggs are laid on the seedling wheat and the larvae are full grown at the ear stage: migration allows the pest to track the northward advance of the growing season. In August, a fourth generation of larvae develop on millet, maize and sometimes rice, in Shandong, Hebei and Liaoning provinces (34–43° N), and sometimes also Jilin and Heilongjiang provinces further north. Finally, in September and October, there is a fifth larval generation on rice in Guangxi, Fujian, Hunan, Jiangxi and Guangdong provinces (21–31° N).

Evidence for a southward migration in autumn is not so strong as for the northward migration in spring. Populations in the fourth-generation area are low in June–July; in the fifth-generation area they are low from April to August, even though temperatures are high enough for around five generations to develop over the summer months (Lin & Chang, 1964). Radar observations in Jilin province in late July and early August of 1985 and 1986 showed movements in several directions, including towards the south (Chen, 1990). Mark-and-capture experiments[1] in 1978–80 confirmed that movements with a southward component occur in autumn (Chen, 1990; Collaboration Group on Armyworm Research in China, unpublished data): in Jilin province in July (two moths), and in

[1] Editors' note: The capture in September 1961 of two apparently marked *M. separata* moths at Wangjiang, Anhui province, 800 km south of a release point in Shandong province (Li *et al.*, 1964) is sometimes cited as firm evidence of a southward migration in autumn. However, in the text of that paper it is stated that the marking on these moths was not as clear as that on the spring recaptures, and that further assessment is required.

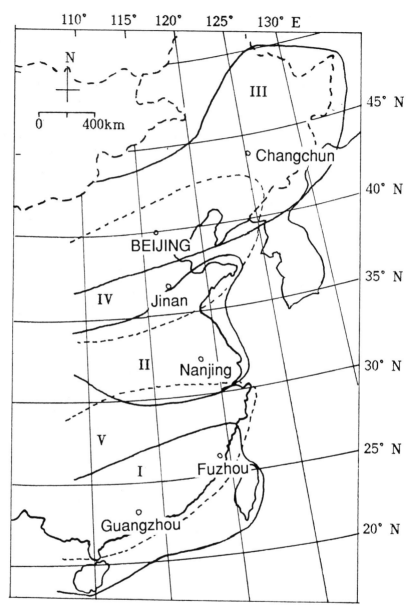

Fig. 4.2. The regions where successive generations of armyworm develop in eastern China each year. Larvae present I, January–March, II, April–May, III, June–July, IV, August–September, V, September–October. After Chen (1979).

Guangdong province in November (one moth), movements of 60–100 km were recorded in directions between southeast and south. Further west in China, a southwestward movement of 80 km in Sichuan province (one moth), and a southward movement of ~1300 km, from Gansu province to Yunnan province (one moth), were recorded in August. (A second moth from the Gansu release flew ~1500 km eastward, to Liaoning province.) Despite these successes, the question of whether moths moving out of northeastern China in autumn reach generation area IV (Fig. 4.2) has still not been answered conclusively by mark-and-capture methods; however, Chinese entomologists generally believe that they do so.

The significance of wind transport for *M. separata* migration was made apparent in an analysis of the dates on which there were sudden increases in the number of moths caught in traps at various locations in northeastern China (Lin *et al.*, 1963; Lin & Hsia, 1963). The increases were presumed to be indicators of influxes of immigrant moths, and were found to follow periods of sustained southwesterly wind. An analysis of wind direction at the standard 950-, 900-, 850-, and 700-hPa levels used by meteorologists showed that winds at the lower levels (950 and 900 hPa, corresponding to ~400 and ~900 m altitude) were most likely to bring the moths to the locations where catches occurred, and it was therefore inferred that the migrations took place at these heights.

The nocturnal behaviour of *M. separata* moths has been studied in the field and the laboratory. Field observations showed that moths copulate, oviposit and fly just above the crops at night, but are completely inactive by day (Chen, 1962). Mass landings on ships far from land occurred at dawn (Hsia *et al.*, 1963). More recently, tethered flight experiments in the laboratory have shown that the moths become increasingly capable of sustained flight during the first 3–4 days after eclosion (Hu & Lin, 1983). Flights of up to 28 h, airspeeds up to 3.7 m s$^{-1}$, and distances covered (in still air) up to 205 km have been recorded; the average airspeed is ~1.5 m s$^{-1}$. When the wind speed is greater than 4 m s$^{-1}$, the moths orientate themselves downwind.

## Observations

In the research described so far, an understanding of the migratory behaviour of *M. separata* was built up by making inferences from indirect observations, e.g. of trap catches or weather conditions. In 1985 it became possible to observe the migrations directly, while they were in progress, using an entomological radar developed by the Jilin Academy of Agricultural Sciences (Chen *et al.*, 1985, 1988). In the following year, a programme of systematic radar observations of the spring migration of *M. separata* into northeastern China commenced. In this section, we summarise the results of observations made over the seven years 1985–1991. A detailed account of the 1986 observations has already appeared (Chen *et al.*, 1989).

### Observation methods

In all years except 1989, the radar was located in the central agricultural plain of Jilin province, near Gongzhuling (43°30' N, 124°50' E) in 1985–88 and near Caijia (43°20' N, 124°40' E) in 1990–91; in 1989 it was operated at Bayuquan (40°20' N, 122°10' E) in Liaoning province, on the shore of the Bohai Sea. Observations were made several times each day and night in 1985, and usually every hour from dusk until dawn in the following years. The radar is of the 'scanning pencil-beam' type, and measurements of migration density and direction, at a series of heights, were made using a 'plan-position indicator' (PPI) display. Observation and analysis methods were adapted from those described by Drake (1981).

In 1986, 1989 and 1991, the radar observations were supplemented by regular measurements of the wind from ground level up to 1500 m, made by tracking a pilot balloon with a theodolite. Profiles of wind speed and direction at 25-m intervals were obtained on 169 occasions, mainly at night. Two or more profiles are available for almost every night the radar was working.

### Results

The radar confirmed that immigration into northeastern China occurred between late May and mid-June, and that it took place at night. A dusk takeoff was frequently observed. Except on one occasion at Bayuquan, migration ceased at dawn, and daytime observations failed to detect migration. The observation of takeoff flights within the destination region demonstrates that the moths make a series of flights over a number of nights. The prevalence of *M. separata* in crops and at lights during this period leaves little doubt that this species was the dominant radar target; in addition, a kite-borne tow-net operated at 100–200 m in 1986 and 1989 collected a total of ten *M. separata* and only two other insects of comparable size. The timing of *M. separata* catches in monitoring traps suggests that no major immigrations occurred outside the periods when radar observations were made.

The radar showed that the moths flew at altitudes up to 1500 m, but mainly between 200 and 1000 m (Fig. 4.3a). Migration occurred at higher altitudes at Bayuquan than in central Jilin province. The migrants always moved downwind. A layer concentration sometimes formed at the altitude where the density was highest, and the moths within the layer often showed some degree of alignment along their (downwind) direction of movement. There was no evidence of alignment near the surface.

Migration was predominantly towards the northeast, on southwesterly winds. Migration densities averaged $\sim 10^{-6}$ m$^{-3}$ at the start and the end of the migration period, and when the wind was from the north. Densities of $\sim 10^{-5}$ m$^{-3}$ were frequent during the peak migration period, reaching $\sim 10^{-4}$ m$^{-3}$ (and occasionally even $\sim 10^{-3}$ m$^{-3}$) in layer concentrations. Fluxes, estimated from migration densities and windspeeds (see below), were highest between 450 and 1000 m, where values were of the order of 1 m$^{-2}$ h$^{-1}$ on peak nights. Nightly overflights, calculated from the densities and speeds at all heights, ranged up to $3.7 \times 10^4$ m$^{-1}$ (Fig. 4.3b), this value being measured at Bayuquan over the night of 3–4 June 1989. Total overflights for the whole observation period were $\sim 0.5 \times 10^4$ m$^{-1}$ in 1986 (nine nights), $\sim 8 \times 10^4$ m$^{-1}$ in 1989 (19 nights), and $\sim 1.5 \times 10^4$ m$^{-1}$ in 1991 (18 nights).

Winds were predominantly southwesterly during these migrations (Fig. 4.3b). There was invariably a strong (>10 m s$^{-1}$) southwesterly wind on the nights of most intense migration, with transport occurring on the southwesterly airflow to the west of the northern Pacific anticyclone and to the southeast of a deep depression over northeast Asia (Fig. 4.4). For the three years with pilot-balloon wind measurements, the wind was southwesterly on between 65% and 70% of observation nights. At Bayuquan, 99% of the overflight occurred during the 68% of nights when the wind blew from the southwest. The overflight for a night with a northwesterly airflow was only 1% of the peak value recorded two nights previously when the wind was from the southwest. At Caijia in 1991, the wind was southwesterly on 67% of nights and northwesterly on the rest; the nights with southwesterly airflow accounted for 92% of the total overflight, with 85% accumulating on just four nights in which the southwesterlies were strong. Nights with northwesterly winds were cool, the surface temperature typically falling to a minimum of 8–10 °C as opposed to 16 °C during southwesterlies.

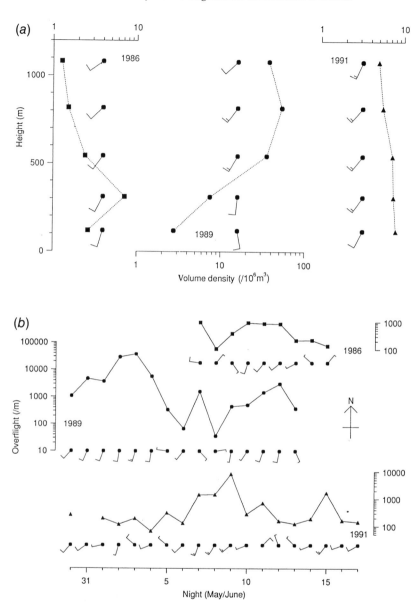

Fig. 4.3. Average winds and migration intensities for the nights of peak armyorm migration in three years. (*a*) Variation of wind and volume density with height. These are averages for the nights of major migration: 7 and 10–13 June in 1986; 30 May–4 June in 1989; and 7–9 and 15 June in 1991. (*b*) Variation from night to night of wind and overflight. Wind values are averages, for the 50–1180-m height interval, of measurements made at ~20.00 LST and either 00.00 or 02.00; overflights are for the whole night.

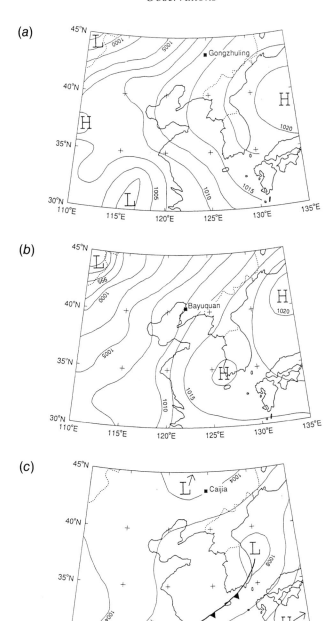

Fig. 4.4. Synoptic charts for nights of intense migration into northeastern China.
(*a*) 12 June 1986 (03.00 LST), (*b*) 2 June 1989 (20.00); and (*c*) 9 June 1991 (09.00).
In (*c*) the situation is slightly complicated by a minor depression over Korea.

An analysis of the nights of strong northeastward migration during the three years 1986, 1989 and 1991 showed that the southwesterly airflow was sheared, with average winds from 170–210° at 100 m but from 210–235° between 400 and 1000 m (Fig. 4.3*a*). Average windspeed increased with height up to ~700 m and fell off above ~1000 m; maximum values were typically 10–15 m s$^{-1}$, with 30 m s$^{-1}$ recorded on one occasion. As the windspeed at the altitude where migration was predominantly occurring was often >10 m s$^{-1}$, the moths' airspeeds will make only a minor contribution to the migration speed; this was confirmed by comparing multiple-exposure PPI measurements of migration speed with balloon measurements of winds. Downwind alignment of the moths was particularly evident in the dense layer concentrations that formed on nights of intense northeastward migration.

## Discussion

These direct observations confirm the conclusions of Lin *et al.* (1963) and Lin & Hsia (1963). They show that migration occurs on sustained southwesterly airflows, predominantly at altitudes between 450 and 1000 m, at speeds of 10–15 m s$^{-1}$. They have also demonstrated that moths fly on a number of successive nights in order to reach their destination, and that the bulk of the immigration into northeastern China takes place on only 3–5 nights each year. Southwesterly winds occur frequently during the immigration period, so that years in which it is not possible for migrants to reach the northeast are rare. The number of immigrants is therefore determined to a considerable extent by the number of moths produced in the source region between 30 and 38° N. These understandings of how *M. separata* moths arrive in northeastern China have been incorporated into procedures for forecasting the intensity of infestation there (Chen, 1979; Crop Pest Forecasting Station of Ministry of Agriculture, 1981).

Migration allows *M. separata* to exploit the rich but temporary habitats that appear in central and northern China each spring and summer. In northeastern China, the seasonal warming in late May and early June is accompanied by (and to some extent due to) the predominance of a warm airflow from the southwest. By flying on successive nights, the immigrant moths are able to cover the ~1000–1500 km that the growing season has advanced northward while they were developing. The larvae of successive generations are therefore able to develop on what appears to be the species' most favourable host: wheat at the seedling to ear stages. In order for this migratory life strategy to be adaptive, moths produced in the temporary northern habitats must move southward in summer or autumn to avoid the lethal temperatures of winter. Such movements occur, and as the return migrants appear to contribute significantly to the overwintering population in the south, it seems likely that they have the effect of maintaining the migration genotype within the population (Han & Gatehouse, 1991).

## References

Cai, X.-M. (1990). [Review of studies on oriental armyworm for 30 years.] In [*Physiology and Ecology of Oriental Armyworm*], ed. C.-S. Lin, R.-L. Chen, X.-Y. Shu, B.-H. Hu & X.-M. Cai., pp. ii–v. Beijing: University of Beijing Press. In Chinese.

Chen, R.-L. (1962). [Discussion of some problems in the study of the regularities of armyworm outbreaks.] *Scientia Agricultura Sinica,* **1962** (7), 12–16. In Chinese.

Chen, R.-L. (1979). [Armyworm.] In [*Handbook of Crop Pest Forecasting*], ed. R.-L. Chen *et al.*, pp. 121–55. Changchun: Jilin People's Press. In Chinese.

Chen, R.-L. (1990). [A model of armyworm migration.] In [*Physiology and Ecology of Oriental Armyworm*], ed. C.-S. Lin, R.-L. Chen, X.-Y. Shu, B.-H. Hu & X.-M. Cai, pp. 322–35. Beijing: University of Beijing Press. In Chinese.

Chen, R.-L., Bao, X.-Z., Drake, V.A., Farrow, R.A., Wang, S.-Y., Sun, Y.-J. & Zhai, B.-P. (1989). Radar observations of the spring migration into northeastern China of the oriental armyworm moth, *Mythimna separata*, and other insects. *Ecological Entomology,* **14**, 149–62.

Chen, R.-L., Bao, X.-Z., Wang, S.-Y., Sun, Y.-J., Li, L.-Q. & Liu, J.-R. (1988). A radar equipment for entomological observation and its test. *Acta Ecologica Sinica,* **8**, 176–82. In Chinese, English summary.

Chen, R.-L., Bao, X.-Z., Wang, S.-Y., Sun, Y.-J., Li, L.-Q., Liu, J.-R., Zhang, D.-K. & Lu, J. (1985). The Gongzhuling entomological radar and observations on migration of armyworm and meadow moths. *Scientia Agricultura Sinica,* **1985** (3), 93. In Chinese.

Chen, R.-L., Leo, C.-S., Liu, T.-Y., Li, M.-C. & Sien, N.-Y. (1965). Studies on the source of the early spring generation of the armyworm, *Leucania separata* Walker in the Kirin Province: II. Studies on the original ground of the immigrant moths. *Acta Phytophylacica Sinica,* **4** (1), 49–58. In Chinese, English summary.

Crop Pest Forecasting Station of Ministry of Agriculture (1981). [Armyworm.] In [*Forecasting Methods of Main Crop Pests*], pp. 165–79. Beijing: Agricultural Press. In Chinese.

Drake, V.A. (1981). *Quantitative Observation and Analysis Procedures for a Manually Operated Entomological Radar*. CSIRO Australia Division of Entomology Technical Paper 19. Melbourne: CSIRO.

Han, E.-N. & Gatehouse, A.G. (1991). Genetics of precalling period in the oriental armyworm, *Mythimna separata* (Walker) (Lepidoptera: Noctuidae), and implications for migration. *Evolution*, **45**, 1502–10.

Hsia, T.-S., Tsai, S.-M. & Ten, H.-S. (1963). Studies of the regularity of outbreak of the oriental armyworm, *Leucania separata* Walker. II. Observations on migratory activity of the moths across the Chili Gulf and Yellow Sea of China. *Acta Entomologica Sinica*, **12**, 554–64. In Chinese, English summary.

Hu, B.-H. & Lin, C.-S. (1983). Experiments on the flight activity of the oriental armyworm moths, *Mythimna separata* (Walker). *Acta Ecologica Sinica*, **3**, 367–75. In Chinese, English summary.

Li, K.-P., Wong, H.-H. & Woo, W.-S. (1964). Route of the seasonal migration of the oriental armyworm moth in the eastern part of China as indicated by a three-year result of releasing and recapturing of marked moths. *Acta Phytophylacica Sinica*, **3**, 101–10. In Chinese, English summary.

Li, M.-C., Chen, R.-L., Liu, T.-Y. & Leo, C.-S. (1965). Studies on the source of the early spring generation of the armyworm, *Leucania separata* Walker, in Kirin Province: I. Studies on the hibernation and problems of the source. *Acta Entomologica Sinica*, **14**, 21–31. In Chinese, English abstract.

Lin, C.-S. & Chang, J.T.-P. (1964). Studies on the regularities of the outbreak of the oriental armyworm (*Leucania separata* Walker). V. A model for seasonal long-distance migration of the oriental armyworm. *Acta Phytophylacica Sinica*, **3**, 93–100. In Chinese, English summary.

Lin, C.-S. & Hsia, T.-S. (1963). Studies of regularity of the outbreak of the oriental armyworm (*Leucania separata* Walker). III. A discussion of the relationship of migratory habit of the oriental armyworm and the concurrent air movement and of its possible mode of migration. *Acta Universitatis Pekinensis*, **9** (3), 291–308. In Chinese, English summary.

Lin, C.-S., Sun, C.-J., Chen, R.-L. & Chang, J.T.-P. (1963). Studies on the regularity of the outbreak of the oriental armyworm, *Leucania separata* Walker. I. The early spring migration of the oriental armyworm moths and its relation to winds. *Acta Entomologica Sinica*, **12**, 243–61. In Chinese, English summary.

Zou, S.-W. (1956). [Review of the armyworm damage and control in the historical record in China.] *Kunchong Zhishi* [*Entomological Knowledge*], **2**, 241–46. In Chinese.

# 5

# Migration of the Oriental Armyworm *Mythimna separata* in East Asia in relation to weather and climate. II. Korea

## J.-H. LEE AND K.-B. UHM

## Introduction

The Oriental Armyworm *Mythimna separata* is an irregular pest of pasture grasses, maize, rice and other grain crops in the Republic of Korea (Table 5.1). Outbreaks occur every few years, most recently in 1981, 1987 and 1990. As pasture acreages are currently being expanded, *M. separata* is expected to become an increasingly serious pest of Korean agriculture. Research on the species has so far been quite limited, especially in comparison with the intensive long-term studies that have taken place in China and Japan.

It is almost certain that *M. separata* does not overwinter in Korea. The eggs, larvae and pupae do not survive the winter in Kyung-gi province (Fig. 5.1) (Choi & Cho, 1976), and even the southernmost extremities of the country (at 34° N) only just reach the boundary zone (32–34° N) for winter survival on the Chinese mainland (Yan 1991). As *M. separata* adults are able to migrate long distances (~1000 km) on the wind (Li *et al.*, 1965; Yan, 1991), it seems likely that at least the early outbreaks each year are initiated by moths migrating across the Yellow Sea from lower-latitude regions of China. Similar northward migrations in spring lead to annual outbreaks in northern China (Lin & Chang, 1964; Li *et al.*, 1965; Chapter 4, this volume). It has been hypothesised (Choi & Cho, 1976) that the flights into Korea take place in the warm sectors of eastward-moving depressions, as has been established for *M. separata* migration into northeastern Japan (Hirai, 1988*a*).

In 1987, a research program to investigate the patterns of occurrence of *M. separata* was initiated by the Rural Development Administration of the Republic of Korea. A network of moth traps extending throughout the country was set up and operated annually from spring to autumn. The temporal and spatial patterns of catches in these traps are presented in

Table 5.1. *Area (in hectares) of* M. separata *outbreaks in Korea, 1987–1991*

|  | Year | | | | |
| Crop | 1987 | 1988 | 1989 | 1990 | 1991 |
| --- | --- | --- | --- | --- | --- |
| Rice | 61.0 | 16.8 | 743.0 | 206.9 | 179.5 |
| Sweetcorn | – | 18.6 | 14.0 | 116.7 | 41.0 |
| Pasture | 1895.0 | 111.9 | 119.6 | 648.4 | 259.0 |
| Others | 562.0 | 95.5 | 180.6 | 232.5 | 56.5 |
| Total | 2518.0 | 242.8 | 1057.2 | 1204.5 | 536.1 |

this chapter, together with an analysis of the synoptic-scale airflows occurring during the periods when moth immigration was inferred to have occurred.

### Materials and methods

Incandescent-light traps for catching *M. separata* moths were located at 151 forecasting units (FUs) (one for each county or city) throughout the Republic of Korea (Fig. 5.1). In 1991, 40 of these FUs were also equipped with black-light (Sankyo, 20W, FL 20 SBLB) traps. All traps were operated between 15 April and 30 September each year from 1987 to 1991, and were cleared each morning. Analysis has been focussed especially on years (1987, 1990 and 1991) with intermediate or relatively high catch totals.

Weather information was provided by the Korean Meteorological Agency (KMA). Synoptic-scale winds at an altitude of ~1500 m were obtained from daily (21.00 LST) 850-hPa pressure-level charts, and the movements of low-pressure areas were read from monthly weather summaries. Regions with low-level jet streams[1] were identified from the KMA wind data with a computer program (Watanabe *et al.*, 1988) developed for forecasting the long-distance migration of planthoppers (Sogawa & Watanabe, 1991; Chapter 18, this volume). The program interpolates upper-wind measurements from observing stations throughout East Asia to produce a grid of wind-direction and speed values.

---

[1] Defined by Watanabe *et al.* (1988) to be winds of over 20 knot (~10.0 m s⁻¹) at altitudes of 1000–3000 m in the warm sector of a depression. Editors' note: the presence of a maximum in the windspeed profile, as implied by the term *jet*, is not specifically required in this definition.

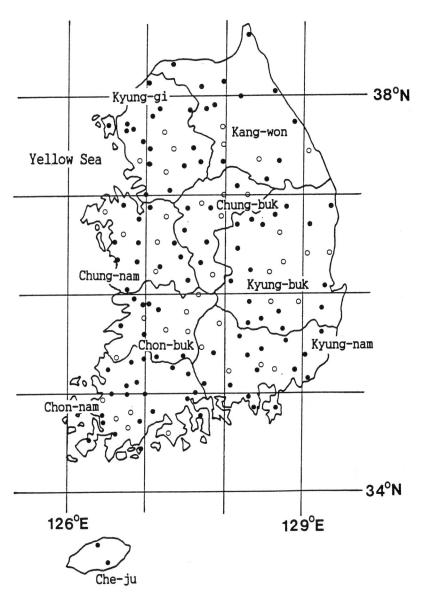

Fig. 5.1. Map of the Republic of Korea, showing location of forecasting units (circles). Incandescent-light traps were present at all sites; in 1991, black-light traps were also operated at the sites marked by open circles.

## Results
### *Temporal and spatial occurrence*

The day-to-day variations of *M. separata* moth catch for the years 1987–1991 are shown in Fig. 5.2. Numbers were relatively high in 1987 and 1990, moderate in 1991, and low in 1988 and 1989. The black-light traps used in 1991 were ~20 times more effective at catching *M. separata* moths than the older design.

Peaks in the catch sequence that were probably the result of immigration events occurred from mid-May to late June. In 1987 there were two small peaks between 21 May and 12 June; in both 1990 and 1991, there was a single peak between 8 and 25 June. A peak starting on 20 June 1987 is of uncertain origin, as it was not associated with airflows from likely source regions in China. Later peaks can be attributed to locally produced moths of the subsequent generation (see below).

The spatial distributions of *M. separata* moth catches during the initial immigration periods in 1987, 1990 and 1991 are shown in Fig. 5.3. The trap data suggest that these immigrants arrive mainly in the central part of the Korean peninsula (between 36°30′ and 38° N), in Kyung-gi, Kangwon, and Chung-nam provinces. Arrival was almost synchronous over the whole country.

### *Relationship of moth immigration to weather*

The 1987–1991 catch sequences suggest that *M. separata* immigration occurs at the same time (mid-May to late June) every year. Identification of synoptic-scale wind systems on which immigration could have occurred has therefore been confined to this period. Three candidate wind systems are shown in Fig. 5.4. In each case, a northeastward-moving depression produced a westerly or southwesterly airflow across the Yellow Sea just before a peak occurred in the trap catches (Fig. 5.2). These airflows, which were identified as low-level jet streams by the computer program of Watanabe *et al.* (1988), could have carried moths to Korea from eastern China. In 1988 and 1989, when trap catches were low, there were no airflows across the Yellow Sea from China during the mid-May to late June candidate immigration period.

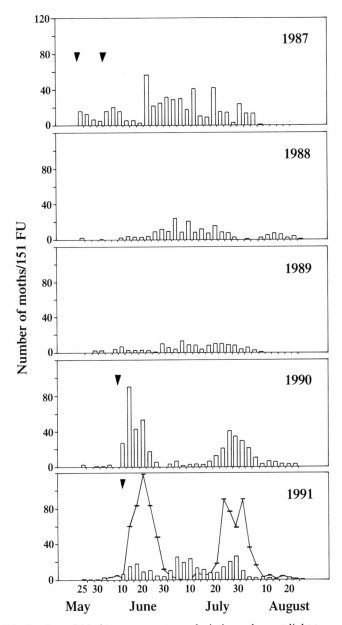

Fig. 5.2. Catches of *Mythimna separata* moths in incandescent-light traps for the seasons 1987–1991 (bars). Values are totals for all 151 forecasting units (FUs), summed in three-day groups. Catches in black-light traps at 40 FUs (1991 only) are also shown (line). Triangles indicate days when depressions passing over or north of Korea produced an airflow across the Yellow Sea from China.

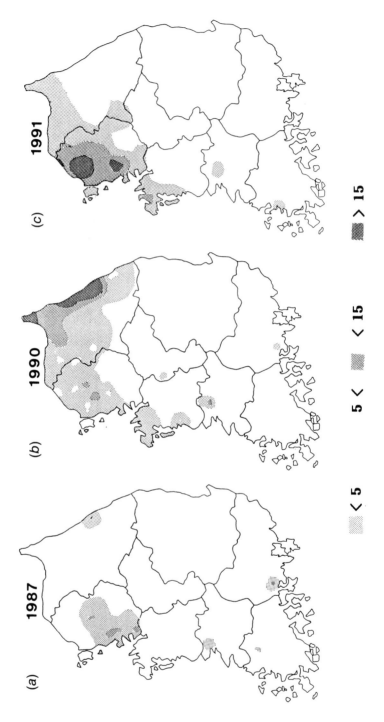

Fig. 5.3. Distribution of *Mythimna separata* catches during the initial immigration periods in (*a*) 1987 (21 May–12 June); (*b*) 1990 (8–25 June); and (*c*) 1991 (8–25 June). Maps prepared using the program *MAPSYS V14* (Lee, Song & Uhm, 1987).

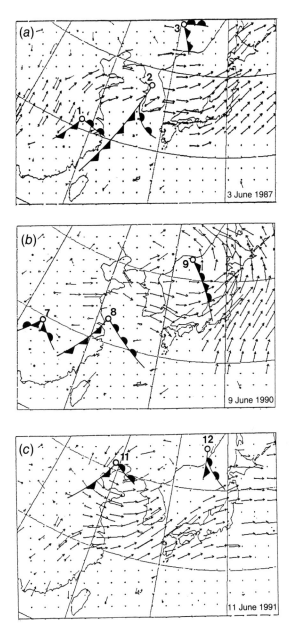

Fig. 5.4. Windfields showing airflows on which immigration from China is thought to have occurred in (a) 1987; (b) 1990; and (c) 1991. Winds interpolated from routine 850-hPa meteorological observations made at 21.00 LST each day; only winds faster than 10 m s$^{-1}$ are shown. Successive daily 21.00 positions of the northeastward-moving depression producing the airflow are also indicated.

## Discussion

### *Temporal and spatial occurrence*

From the five years of data available, it appears that the number of *M. separata* moths caught in May–June may be a useful predictor of damaging outbreaks later in the season. In the years when catches during this initial period were highest (1987 and 1990), severe infestation of pasture grasses followed.

Immigration started much earlier in 1987 than in the other years. Similar early invasions of *M. separata* moths occurred in 1974, when large numbers were caught between 26 and 30 May (Choi & Cho, 1976), and 1981, when the catches were in early May (K.-B. Uhm, unpublished data). In the last case, ~1200 ha of grass pastures and rice in Kyung-gi, Kang-won and Chung-nam provinces (which had all received moths in the initial invasion) were seriously infested by larvae in June. The pattern of damage occurrence in 1987 was similar: these three provinces were again infested with larvae in June.

The peaks in the trap catches in July (Fig. 5.2) can be attributed to local emergence of the subsequent generation. The peak dates are consistent with development periods (egg to adult emergence) of 55 and 40 days at 20 and 25 °C, respectively (Hirai & Santa, 1983). Mean daily air temperatures in June and July are above 20 °C in the outbreak areas. For reasons that are not yet understood, larvae of the subsequent generation (initiated by these locally emergent moths) occur only at low densities (J.-H. Lee & K.-B. Uhm, unpublished data). Consequently, damage by *M. separata* in Korea seems to be due almost entirely to the first generation of larvae, i.e. those hatching from the moths invading in May and June.

Moth invasions seem to be most intense in the northern provinces of Kyung-gi and Kang-won. Historically, damage to crops and pastures has been confined mainly to this area and the adjacent Chung-nam province. It is not known whether these outbreaks extended into North Korea. An outbreak occurred further south (in Chon-nam province) in 1974 (Choi & Cho, 1976). Moths were caught in southwestern Korea during the present study, though in relatively low numbers.

### *Relationship of moth immigration to weather*

The hypothesis of Choi & Cho (1976) that *M. separata* moths reach Korea on the warm-sector winds of eastward-moving depressions has largely been confirmed by this study. Immigration occurs most frequently on strong southwesterly or westerly winds, with windspeeds at 1500 m being typically >10 m s$^{-1}$ and temperatures in the range 18–20 °C. The depressions with which these winds are associated usually pass over the northern Yellow Sea or northeastern China (Fig. 5.4a,c), but are occasionally centred further south (Fig. 5.4b). In the latter case, migration presumably occurs on the cooler airflow behind the cool front rather than in the warm sector.

The synoptic weather patterns associated with transport of *M. separata* moths to Korea appear similar to those associated with transport to Japan (Oku & Kobayashi, 1974; Hirai, 1982; Hirai et al., 1985; Hirai, 1988a,b). *M. separata* moths appeared suddenly in northern Japan on and immediately after 6 June 1987 (Hirai, 1988b; Chapter 6, this volume), which coincides with the second peak in the Korean catches (Fig. 5.2). The origin of these immigrants was either China or southwestern Japan, and they were transported on airflows associated with either the depression shown in Fig. 5.4a or a subsequent one that passed over from 5–8 June (Chapter 6, this volume). As in Korea, an outbreak followed later in the summer.

An insect migration can be divided into 'takeoff', 'transmigration' and 'landing' phases, with the last two determining the region where an invasion occurs (Oku & Kobayashi, 1974). *M. separata* moths have been reported as landing in cyclonic centres, at cold fronts, and in thunderstorms (Lin, 1963). In the two years of strongest immigration, 1987 and 1990, cyclonic centres passed through and brought considerable rain to the central Korean peninsula at the time the invasions occurred (Fig. 5.4a,b). Average rainfall was significantly higher in the invasion area (Kyung-yi province) than in the south of the country (Chon-nam province) in late May 1987 (80 v. 11 mm) and early June 1990 (30 v. 15 mm), but only marginally so in early June 1987 (116 v. 104 mm).

Topographical features may also have had an influence on the moths' migrations. Outbreaks in 1987 and 1991 were concentrated in the western coastal lowlands, but in 1990 many moths apparently crossed the inland mountains to reach the east coast (Fig. 5.3).

### Source region and migration route

There are no significant sources of *M. separata* moths within Korea. The major source region for the May–June invasions is most likely to be eastern China (Fig. 5.5), because (a) spring armyworm populations are usually very large (larval densities of 20–50 m$^{-2}$) there (Hirai *et al.*, 1985); (*b*) adults from these populations migrate to northeastern China (Li, Wong & Woo, 1964; Chen *et al.*, 1965) and northern Japan (Oku & Kobayashi, 1974; Hirai *et al.*, 1985) during the same period; and (*c*) *M. separata* moth invasions are associated with airflows from this region.

A single migration route can account for invasions of both Korea and northern Japan: the moths originate in eastern China at 30–34° N, cross the northern Yellow Sea, pass over the Korean peninsula at 37–38° N, and continue to northern Japan at 40–45° N (Fig. 5.5). The centres of the depressions that produce the moth-transporting winds follow much the same path. This route can be seen as a southern variant of that which carries moths from the same source area into northeastern China at exactly the same time of year (Chapter 4, this volume). These more regular flights, which probably pass over the Bohai Sea, are associated with depressions that follow tracks well to the north of those shown in Fig. 5.4.

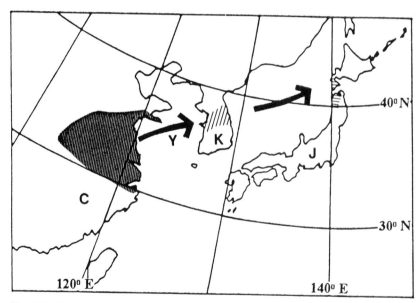

Fig. 5.5. Schematic representation of the migration route which brings *M. separata* moths from source areas in eastern China (dark hatching) to invasion areas in Korea and northern Japan (light hatching). C, China; J, Japan; K, Korea; Y, Yellow Sea.

# References

Chen, R.-L., Leo, C.-S., Liu, T.-Y., Li, M.-C. & Sien, N.-Y. (1965). Studies on the source of the early spring generation of the armyworm, *Leucania separata* Walker in the Kirin province. II. Studies on the original ground of the immigrant moths. *Acta Phytophylacica Sinica*, **4**, 49–58. In Chinese, English abstract.

Choi, K.-M. & Cho, U.-H. (1976). [Studies on the factors of occurrence of the oriental armyworm and development of its artificial diet.] In [*1975 Annual Research Report*], pp. 103–35. Suwon, Korea: Agricultural Science Institute. In Korean.

Hirai, K. (1982). Outbreaks of the armyworm, *Pseudaletia separata* Walker, in and around west Japan in early summer of 1981, with the analyses of weather maps involved in the moth's displacement. *Kinki Chugoku Nogyo Kenkyu [Kinki Chugoku Agricultural Research]*, **64**, 66–8. In Japanese.

Hirai, K. (1988*a*). Sudden outbreaks of the armyworm, *Pseudaletia separata* Walker and its monitoring systems in Japan. *Japan Agricultural Research Quarterly*, **22**, 166–74.

Hirai, K. (1988*b*). Possible source and route of migration of the armyworm, *Pseudaletia separata* Walker, invaded to Tohoku District in June of 1987. *Annual Report of the Society of Plant Protection of North Japan*, **39**, 52–7. In Japanese, English abstract.

Hirai, K., Miyahara, Y., Sato, M., Fujimura, T. & Yoshida, A. (1985). Outbreaks of the armyworm, *Pseudaletia separata* Walker (Lepidoptera: Noctuidae) in northern Japan in the mid-July of 1984, with analysis of weather maps involved in the moths' displacement. *Japanese Journal of Applied Entomology and Zoology*, **29**, 250–3. In Japanese, English abstract.

Hirai, K. & Santa, H. (1983). Comparative physio-ecological studies on the armyworms, *Pseudaletia separata* Walker and *Leucania loreyi* Duponchel (Lepidoptera: Noctuidae). *Bulletin of the Chugoku National Agricultural Experiment Station E*, **21**, 55–101. In Japanese, English summary.

Lee, M.-H., Song, Y.-H. & Uhm, K.-B. (1987). [Studies on the computerization in forecasting insect pests.] In [*1986 Annual Research Report*], pp. 355–72. Suwon, Korea: Agricultural Science Institute. In Korean.

Li, K.-P., Wong, H.-H. & Woo, W.-S. (1964). Route of the seasonal migration of the oriental armyworm moth in the eastern part of China as indicated by a three-year result of releasing and recapturing of marked moths. *Acta Phytophylacica Sinica*, **3**, 101–10. In Chinese, English summary.

Li, M.-C., Chen, R.-L., Liu, T.-Y. & Leo, C.-S. (1965). Studies on the source of the early spring generation of the armyworm, *Leucania separata* Walker, in Kirin province. I. Studies on the hibernation and problems of the source. *Acta Entomologica Sinica*, **14**, 21–31. In Chinese, English summary.

Lin, C.-S. (1963). Studies on the regularity of the outbreak of the oriental armyworm (*Leucania separata* Walker). IV. An analysis of meteoro-physical conditions during the descending movement of the long-distance migration of the oriental armyworm moth. *Acta Phytophylacica Sinica*, **2**, 111–22. In Chinese, English summary.

Lin, C.-S. & Chang, J.T.-P. (1964). Studies on the regularities of the outbreak of the oriental armyworm (*Leucania separata* Walker). V. A model for seasonal long-distance migration of the oriental armyworm. *Acta Phytophylacica Sinica*, **3**, 93–100. In Chinese, English summary.

Oku, T. & Kobayashi, T. (1974). Early summer outbreaks of the oriental armyworm, *Mythimna separata* Walker, in the Tohoku district and possible causative factors (Lepidoptera: Noctuidae). *Applied Entomology and Zoology*, **9**, 238–46.

Sogawa, K. & Watanabe, T. (1991). Redistribution of the rice planthoppers and its synoptic monitoring in East Asia. In *Migration and Dispersal of Agricultural Insects*, pp. 215–28. Tsukuba, Japan: National Institute of Agro-Environmental Sciences.

Watanabe, T., Seino, H., Kitamura, C. & Hirai, Y. (1988). Computer program for forecasting the overseas immigration of long-distance migratory rice planthoppers. *Japanese Journal of Applied Entomology and Zoology*, **32**, 82–5. In Japanese, English abstract.

Yan, Y.-L. (1991). Armyworm (*Mythimna separata*) migration and outbreak forecast in China. In *Migration and Dispersal of Agricultural Insects*, pp. 31–9. Tsukuba, Japan: National Institute of Agro-Environmental Sciences.

# 6

# Migration of the Oriental Armyworm *Mythimna separata* in East Asia in relation to weather and climate. III. Japan

K. HIRAI

## Introduction

The Oriental Armyworm *Mythimna separata* is a serious pest of gramineous crops and pastures in Asia and New Zealand (Sharma & Davies, 1983). In Japan, outbreaks occur every few years, mainly along the western (Sea of Japan) coasts of Hokkaido and Honshu and in western Kyushu (Fig. 6.1). Outbreaks affecting 200 ha or more of crops occurred in eight of the 31 years 1958–1988: in 1960, 1971, 1972, 1978, 1981, 1982, 1984 and 1987. The most serious outbreaks (over 20 000 ha affected) were in 1960, 1971 and 1987 (Oku & Kobayashi, 1977; Hirai, 1988*a*). These outbreaks are of significant economic importance and some required large-scale insecticidal control.

Since 1941, insect pests of rice (especially planthoppers (Delphacidae) and leafhoppers (Deltocephalidae)) have been monitored in Japan with a nationally operated light-trap network. The network extends throughout the country, with four or five traps in each of the 46 prefectures. *M. separata* is caught in these traps, but only in small numbers: the incandescent lamps used do not attract this species efficiently. A smaller network of food-lure traps (containing a mixture of sake lees, water, brown sugar and vinegar in the ratios 13 : 32 : 5 : 1 by weight) has therefore also been set up specifically to monitor *M. separata*. It extends to just four prefectures: Hokkaido, Aomori and Akita in the north and Kagoshima in the south.

In 1983, a research programme aimed at developing a method for forecasting *M. separata* outbreaks was initiated by the Agriculture, Forestry and Fisheries Research Council of Japan. This recent research has confirmed earlier findings on the key role played by migration in the initiation and spread of outbreaks. The conditions leading to armyworm outbreaks, the role of migration in initiating them, and the methods being developed for forecasting them, are described in this chapter.

117

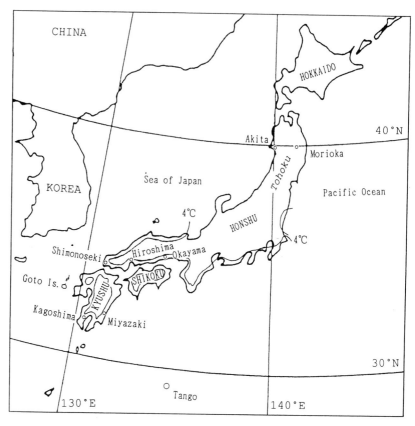

Fig. 6.1. Map of Japan showing localities named in text and 4 °C January average isotherm.

## *M. separata* in Japan

### *Overwintering*

The primary factor limiting *M. separata* populations in Japan appears to be winter mortality. Overwintering, which occurs as larvae or pupae at the base of gramineous pasture plants, is possible only in the warmer climates of southern Kyushu, southern Shikoku, and the southern half of Honshu along the Pacific Ocean coast (Fig. 6.1) (Nagano, 1974; Tanaka, 1976; Hirai & Santa, 1983). The overwintering area corresponds approximately with the region where the January average temperature exceeds 4 °C. Larval densities of $1-2$ m$^{-2}$ in pastures are typical in this area in winter (Hirai & Santa, 1983). Overwintering larvae have not been found in regions where there is continuous lying snow during winter, or where autumn conditions are too cool for crops and pastures to develop a thick vegetation cover before growth ceases (K. Hirai, unpublished data). In China, where overwintering also occurs at the base of plants, survival of the perhaps more numerous larvae can be significant in regions where the January average temperature exceeds 0 °C, i.e. south of ~33° N (Li, Wong & Woo, 1964).

### Flight and migration

*M. separata* adults emerge mainly during the 2 h following dusk (Hirai, 1984). Because the teneral period for this species is ~11–14 h (Hirai & Santa, 1983), most newly emerged adults will be unable to fly during the night of emergence. Active flight commences at dusk on the evening of the day following emergence, when the moths take off with a characteristic ascending flight and move downwind (Hirai, 1991). Similar flights occur on at least the four subsequent nights, unless it is raining. Laboratory flight-mill experiments (Hill and Hirai, 1986) suggest that flight may be continued until, but not usually after, dawn; however, observations of moths released over a lake during the daytime (K. Hirai, unpublished data) and from a ship at sea (Asahina & Turuoka, 1969) indicate that moths continue flying in daylight when over water.

The migrant status of *M. separata* was first established in China in the early 1960s (see e.g. Li *et al.*, 1964; Chapter 4, this volume). The capture of armyworm moths on a weather ship at location 'Tango', almost 500 km south of Japan in the Pacific Ocean (Asahina & Turuoka, 1969; Fig. 6.1), drew attention to the possibility that moths originating in China might reach Japan. It has now been established that long-distance migration occurs on the wind, and that armyworm moths are sometimes transported to Japan from eastern China on the southwesterly airflows in the warm sector of an eastward-moving low-pressure system (i.e. a temperate depression) (Oku, 1983; Hirai, 1988*b*).

### Breeding

The females oviposit on dry or partly dry gramineous plants; frequently used sites include those between the stem and the sheaf of a dried and wilted leaf, and among the corrugations of dead leaves. The neonate larvae move to fresh leaves and feed on them, eventually pupating in the soil or among larval frass at the base of the plant (Tanaka, 1976; K. Hirai, unpublished data).

On arrival, immigrant females have slender abdomens and undeveloped ovaries, but are sometimes already mated, though with no more than one spermatophore. They first fly to and feed on nectar sources, e.g. *Acacia* spp. and *Amorpha fruticosa*, and then mature, mate and commence egg-laying (Hirai, 1988*a*).

## Phenology and spread of populations

In non-outbreak years, *M. separata* moths occur in Japan in small numbers between late-March and October, mainly in maize and rice fields and in pastures. In these years *M. separata* is less numerous than the non-migratory armyworms *Leucania loreyi* and *L. striata* (which predominate in western Japan) and *L. pallens* (which predominates in northern Japan) (Hirai & Santa, 1983; Hirai, 1986).

Moths first appear each year in the southernmost overwintering areas (e.g. around Kagoshima in southern Kyushu), at the end of March. Captures in food-lure traps in this region increase gradually, suggesting local emergence rather than immigration from lower latitudes. By early April moths are being trapped in small numbers in northern Kyushu and the far west of Honshu; this may also be due to local emergence, or to immigration from further south. By the end of April moths have appeared along the entire Sea of Japan coast of Honshu, and by early May they have reached western Hokkaido; i.e. their range has extended well beyond the region where overwintering is possible (Fig. 6.2). Increases in trap catches in the northern areas follow periods of warm southwesterly winds which persist for 1–2 days as temperate depressions move from southwest to northeast between April and May. This gradual northward advance of adults apparently occurs primarily by migration, either from overwintering areas in southern Japan or from ~27–32° N in eastern China, where a large-scale emergence occurs between late March and mid-April (Li *et al.*, 1964).

In outbreak years, major egg-lays first occur in late April and continue until early August; major damage occurs between late June and mid-September. The initial infestations, in late June and early July, are mainly in pastures and wheat crops along the Sea of Japan side of the central mountain range of Honshu and in inland areas of the Tohoku region (northern Honshu): regions in which the species cannot overwinter. Localized outbreaks occur simultaneously along the Sea of Japan coast,

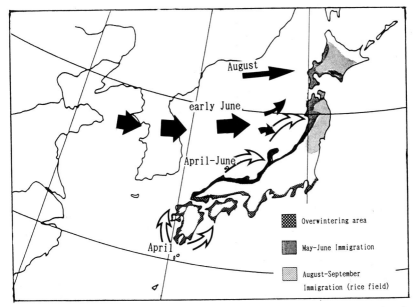

Fig. 6.2. Map of Japan showing *Mythimna separata* overwintering and outbreak areas, and the principal migration routes and times. Open arrows, migration within Japan; solid arrows, immigration from continental East Asia.

and in 1981 and 1987 there were also simultaneous outbreaks in Korea (Hirai, 1982, 1988*a,b*; Chen *et al.*, 1991; Uhm & Lee, 1991; Chapter 5, this volume). These outbreaks appear to be initiated mainly by immigrants from eastern China (Fig. 6.2), with perhaps some immigration from southern Japan also contributing. The emerging moths initiate further larval infestations in rice and maize crops and in pastures during late August and September; these extend to the eastern (Pacific Ocean) coast of Tohoku.

A second, localised, type of outbreak occurs in August and September in rice paddies and maize fields in those parts of the western half of Honshu where the species can overwinter. These outbreaks follow winters that are warmer and dryer than usual (Tamura, 1942), and are apparently initiated by populations overwintering in Japan.

### The origin of outbreaks

A variety of factors have been identified as leading to *M. separata* outbreaks, including both flood (Nawa, 1916; Mukaigawa, 1918) and below-average rainfall (Tamura, 1942; Oumori, 1960; Koyama, 1970). Flooding is now known not to be essential (Hasegawa and Watanabe, 1958; Hirai, 1982; Oku, 1983), and it is clear that major outbreaks follow heavy egg-laying by immigrant moths, a hypothesis first proposed by Okada (1948) and Kato (1948).

The role of immigrants in initiating Japanese outbreaks was first established in the north of the country (Fujita, Toki & Fujimura, 1972), where a relationship between immigration events and the movement of low-pressure areas across the Sea of Japan from eastern China was soon identified (Oku & Kobayashi, 1974). These immigrations occur primarily during two periods: from late May to early June, and during August. An analysis of the incidence and track of the depressions suggests that moths arriving during the first period originate in China at ~36–39° N (i.e. in Shandong and Hebei provinces), while those in the second period come from the southern part of northeastern China (Liaoning and Jilin provinces) at 40–44° N (Oku, 1983). The speeds of the transporting winds, estimated from surface or 850-hPa weather charts, were in the range 5–17 m s$^{-1}$ (18–60 km h$^{-1}$).

Such immigrations can result in outbreaks occurring simultaneously over an area extending from Kyushu in the south to Hokkaido in the north (Hirai, 1982, 1988*a,b*; Oku, 1983; Hirai *et al.*, 1985). The outbreaks of 1954, 1960, 1961, 1971, 1972, 1981 and 1987 followed immigrations during the first (late May to early June) migration period (Moritsu & Hamasaki, 1963; Hirai, 1982, 1988*b*; Oku, 1983), while those of 1953 and 1962 followed immigrations during the second (August) period (Oku, 1983).

### The 1987 outbreak: a case study

The winter of 1986–87 was warmer than normal, and densities of over-wintering larvae averaged 3.5 m$^{-2}$, approximately double that in a normal year (Hirai, 1988*b*). Adults appeared in southern Kyushu in late March, and the first outbreak occurred in a wheat crop in the Goto Islands to the northwest of Kyushu at the end of May. This was followed in late June and early July by outbreaks along the whole of the Sea of Japan side of Honshu, and in Hokkaido. In late August and early September there was a second series of outbreaks in rice, maize and pastures over a wide area of northern Japan. The season closed with a further series of infestations, affecting pastures, in October.

The initial infestation of northern Japan appears to have been due to moths immigrating from both eastern China and Kyushu in April and May, during brief (1–2 days) spells of warm (maximum temperature >25 °C) weather that occur as depressions move eastwards over Honshu. Starting in April 1987, a succession of depressions moved eastwards across the Sea of Japan, and on each occasion small numbers of *M. separata* moths were caught during the next few days in sex-pheromone or food-lure traps in the regions where outbreaks subsequently occurred. Catches in a food-lure trap in northern Honshu in mid-June were 250 times larger than during the previous (non-outbreak) year. The warm airflow associated with a depression that passed over between 5 and 8 June (Fig. 6.3) appears to have been particularly important, and was noteworthy for carrying moths up to 100 km inland from the Sea of Japan coast. Catches in food-lure traps in Morioka, on the Pacific Ocean side of Honshu, between 8 and 17 June suggest that temperatures (18 °C at 1000 m) were sufficiently high on this occasion for moths to be carried across the 1000-m central mountain chain.

Moths appearing in northern Japan in late July and early August are believed to have been the locally produced offspring of the immigrants that arrived in June (Hirai, 1988*a*). These moths gave rise to a second wide-area infestation between mid-August and early September, in which rice, maize and pastures were severely damaged. Moths of this generation emerged between mid-September and early October and initiated a further larval infestation in pastures, at densities of up to 0.7 m$^{-2}$, during October. An increase in the size of moth catches in western Japan during October suggests that there may have been a return migration towards the south (Hirai, 1988*a,b*).

### Forecasting armyworm outbreaks in Japan

In Japan, formal forecasts of agricultural insect pests are issued by the Division of Plant Protection of the Ministry of Agriculture, Forestry and Fisheries. These short-term forecasts are based on reports from the plant-disease and insect-control stations in each prefecture. It would be desirable to issue also a longer-term forecast of early summer armyworm outbreaks, so that a decision can be made on whether an armyworm monitoring and survey programme is needed during the coming summer; in non-outbreak years populations are so low that this is not justified. Such an initial forecast would need to be issued by April, before immigration has occurred; it should ideally be based on information about densities in likely source regions, and on the likelihood of migrants from

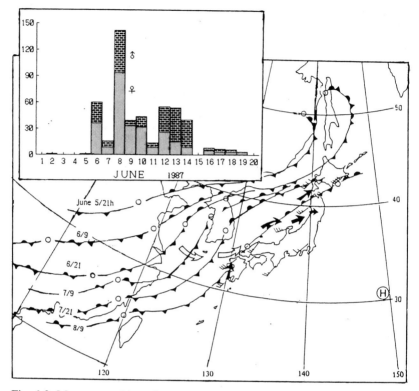

Fig. 6.3. Movement of low air pressure in association with mass immigration of *M. separata*, 5–8 June 1987. Frontal position shown at 09.00 and 21.00 LST each day. Arrows indicate a likely migration route. H on the Pacific stands for a high atmospheric pressure. The bar chart shows the number of adults captured each night in a food-lure trap at Akita. Reproduced from Hirai (1988*a*), with permission.

these regions being transported to Japan. However, overwintering populations of *M. separata* in Japan are at such low densities that surveying them is impracticable and is not attempted, and information from overwintering areas and source regions in China has not been readily available. Research has therefore been focussed on predicting overwintering success from meteorological conditions in the Japanese overwintering areas. This approach has proved effective, even though it is now recognised that the immigrants are more likely to be derived from populations that overwintered in southern China. The method probably works because winter severity in the two regions may be correlated. (A significant correlation has been established between winter severity in each of these regions and winter severity in Shanghai (eastern China) (Hirai,

1990), so a correlation between the two regions themselves is likely.)

It has been found that Japanese armyworm outbreaks occur in years when western Japan experiences higher than average temperatures and lower than average rainfall in January and February. Experimental forecasts, which are being issued informally by the author, are based on an index of wet-cold conditions[1] for February: this has below-average values in outbreak years (Fig. 6.4; Hirai, 1990). The only historical outbreak that this method would have failed to predict is that of 1972, which followed a wet, but exceptionally warm, winter. It is hypothesised (but not yet established) that the number of locations in western Japan with a below-average wet-cold index may provide an indication of the size of the outbreak.

The following scheme is proposed for using these forecasts operationally if this is considered appropriate at some future date. In years when winter temperatures indicate that an outbreak is likely to follow, a watch would be kept between mid-April and June for wind systems likely to carry moths from China to Japan. Suitable depressions would be identified and tracked on standard surface-synoptic charts. Mass immigration would be expected when warm southwesterly winds blow over northern Japan for several days, with temperatures of >20 °C and little rainfall. If no such wind systems occur, the outbreak alert would be cancelled. (A lack of suitable transporting winds apparently accounts for the absence of an outbreak in 1992, when overwintering conditions were favourable.) Once the immigrants have started to arrive, updated forecasts would be issued based on the number of moths caught in monitoring traps in potential outbreak areas. Surveys for larvae would need to be made three weeks after any significant catches, to ensure early identification of infested crops and timely control. If this were not done, control of the subsequent generation between mid-August and early September would be necessary to prevent damage to rice, maize and pastures.

At present, monitoring for immigrant moths relies mainly on a network of food-lure traps that has been set up on the Sea of Japan side of northern Japan (Hirai, 1987). The traps are placed in grass pastures of area >1 ha, in sheltered sites open to the west or southwest. Results from seven years of operation indicate that catches of more than 16 female moths per trap-night are followed about one month later by damaging infestations in nearby crops and pastures (Hirai, 1988*a*). Immigrants are also readily

---

[1] $R/T_{av}$, where $R$ is the rainfall for the month (in mm) and $T_{av}$ is the average temperature for the month (in °C).

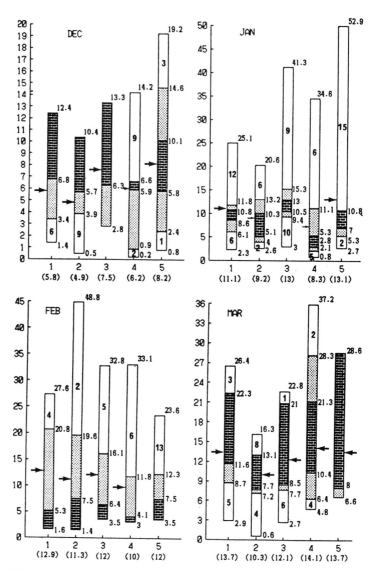

Fig. 6.4. Ranges of the index of wet/cold conditions for the months December through March for five stations in the *Mythimna separata* overwintering area of western Japan. Ranges calculated from observations over 31 years (1957–1988); average values (arrowed) appear in parentheses below the station number. Key to stations: 1, Hiroshima; 2, Okayama; 3, Shimonoseki; 4, Miyazaki; 5, Kagoshima. The shading shows the range of the index for the six outbreak years (1960, 1971, 1978, 1981, 1984 and 1987), and the darker shading the range for the two years with widespread outbreaks (1960, 1987). Reproduced from Hirai (1990), with permission

caught in black-light traps (Hirai, 1986, 1988*a*; Chapter 5, this volume), which have a much higher catching efficiency for *M. separata* than the incandescent-lamp traps of the national pest-monitoring network. (A pheromone trap has also been developed (Takahashi *et al.*, 1979), but because the synthetic lure is less attractive than pheromone from a calling female, it is ineffective at the high moth densities that follow a mass immigration.)

*M. separata* regularly migrates across international boundaries, and outbreaks in one country are often initiated by moths that have developed in another. Regular exchanges of current information on armyworm populations between the various national monitoring programmes would greatly enhance the forecasting and control capabilities of all affected states.

## References

Asahina, S. & Turuoka, Y. (1969). Records of the insects visited a weather-ship located at the Ocean weather station 'Tango' on the Pacific. III. *Kontyu*, **37**, 290–304. In Japanese, English summary.

Chen, G., Jian, H., Ren, H.-F., Luo, L.-W., Zhou, S.-W., Zhang, Z.-L., Lei, S.-F. & Chen, B.-Q. (1991). A 15–year study on the natural control of *Mythimna separata* (Lep.: Noctuidae) by *Ovomermis sinensis* [Nematoda: Mermithidae]. *Chinese Journal of Biological Control*, **7**, 145–50. In Chinese, English abstract.

Fujita, K., Toki, A. & Fujimura, T. (1972). [Outbreaks of *Mythimna separata* in uncultivated fields.] *Annual Report of the Society of Plant Protection of North Japan*, **23**, 135. In Japanese.

Hasegawa, T. & Watanabe, K. (1958). [Outbreaks of *Mythimna separata* in Akita Prefecture in 1957 and their control.] *Annual Report of the Society of Plant Protection of North Japan*, **9**, 73–4. In Japanese.

Hill, M.G. & Hirai, K. (1986). Adult responses to larval rearing density in *Mythimna separata* and *Mythimna pallens* (Lepidoptera: Noctuidae). *Applied Entomology and Zoology*, **21**, 191–202.

Hirai, K. (1982). Outbreaks of the armyworm, *Pseudaletia separata* Walker, in and around west Japan in early summer of 1981, with the analyses of weather maps involved in the moth's displacement. *Kinki Chugoku Nogyo Kenkyu* [*Kinki Chugoku Agricultural Research*], **64**, 66–8. In Japanese.

Hirai, K. (1984). Migration of *Pseudaletia separata* Walker (Lepidoptera: Noctuidae): considerations of factors affecting time of taking-off and flight period. *Applied Entomology and Zoology*, **19**, 422–9.

Hirai, K. (1986). Light and molasses attraction of some noctuid moths. *Tohoku Kontyu* [*Northeast Insects (Japan)*], **24**, 3–5. In Japanese.

Hirai, K. (1987). Live-capture trap for insects. *Kongetsu no Nogyo* [*Agriculture of this Month (Tokyo)*], **31** (7), 92–6. In Japanese.

Hirai, K. (1988*a*). Sudden outbreaks of the armyworm, *Pseudaletia separata* Walker and its monitoring systems in Japan. *Japan Agricultural Research Quarterly*, **22**, 166–74.

Hirai, K. (1988*b*). Possible source and route of migration of the armyworm, *Pseudaletia separata* Walker, invaded to Tohoku District in June of 1987. *Annual Report of the Society of Plant Protection of North Japan*, **39**, 52–7. In Japanese, English abstract.

Hirai, K. (1990). Relationship of weather in overwintering areas to outbreaks of the armyworm, *Pseudaletia separata* Walker (Lepidoptera: Noctuidae). *Japanese Journal of Applied Entomology and Zoology*, **34**, 189–98. In Japanese, English abstract.

Hirai, K. (1991). Ascending flights for migration of the oriental armyworm, *Pseudaletia separata*. *Entomologia Experimentalis et Applicata*, **58**, 31–6.

Hirai, K., Miyahara, Y., Sato, M., Fujimura, T. & Yoshida, A. (1985). Outbreaks of the armyworm, *Pseudaletia separata* Walker (Lepidoptera: Noctuidae) in northern Japan in the mid-July of 1984, with analysis of weather maps involved in the moths' displacement. *Japanese Journal of Applied Entomology and Zoology*, **29**, 250–3. In Japanese, English abstract.

Hirai, K. & Santa, H. (1983). Comparative physio-ecological studies on the armyworms, *Pseudaletia separata* Walker and *Leucania loreyi* Duponchel (Lepidoptera: Noctuidae). *Bulletin of the Chugoku National Agricultural Experiment Station E*, **21**, 55–101. In Japanese, English summary.

Kato, M. (1948). [A note on god insect.] *Shinkontyu* [*New Insects (Japan)*], **1**, 31. In Japanese.

Koyama, J. (1970). [Outbreaks of *Mythimna separata* in 1969 in Tohoku District and the cause.] *Annual Report of the Society of Plant Protection of North Japan*, **21**, 45. In Japanese.

Li, K.-P., Wong, H.-H. & Woo W.-S. (1964). Route of the seasonal migration of the oriental armyworm moth in the eastern part of China as indicated by a three-year result of releasing and recapturing of marked moths. *Acta Phytophylacica Sinica*, **3**, 101–10. In Chinese, English summary.

Moritsu, M. & Hamasaki, S. (1963). An outbreak of *Leucania separata* Walker on Italian rye grass. *Odokon-Chugoku* [*Japanese Journal of Applied Entomology and Zoology, Chugoku Branch*], **5**, 1. In Japanese.

Mukaigawa, Y. (1918). [Outbreaks of *Mythimna separata*.] *Kontyu Sekai* [*Insect World (Japan)*], **22**, 73–4. In Japanese.

Nagano, M. (1974). Forecast of outbreak period of the armyworm, *Leucania separata* Walker in grassland. *Bulletin of the Nagasaki Agricultural and Forestry Experiment Station (Section of Agriculture)*, **2**, 38–45. In Japanese.

Nawa, U. (1916). Outbreaks of *Mythimna separata* in rice fields. *Kontyu Sekai* [*Insect World (Japan)*], **20**, 365–8, 403–11. In Japanese.

Okada, I. (1948). [God insect.] *Shinkontyu* [*New Insects (Japan)*], **1**, 24–5. In Japanese.

Oku, T. (1983). Annual and geographical distribution of crop infestation in northern Japan by the oriental armyworm in special relation to the migration phenomena. *Miscellaneous Publication of the Tohoku National Agricultural Experiment Station*, no. 3, 1–49. In Japanese, English summary.

Oku, T. & Kobayashi, T. (1974). Early summer outbreaks of the oriental armyworm, *Mythimna separata* Walker, in the Tohoku district and possible causative factors (Lepidoptera: Noctuidae). *Applied Entomology and Zoology*, **9**, 238–46.

Oku, T. & Kobayashi, T. (1977). The oriental armyworm outbreaks in Tohoku district, 1960, with special reference to the possibility of mass immigration from China. *Bulletin of the Tohoku National Agricultural Experiment Station*, **55**, 105–25. In Japanese, English summary.

Oumori, H. (1960). [Outbreaks of *Mythimna separata* in 1960.] *Shokubutsu Boeki [Plant Protection (Japan)]*, **14**, 543–6. In Japanese.

Sharma, H.C. & Davies, J.C. (1983). *The Oriental Armyworm,* Mythimna separata *(Wlk.) Distribution, Biology and Control: a Literature Review.* Centre for Overseas Pest Research, ODA Miscellaneous Report 59. London: Centre for Overseas Pest Research.

Takahashi, S., Kawaradani, M., Sato, Y. & Sakai, M. (1979). Sex pheromone components of *Leucania separata* Walker and *Leucania loreyi* Duponchel. *Japanese Journal of Applied Entomology and Zoology*, **23**, 78–81. In Japanese, English abstract.

Tamura, I. (1942). [Outbreaks of *Mythimna separata* Walker and their control.] *Nogyo Oyobi Engei [Agriculture and Horticulture (Japan)]*, **17**, 1429–34. In Japanese.

Tanaka. A. (1976). [Ecology of the armyworm, *Mythimna (Leucania) separata* (Walker).] *Shokubutsu Boeki [Plant Protection (Japan)]*, **30**, 431–7. In Japanese.

Uhm, K.-B. & Lee, J.-H. (1991). Forecasting of occurrence of rice pests in Korea. In *Migration and Dispersal of Agricultural Insects*, pp. 155–65. Tsukuba, Japan: National Institute of Agro-Environmental Sciences.

# 7

# Insect migration in an arid continent. I. The Common Armyworm *Mythimna convecta* in eastern Australia

## G. McDONALD

## Introduction

There is now strong evidence that the arid zones of Australia, the driest continent on Earth, play a vital role in the annual life cycle of several of the country's most significant insect pests. This chapter provides an overview of recent findings on the migratory behaviour of the Common Armyworm *Mythimna convecta*, and of how it leads to outbreaks in the agricultural zones following population buildup in the dry inland. The following two chapters provide similar treatments for the Native Budworm *Helicoverpa punctigera* and the Australian Plague Locust *Chortoicetes terminifera*.

Australia's climate is characterised by relatively low and irregular rainfall: more than 80% of the continent receives less than 600 mm of rain annually (Fig. 7.1a,b). There are major interannual fluctuations in rainfall across the continent which are attributable to the El Niño / Southern Oscillation (ENSO) phenomenon (Nicholls, 1991). The most predictable rainfalls occur in the coastal areas, and regularity and quantity diminish rapidly inland (Division of National Mapping, 1982). The southern part of the continent receives most of its rain in winter in conditions associated with mid-latitude frontal depressions whose centres move eastwards at a latitude of about 40° S. The tropical and subtropical climate zones cover the northern half of Australia and are characterised by rainfall occurring mainly in summer. However, much of the inland of the continent is arid. It is neither far enough north to receive regular monsoon rainfall, nor sufficiently far south to be often visited by rain-bearing cold fronts. The eastern coastal zone receives rain from the moist air that flows from the Pacific Ocean and is forced to rise (orographic lifting) by an eastern fringe of highlands, the Great Dividing Range, and, in summer, by a heated land mass. The range is situated within about

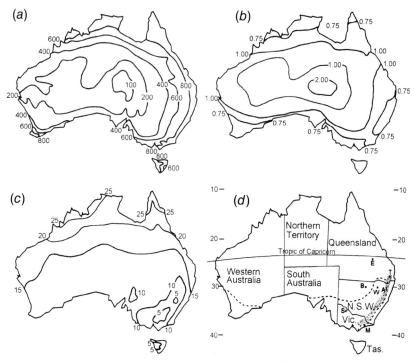

Fig. 7.1. Climatic and political regions of Australia. (*a*) Annual median rainfall; (*b*) rainfall variability; (*c*) average daily temperatures in mid-winter (July); (*d*) political boundaries and trap sites (Fig. 7.3). N.S.W., New South Wales; Vic., Victoria; Tas., Tasmania; E, Emerald; T, Toowoomba; B, Bourke; W, Wee Waa; A, Armidale; S, Swan Hill; M, Maffra. The dashed line indicates the approximate southern limit of the tropical grasslands and the northern limit of the temperate grasslands; the stippling indicates the Great Dividing Range. The measure of rainfall variability used in (*b*) is the ratio $(P_{90}-P_{10})/P_{50}$, where $P_N$ is the $N$-percentile rainfall, i.e. the rainfall not exceeded in $N\%$ of years (Linacre & Hobbs, 1977).

100 km of the eastern and southeastern coast of Australia, and separates the fertile, high-rainfall, coastal fringe from the progressively drier inland. Cropping occurs principally just inland of the Range, whilst grazing extends even further inland.

*Mythimna convecta* is the most widely distributed armyworm species in Australia. It has long been known to damage cereal and grass crops (especially barley) throughout temperate and subtropical mainland Australia (Greenup, 1970). Its occurrence and habits beyond the cropping zone have been the subject of a major ecological study, completed in 1990, that extended throughout the subtropical and temperate eastern mainland (McDonald, 1991*a*). Adult and larval incidence and abundance were monitored over the entire study area throughout the years 1986–89 by a network of food-based lure ('FE') traps (McDonald, 1990*a*) and a series of surveys. This chapter, which is based largely on the study results, presents an overview of current knowledge of the species' distribution and movements, and the biological and meteorological mechanisms that appear to influence migration.

## Evidence of distribution changes

The annual distribution of *M. convecta* within eastern Australia showed a marked seasonal shift, apparently in response to changing temperature and rainfall regimes and their effects on subtropical and temperate grasslands (Fig. 7.1*d*). This seasonal cycle of distribution is shown in Fig. 7.2; its main features in each of the four seasons are described below.

### *Autumn*

In autumn (March–May), moth numbers were low throughout the study area (e.g. Fig. 7.3) while larval densities were often high in the subtropical regions. The largest populations of larvae observed in this study were consistently found in late autumn in northern New South Wales (NSW) and southern Queensland, east of the 350-mm rainfall isohyet and west of the Range. The age distributions of the autumn populations, unlike at other times of the year, were generally narrow and suggested a synchronised egg-lay. However, very high larval densities of *M. convecta* are also known to occur on the north coast of NSW in the February–May period (Greenup, 1970; Rand & Wright, 1978). Likewise, *M. convecta* moths collected at numerous sites on Queensland's east coast were caught mostly during the autumn months (Common, 1965). In the southern regions, where pastures are still dry in March, larvae were usually not found until late April or May, by which time pasture germination and/or regrowth had been stimulated by the first autumn rains. In these temperate pastures, mixed populations of *M. convecta* and the Southern Armyworm *Persectania ewingii* (also a pest species) were recorded in the young growth, although *M. convecta* densities were always low. The autumn surveys did not include all the potential breeding areas; for example, only a few sites were sampled on the eastern coast of NSW and Queensland.

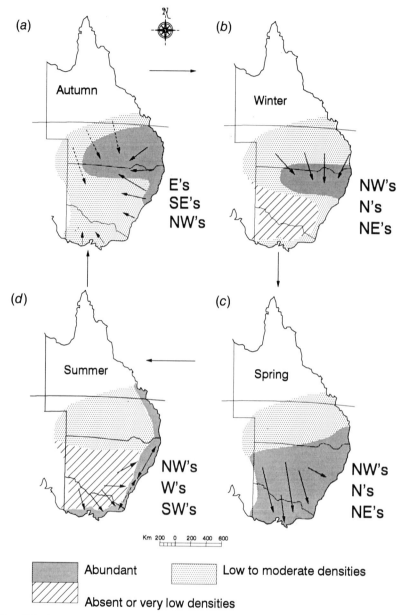

Fig. 7.2. Seasonal shifts (*a–d*) in the distribution of *Mythimna convecta* across mainland eastern Australia, reflecting changes in the availability of green hosts. Major directions of moth migration and associated wind directions necessary to explain the shifts in distribution are indicated. Broken arrows indicate less probable or less frequent movements.

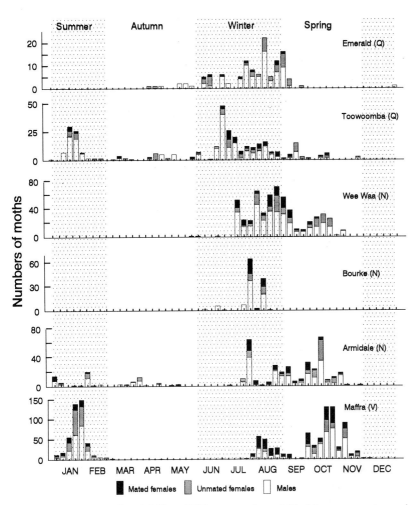

Fig. 7.3. The seasonal variability of FE-trap catches of *Mythimna convecta*, and the reproductive status of the moths caught, at selected trapping sites (see Fig. 7.1*d*) throughout eastern Australia in 1988. (From McDonald (1991*a*)).

### Winter

During winter (June–August), *M. convecta* larvae disappear from the colder southern regions of eastern Australia, except for some coastal areas which do not experience the most extreme temperatures (Fig. 7.1*c*). Elsewhere, the species remains common: larvae were frequently found in northern NSW and relatively high moth catches were recorded in traps throughout southern Queensland, central and northern NSW, and occasionally coastal Victoria (e.g. Fig. 7.3). Winter incidence was most regular, and larval densities highest, at latitudes north of ~34° S.

There is no evidence that winter breeding in the subtropical grasslands of Queensland is less significant than that in the temperate grasslands of northern and central NSW, but it seems likely that this is so. In contrast to the actively growing temperate grasslands, the subtropical grasses are quiescent in this season, and most species are also frost-susceptible (Tothill & Hacker, 1973). Burning of the green foliage by frosts and the absence of new winter growth is likely to diminish both habitat and food quality for *M. convecta* larvae. The larvae themselves are also frost-susceptible, but survival may be enhanced by the dense form of sub-tropical grass tussocks. Where small amounts of green growth persist, larvae may be found embedded and protected in the thick crown of tussock grasses after frosts.

The occurrence of *M. convecta* throughout winter in much of the northern inland has been overlooked by previous workers. Greenup (1970) believed that winter populations were largely confined to the northern coastal and mid-coastal regions of NSW and that these regions were the likely sources for spring infestations. In contrast, Goodyer (1983) suggested that the several overlapping generations of *M. convecta* occur only between spring and autumn, and that the species 'overwinters' as larvae, or as pupae (Smith & Caldwell, 1947). The present study has established that there is no diapause; development through the year is continuous, as exemplified by the mid-winter emergence of large numbers of moths (e.g. Fig. 7.3), and only slows during winter (McDonald, 1990*b*).

As predicted by Smith (1984) on the basis of laboratory studies, survival of *M. convecta* larvae in Victoria during winter is low. In southern Victoria, larvae were absent or below detection thresholds. This was consistent with a previous study in which the only winter collections of *M. convecta* were a few isolated larvae found in the northeast of the state (McDonald & Smith, 1986). The major cause of low larval survival in these southern areas may be less related to the impact of lethally low temperatures (Smith, 1984) than to the indirect effects of protracted development and natural hazards (McDonald, 1991*a*).

### *Spring*

*M. convecta* is most widely distributed during spring (September–November). Throughout most of its range, the species attains its pest status from larval feeding activity in pastures and winter cereals from late winter to early summer, but mostly in spring. Crop damage occurs progressively later in the season, moving towards the south and east. This trend was most clearly demonstrated by the development of a major outbreak of *M. convecta* during winter and spring 1983 (McDonald, Bryceson & Farrow, 1990). The outbreak had its origins in southwestern Queensland during late summer, and there was a gradual latitudinal spread southwards into Victoria and Tasmania, concluding ten months later. Generally, the earliest outbreaks in eastern Australia are reported from June to August in irrigated cereals of central Queensland, particularly Emerald (23°31′ S, 148°10′ E). In southern Queensland, cereals are damaged from August to October (Broadley, 1979), while damage to other gramineous crops may continue into summer (Passlow, 1952). Localised breeding in the northern inland diminishes over the October to December period, and may be directly linked to food and habitat availability. The food quality of the northern temperate grasslands of northern NSW and South Australia deteriorates rapidly over spring as the temperate grasses mature and dry off from early September to October. In addition, the new season's growth of subtropical summer-growing perennials is delayed until the mid-summer rains. In Victoria, there is little evidence for larval activity of *M. convecta* before October (McDonald & Smith, 1986).

### Summer

Most summer activity of *M. convecta*, as measured by moth numbers in the FE traps and presumably resulting in breeding, was confined to the coastal areas of southeastern Australia; only small and occasional catches of moths were recorded in the inland. Except for Victoria, there were few surveys west of the Great Dividing Range during summer, and hence the extent of *M. convecta* breeding in these areas was not adequately assessed. Nevertheless, it can be inferred that breeding in southern NSW and Victoria at this season is unlikely, as the temperate grass pastures of these areas invariably dry off in early summer. In contrast, some limited summer breeding may occur in the grasslands of the northern subtropics, where over 80% of annual grassland production results from summer-growing species (Christie, 1979). This would provide a source of moths for subsequent autumn infestations of the inland: isolated areas of late-summer breeding in southwestern Queensland appeared to be important in the development of the major outbreak of *M. convecta* in 1983

(McDonald *et al.*, 1990). Summer breeding in the subtropics is probably limited to periods when several falls of rain occur over 6–8 weeks, because grass growth after only one rainfall is short-lived due to the high rates of evaporation. Extreme temperatures in the inland may result in high larval mortality, but do not preclude survival, as the species has a relatively high thermal optimum for development (33–34 °C) (McDonald, 1990*b*). In fact *M. convecta*, an essentially tropical species, has presumably evolved this high thermal optimum to permit rapid development in relatively transient grasslands.

The summer populations in Victoria and coastal NSW may be grouped into two categories: those that arise from egg-laying in November and early December (early summer), and those from egg-laying in late January and February (late summer) (Smith & McDonald, 1986). The early-summer moths are probably migrants from northern regions, the result of late-winter or spring generations in central or southern NSW. Their hosts in Victoria include the late-maturing crops and pastures in the northeast and south of the state. The late-summer generation is probably the progeny of local spring populations. Their hosts are largely restricted to environments with moderate temperatures and high rainfall, particularly coastal and near-coastal regions. Such areas include the NSW north coast and southeastern Victoria. Moderate to high densities of *M. convecta* larvae were also reported in February 1988 in green *Lolium* sp. pastures on the NSW northern tablelands where summer rainfall is significant and where temperatures are moderated by altitude (Del Socorro, 1991).

**Evidence for migration**

The most compelling direct evidence in support of *M. convecta* migration at altitude was recorded by Del Socorro (1991) and Gregg *et al.* (1994) who collected moths from an upwards-facing light trap mounted on a tower on a mountaintop (Point Lookout, at 30°30′ S 152°24′ E) in the northern tablelands of NSW. Most females caught were unmated, as is often found with migrant insects (Johnson, 1969). Trajectory analyses from these catches indicated that the moths were migrating in a range of directions, notably including northward and westward movements from and along the adjacent coastal regions. Indirect evidence of migration arose from an FE-trap and radar study at a *M. convecta* emergence site in southeastern NSW in October 1987 (McDonald 1991*a*; V.A. Drake, unpublished data). The study associated local emergence and boundary-layer flight activity of high densities of *M. convecta* with the rapid direct ascent and high altitude (1–2 km) migration of *M. convecta*-like noctuids predominantly moving to the south or southeast. The population size declined considerably over the 30-day observation period. Most females caught at the site were unmated, and unlike reproductively active populations observed at other sites, the proportion of mated females did not increase throughout the study, implying that newly emerged moths were moving away from the site. Other evidence for migration was provided implicitly from the trap records compiled from the present study of distribution. In particular, migration was inferred from many of the moth catches that simultaneously peaked at geographically isolated sites after the passage of synoptic events known to promote migration (McDonald, 1991*a*).

### Permanent and temporary ranges and the migrations that link them

The three seasonal constraints confronted by *M. convecta* within each year are: low winter temperatures in the southern, temperate areas; high summer temperatures in the northern, subtropical areas, particularly inland; and the drying out of most non-coastal grasslands from mid-spring to autumn (Fig. 7.2). It is probably the third constraint, the rapid deterioration of habitats and food quality from mid-spring in the north to autumn in the south, that most seriously affects the permanent range. Over this period, the subtropical ($C_4$) grasses, including those that dominate the northern range and the eastern and southern coastal fringes, will grow vigorously with rains and thus provide potential year-round habitat (Fitzpatrick & Nix, 1970). However, common grass hosts in the inland temperate zone tend not to respond to summer rain and hence cannot support breeding populations.

The temperature constraints are less likely to influence the permanent range directly; laboratory studies suggest that intermittent temperature extremes would not generally be lethal (McDonald, 1990*b*). Therefore, *M. convecta* larvae that pupate before mid-winter could probably survive the southern winters. Similarly, it appears likely that under the hot arid conditions of the northern inland summer, the species may survive in areas where, after good rains, grasses become palatable and provide sufficient protection from the heat and natural enemies. In both situations, mortality is probably much higher than in populations found in areas where the temperatures are less extreme, but still some insects may be expected to survive.

If these hypotheses on survival are correct, then small permanent populations could inhabit the entire northern range of the species together with pockets in the eastern and southern coastal belts (Fig. 7.4); the remaining (southern, non-coastal) range would be occupied only temporarily.

According to this model of seasonal distribution, *M. convecta* is most widely distributed in autumn and spring, and least widely in summer (Fig. 7.2). Although the summer range may include the inland subtropics, in practice most breeding is likely to occur in the high-rainfall coastal regions. This suggests two important periods of migration or migratory phases: (*a*) autumn, with movement into the inland as temperatures decrease; and (*b*) late winter, spring and summer, with movement to the south and eventually out of the inland as temperatures increase. Southward and eastward movements are undoubtedly numerically dominant, but limited-range northward and westward movements, via the east coast, are also probable and provide options for various migration loops (Fig. 7.4).

### Autumn movement into the inland

Reinvasion and/or colonisation of the inland grasslands during autumn (Fig. 7.2*a*) is a critical process in *M. convecta*'s annual cycle of population growth and displacement. Autumn is the only season during which there is at least a reasonable chance of rain throughout the species' entire range. This rain, coupled with declining evaporation rates in the subtropics and declining temperatures and day-length in the temperate zone, induces vigorous grass growth (Fitzpatrick & Nix, 1970). The resulting conditions appear optimal for breeding provided moths from summer generations are able to locate the areas of rain. For example, in northern NSW and southern Queensland, synchronous outbreaks occurred over vast areas after heavy rain in the three successive autumns of 1988–90. Although there are various meteorological processes, particularly wind convergence, that may concentrate moths at rain-affected sites (see e.g. Pedgley *et al.*, 1982; Rose *et al.*, 1987; Drake & Farrow, 1989), the specific mechanisms responsible for transporting and possibly concentrating *M. convecta* moths into such an immense outbreak zone are not understood.

There are several potential source regions of autumn migrants. The arid inland may support small, low-density populations in favourable patches of habitat throughout summer and thus act as a source of moths.

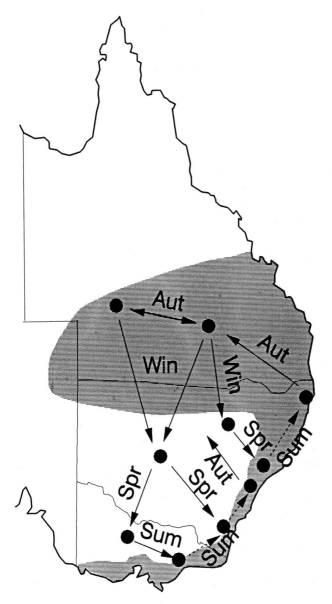

Fig. 7.4. Probable permanent range of *Mythimna convecta* throughout mainland eastern Australia in average years (stippled). In individual years, the range may be less continuous than shown here. The arrows show the proposed eastward and southward paths of migration, and examples of potential return loops, highlighting the significant role of the eastern coastal fringes.

A similar process has been suggested for the African Armyworm *Spodoptera exempta* in East Africa, although in this case the region where populations are believed to persist through the dry season is located near the coast (Pedgley *et al.*, 1989). The grasslands on the northeastern and southeastern coast of Australia, where the high annual rainfall provides good conditions for year-round breeding (Fig. 7.2*a*), are also probably important sources. These coastal regions are the only areas in eastern Australia in which *M. convecta* adults (McDonald, 1991*a*) and larvae (Greenup, 1970; Rand & Wright, 1978) are readily found during summer and autumn. They also provide the species with a corridor through which the descendants of the southern immigrants may regain access to the northern inland, possibly through successive generations and northern and southern migratory loops (e.g. Fig. 7.4). While the large southward migrations appear to be synchronised around major unidirectional wind systems, there is less evidence that return movement of moths from the coastal regions (north and south) to the inland occurs on the same scale. However, Gregg *et al.* (1994) have shown, using backtracks of moth trajectories based on upper-wind data, that migrations of *M. convecta* from the eastern coastal region to the north and northwest may be common during late summer and autumn. There are various wind systems that would assist the westward and northward migrations of coastal emigrants. These include easterly winds, generated by stationary high-pressure cells over central Australia (Baines, 1990), temporary sea-breeze effects that extend well inland (Drake, 1982), or winds around tropical troughs and depressions (Farrow, 1990). The autumn 'return migration' of the Bogong Moth *Agrotis infusa* from the eastern alps to the western plains of NSW and Victoria would also occur on these winds (Common, 1954).

Immigrant females arriving in the previously parched inland in autumn face a dilemma. The immediate prospects of locating green host plants are relatively poor. In the subtropics, there is some chance that previous summer rains may have maintained grass growth, but effective lush growth, which is most conducive to successful breeding, is not often sustained. No summer growth is expected in the temperate regions. Despite this, the largest outbreaks observed in this study occurred in autumn, but immediately after prolonged heavy rain. Autumn incursions of *M. convecta* reported in the literature also appeared to be initiated by heavy rain (Greenup, 1970; Rand & Wright, 1978), as is the case with *S. exempta* in Africa (see e.g. Pedgley *et al.*, 1989). Subtropical grasses can produce new growth within days of heavy autumn rain. Thus, the preparedness or even preference of immigrant *M. convecta* females to lay eggs in dried grasses will ensure that the young larvae have maximum exposure to new foliage, which may have appeared even before the eggs hatch (McDonald, 1991*b*). It is not clear whether the females persist with egg-laying in dried grasses in the absence of rain.

### Late winter to summer movement out of the inland

The decline of grassland habitats commences in winter in the north (with frosting of the quiescent subtropical grasses), and gradually progresses eastwards and southwards (with senescence of the temperate species), concluding in the NSW coastal region, southern Victoria, and Tasmania in mid- to late summer. This process displaces a succession of generations through southern Queensland and NSW to the southern and coastal regions (Fig. 7.2*b–d*). Once the grasses have dried in each area, the prospects of moths finding suitable habitat locally are usually low (although occasionally small patches of favourable habitat may persist through summer). Hence, southward or eastward migrations between each generation improve the chances of some moths being transported into grasslands undergoing vigorous growth. The migration distances may be substantial (500–1500 km) during spring, particularly for moths emerging in the extreme inland areas (Fig. 7.2*c*). During late summer, the last moths from the spring generation emerge in the southern inland areas and need only migrate relatively small distances to coastal or highland habitats (Fig. 7.2*d*). At the end of summer, therefore, the populations of *M. convecta* in the coastal regions of NSW and Victoria may consist of locally emerged individuals together with those that have travelled from distant regions of southern Queensland, northern South Australia, NSW and Victoria.

These migrations are greatly assisted by the prevailing wind systems. Pre-frontal northwesterly airflows provide the strongest and most frequently occurring means of long-distance transportation (Drake & Farrow, 1989). They occur across southern Australia especially from autumn through to spring, and are perhaps most developed during spring. These airflows are highly effective at moving moths into the more productive grasslands of the south or east coasts, as shown for other Australian noctuids, including *Agrotis* spp. (Drake *et al.*, 1981; Common, 1954), *H. punctigera* (Fox, 1978; Chapters 8 and 9, this volume) and *P. ewingii* (Helm, 1975; Drake *et al.*, 1981). In late summer, most successful emigrants from the southern inland presumably use westerly and southwesterly winds, which at this time of year are not so cold as to suppress flight, to migrate to the eastern and northeastern coastal regions.

## A general perspective on migration of *M. convecta* and other armyworm species

### M. convecta – *obligate migrant?*

Migration is an integral component of the life-history strategy of *M. convecta*. The dispersal of young moths between regions and seasons allows the species to recolonise newly formed habitats (produced by rainfall) in a largely ephemeral environment. Similarly, the species may follow seasonally available habitats, as most grasslands do not persist in a suitable form for more than one generation. The exceptions are the grass pastures of the coastal region of eastern Australia, which in many years remain green continuously, and the temperate grasslands. In the latter, at least two generations are possible before the grasses senesce, one commencing in autumn and the second in winter. This raises the question of the need to migrate in all situations. If *M. convecta* relied solely on environmental cues (increasing photoperiod and temperature, decreasing food quality) to induce migration, the winter migrations that were apparent would not have occurred. In addition, the high incidence of mated females in the winter and spring trap catches, relative to those in summer catches, may suggest that the former populations consisted of relatively fewer emigrants. Therefore, it is unlikely that all individuals are obligate migrants. Nonetheless, at least a proportion of a *M. convecta* population appears to migrate regardless of the environment, suggesting the existence of a migration genotype with variable expression, as appears to be the case with *S. exempta* (Parker & Gatehouse, 1985) and *M. separata* (Han & Gatehouse, 1991a). The migratory strategy of *M. convecta* is typical of many insect species that inhabit uncertain environments and that have no form of diapause (Dingle, 1982; Gatehouse, 1986).

An extended pre-reproductive period may assist migrating *M. convecta*. If migration is primarily confined to the pre-reproductive period (Kennedy, 1961), any delay in reproductive maturity would add substantially to the time available to locate the increasingly scarce habitats. Reproductive delays in *M. convecta* are induced by long day-length, or a combination of environmental factors probably including the decline in food quality and larval crowding (McDonald & Cole, 1991). Similar strategies have been proposed for other *Mythimna (Pseudaletia)* spp. In

North America, the combined effect of short photoperiod and low temperature in autumn and winter extends the pre-reproductive period of *M. unipuncta* (Delisle & McNeil, 1987), and allows this species more time to migrate south to the tolerable winter-temperature regimes of Central America (McNeil, 1987). In Asia, the pre-reproductive period of *M. separata* is genetically controlled, providing a mechanism for the genetic regulation of migratory capacity (Han & Gatehouse, 1991*a*). There is also some evidence that the declining photoperiods of autumn induce extended pre-reproductive periods, providing further opportunities for southward migrations before the onset of the harsh winter (Han & Gatehouse, 1991*a,b*). In both of these northern-hemisphere species, photoperiod appears to be one of the most important environmental cues used to anticipate the approaching hazardous winter, allowing escape through emigration. Deterioration in habitat per se has not been implicated. In contrast, increasing photoperiod and probably declining food quality appear to be used by *M. convecta* as cues for emigration to escape the onset of a hazardous summer.

### *Temporary ranges*

Exotic grasses and crops are the most common habitats of *M. convecta* within its temporary (temperate) range, suggesting that its capacity to readily expand its range to the higher latitudes has been greatly aided by agricultural practices. Similarly, the two closely related northern-hemisphere species *M. separata* and *M. unipuncta* have permanent distributions in the south of their ranges, and extend their range northwards during spring and summer into temporary habitats that have been expanded through agriculture (Li *et al.*, 1965; Fields & McNeil, 1984). Neither of these species can survive the subzero temperatures of late autumn and winter in the north and consequently their distributions contract to the lower latitudes. The response of *M. convecta* to seasonal conditions differs from these species in two significant ways. Firstly, the distribution of *M. convecta* is more likely to contract into the lower latitudes (north) by habitat decline in summer and early autumn than by temperature extremes in winter; and secondly, any excursions into the higher latitudes (south) cannot be regarded as non-adaptive (Dingle, 1982) or 'dead-end', even if long-distance return migrations do not occur. This is because southern and eastern coastal habitats provide permanent refugia in contrast to the otherwise parched conditions of the inland temperate grasslands. With the advent of autumn rains, these populations need only move relatively small distances inland (50–200 km), to the north and west, to colonise rejuvenated pastures (Figs. 7.2a, 7.4).

### Return migrations

The question of return migrations of noctuids, involving movement from higher to lower (tropical) latitudes, has been a contentious issue largely because there has been far less firm evidence for this type of movement than for the spring migrations to higher latitudes (Gatehouse, 1986). Rabb & Stinner (1979) proposed that the spring movement to higher latitudes of pest insects such as armyworm moths was unidirectional, and was a chance exploitation of temporary habitats. This was termed the 'Pied Piper' phenomenon, because it was thought to represent a random dispersal of the insects into regions where overwintering was impossible. Walker (1980) suggested that this presents an evolutionary dilemma, as selection would work against those moths moving into the temperate latitudes. He proposed a model that includes a return equatorward migration, which provides a definite evolutionary advantage. There is some evidence for a return migration of *M. separata* moths in China (Chapter 4, this volume). Han & Gatehouse (1991a) suggest that the spring migrations into northern China create a genetic cline across latitude, thereby ensuring that most moths in the autumn generation have the genetic constitution to undertake the return southward journey. There is also circumstantial evidence in support of regular annual return migrations for *M. unipuncta* in North America (Delisle & McNeil, 1987; McNeil, 1987; Chapter 2, this volume). Johnson (1987) argues that while return migrations of *Spodoptera frugiperda* from North to Central America probably do occur, these are not strictly necessary, because migration is adaptive even for populations remaining in the tropics, and this will provide enough selection pressure to maintain the migration genotype. For *M. convecta*, return migrations to the north and west of varying magnitudes seem plausible. While the necessary winds for return migration, the easterlies, southeasterlies and southerlies, are common during the warmer months of January–March in southeastern Australia (Symmons, 1986; Gregg *et al.*, 1994), they tend to be weaker and cooler, and not as favourable for long-range migration (Drake & Farrow, 1985). However, in the south, moths redistributing eastwards or northwards from southeastern coastal regions need only travel 50–200 km to colonise rejuvenated grasslands in autumn. If these movements occur regularly, then there should be no evolutionary dilemma for *M. convecta*.

The population dynamics of *M. convecta* result from a life strategy of migration between favourable but often unpredictable environments. It is typical of the major armyworm pests of the world, including *M. separata* (Li *et al.*, 1965), *M. unipuncta* (Fields & McNeil, 1984), *S. exempta* (Rose *et al.*, 1987) and *S. frugiperda* (Johnson, 1987). The mostly arid and highly unpredictable nature of the Australian environment provides only limited opportunities for *M. convecta* populations to grow rapidly, but the species appears to be well adapted to capitalise on short-lived, favourable conditions.

## Acknowledgements

I greatly appreciate the efforts of Mr A.M. Smith and Dr P.M. Ridland (Department of Agriculture, Victoria), Dr V.A. Drake (The University of New South Wales, Canberra) and Drs D.E. Pedgley and D.R. Reynolds (Natural Resources Institute, UK) for comments on earlier drafts of this manuscript. The original work on *M. convecta* was funded by the Rural Credits Development Corporation and the Barley Research Council.

## References

Baines, P.G. (1990). What's interesting and different about Australian meteorology? *Australian Meteorological Magazine*, **38**, 123–41.

Broadley, R.H. (1979). Armyworms in South Queensland field crops. *Queensland Agricultural Journal*, **105**, 433–43.

Christie, E.K. (1979). Eco-physiological studies of the semiarid grasses *Aristida leptopoda* and *Astrebla lappacea*. *Australian Journal of Ecology*, **4**, 223–8.

Common, I.F.B. (1954). A study of the ecology of the adult bogong moth, *Agrotis infusa* (Boisd.) (Lepidoptera: Noctuidae), with special reference to its behaviour during migration and aestivation. *Australian Journal of Zoology*, **2**, 223–63.

Common, I.F.B. (1965). The identity and distribution of species of *Pseudaletia* (Lepidoptera: Noctuidae) in Australia. *Journal of the Entomological Society of Queensland*, **4**, 14–17.

Delisle, J. & McNeil, J.N. (1987). The combined effect of photoperiod and temperature on the calling behaviour of the true armyworm, *Pseudaletia unipuncta*. *Physiological Entomology*, **12**, 157–64.

Del Socorro, A.P. (1991). *Ecology of armyworms, particularly the common armyworm, in north-eastern New South Wales*. MRurSci thesis, University of New England, Armidale, Australia.

Dingle, H. (1982). Function of migration in the seasonal synchronisation of insects. *Entomologia Experimentalis et Applicata*, **31**, 36–48.

Division of National Mapping (1982). *Atlas of Australian Resources (Third Series)*, vols 1–4, ed. T. Plumb, pp. 4–8. Canberra: Australian Government Publishing Service.

Drake, V.A. (1982). Insects in the sea-breeze front at Canberra, a radar study. *Weather*, **37**, 134–43.

Drake, V.A. & Farrow, R.A. (1985). A radar and aerial-trapping study of an early spring migration of moths (Lepidoptera) in inland New South Wales. *Australian Journal of Ecology*, **10**, 223–35.

Drake, V.A. & Farrow, R.A. (1989). The 'aerial plankton' and atmospheric convergence. *Trends in Ecology and Evolution*, **4**, 381–5.

Drake, V.A., Helm, K.F., Readshaw, J.L. & Reid, D.G. (1981). Insect migration across Bass Strait during spring: a radar study. *Bulletin of Entomological Research*, **71**, 449–66.

Farrow, R.A. (1990). Flight and migration in acridoids. In *Biology of Grasshoppers*, ed. R.F. Chapman & A. Joern, pp. 227–314. New York: John Wiley & Sons.

Fields, P.G. & McNeil, J.N. (1984). The overwintering potential of true armyworm, *Pseudaletia unipuncta* (Lepidoptera: Noctuidae), populations in Quebec. *The Canadian Entomologist*, **116**, 1647–52.

Fitzpatrick, E.A. & Nix, H.A. (1970). The climatic factor in Australian grassland ecology. In *Australian Grasslands*, ed. R.M. Moore, pp. 3–26. Canberra: ANU Press.

Fox, K.J. (1978). The transoceanic migration of Lepidoptera to New Zealand – a history and a hypothesis on colonisation. *New Zealand Entomologist*, **6**, 368–80.

Gatehouse, A.G. (1986). Migration in the African armyworm *Spodoptera exempta*: genetic determination of migratory capacity and a new synthesis. In *Insect Flight: Dispersal and Migration*, ed. W. Danthanarayana, pp. 128–44. Heidelberg: Springer-Verlag.

Goodyer, G.J. (1983). *Armyworm Caterpillars*, Agfact AE.15. Sydney: NSW Department of Agriculture.

Greenup, L.R. (1970). Ecological studies of the common armyworm *Pseudaletia convecta* Walk. MSc thesis, The University of New South Wales.

Gregg, P.C., Fitt, G.P., Coombs, M. & Henderson, G.S. (1994). Migrating moths (Lepidoptera) collected in tower-mounted light traps in northern New South Wales, Australia: influence of local and synoptic weather. *Bulletin of Entomological Research*, **84**, 17–30.

Han, E.-N. & Gatehouse, A.G. (1991*a*). Genetics of precalling period in the oriental armyworm, *Mythimna separata* (Walker) (Lepidoptera: Noctuidae), and implications for migration. *Evolution*, **45**, 1502–10.

Han, E.-N. & Gatehouse, A.G. (1991*b*). Effect of temperature and photoperiod on the calling behaviour of a migratory insect, the oriental armyworm *Mythimna separata*. *Physiological Entomology*, **16**, 419–27.

Helm, K.F. (1975). Migration of the armyworm *Persectania ewingii* moths in spring and the origin of outbreaks. *Journal of the Australian Entomological Society*, **14**, 229–36.

Johnson, C.G. (1969). *Migration and Dispersal of Insects by Flight*. London: Methuen.

Johnson, S.J. (1987). Migration and the life history strategy of the fall armyworm, *Spodoptera frugiperda* in the western hemisphere. In *Recent*

*Advances in Research on Tropical Entomology*, ed. M.F.B. Chaudhury, pp. 543–9. *Insect Science and its Application*, vol. 8, Special Issue. Nairobi: ICIPE Science Press.

Kennedy, J.S. (1961). A turning point in the study of insect migration. *Nature*, **189**, 785–91.

Li, M.-C., Chen, R.-L., Liu, T.-Y. & Leo, C.-S. (1965). Studies on the source of the early spring generation of the armyworm, *Leucania separata* Walker, in Kirin province. I. Studies on the hibernation and problems of the source. *Acta Entomologica Sinica*, **14**, 21–31. In Chinese, English summary.

Linacre, E. & Hobbs, J. (1977). *The Australian Climatic Environment*. Brisbane: John Wiley & Sons.

McDonald, G. (1990*a*). A fermentation trap for selectively monitoring activity of *Mythimna convecta* (Walker) (Lepidoptera: Noctuidae). *Journal of the Australian Entomological Society*, **29**, 107–8.

McDonald, G. (1990*b*). Simulation models for the phenological development of *Mythimna convecta* (Walker) (Lepidoptera: Noctuidae). *Australian Journal of Zoology*, **38**, 649–63.

McDonald, G. (1991*a*). The phenology, ecology and movement of the common armyworm, *Mythimna convecta* (Walker) (Lepidoptera: Noctuidae), in eastern Australia. PhD thesis, La Trobe University, Bundoora, Australia.

McDonald, G. (1991*b*). Oviposition and larval dispersal of the common armyworm, *Mythimna convecta* (Walker) (Lepidoptera: Noctuidae) in eastern Australia. *Australian Journal of Ecology*, **16**, 385–93.

McDonald, G., Bryceson, K.P. & Farrow, R.A. (1990). The development of the 1983 outbreak of the common armyworm, *Mythimna convecta*, in eastern Australia. *Journal of Applied Ecology*, **27**, 1001–19.

McDonald, G. & Cole, P.G. (1991). Factors influencing oocyte development in *Mythimna convecta* (Lepidoptera: Noctuidae) and their possible impact on migration in eastern Australia. *Bulletin of Entomological Research*, **81**, 175–84.

McDonald, G. & Smith, A.M. (1986). The incidence and distribution of the armyworms *Mythimna convecta* (Walker) and *Persectania* spp. (Lepidoptera: Noctuidae) and their parasitoids in major agricultural districts of Victoria, south-eastern Australia. *Bulletin of Entomological Research*, **76**, 199–210.

McNeil, J.N. (1987). The true armyworm, *Pseudaletia unipuncta*: a victim of the Pied Piper or a seasonal migrant? In *Recent Advances in Research on Tropical Entomology*, ed. M.F.B. Chaudhury, pp. 591–7. *Insect Science and its Application*, vol. 8, Special Issue. Nairobi: ICIPE Science Press.

Nicholls, N. (1991). The El Niño / Southern Oscillation and Australian vegetation. *Vegetatio*, **91**, 23–36.

Parker, W.E. & Gatehouse, A.G. (1985). Genetic factors controlling flight performance and migration in the African armyworm moth, *Spodoptera exempta* (Walker) (Lepidoptera: Noctuidae). *Bulletin of Entomological Research*, **75**, 49–63.

Passlow, T. (1952). Armyworm control in cereal crops. *Queensland Agriculture Journal*, **75**, 242–4.

Pedgley, D.E., Page, W.W., Mushi, A., Odiyo, P., Amisi, J., Dewhurst, C.F., Dunstan, W.R., Fishpool, L.D.C., Harvey, A.W., Megenasa, T. &

Rose, D.J.W. (1989). Onset and spread of an African armyworm upsurge. *Ecological Entomology*, **14**, 311–33.

Pedgley, D.E., Reynolds, D.R., Riley, J.R. & Tucker, M.R. (1982). Flying insects reveal small-scale wind systems. *Weather*, **37**, 295–306.

Rabb, R.L. & Stinner, R.E. (1979). The role of insect dispersal and migration in population processes. In *Radar, Insect Population Ecology, and Pest Management*. NASA Conference Publication 2070, ed. C.R. Vaughn, W.W. Wolf & W. Klassen, pp. 3–16. Wallops Island, Virginia: NASA Wallops Flight Centre.

Rand, J.R. & Wright, W.E. (1978). Control of the armyworms *Spodoptera mauritia* and *Pseudaletia convecta* in pastures on the north coast of New South Wales. *Australian Journal of Experimental Agriculture and Animal Husbandry*, **18**, 249–57.

Rose, D.J.W., Dewhurst, C.F., Page, W.W. & Fishpool, L.D.C. (1987). The role of migration in the life system of the African armyworm *Spodoptera exempta*. In *Recent Advances in Research on Tropical Entomology*, ed. M.F.B. Chaudhury, pp. 561–9. *Insect Science and its Application*, vol. 8, Special Issue. Nairobi: ICIPE Science Press.

Smith, A.M. (1984). Larval instar determination and temperature-development studies of immature stages of the common armyworm, *Mythimna convecta* (Walker) (Lepidoptera: Noctuidae). *Journal of the Australian Entomological Society*, **23**, 91–7.

Smith, A.M. & McDonald, G. (1986). Interpreting ultraviolet-light and fermentation trap catches of *Mythimna convecta* (Walker) (Lepidoptera: Noctuidae) using phenological simulation. *Bulletin of Entomological Research*, **76**, 419–31.

Smith, J.H. & Caldwell, N.E.H. (1947). Armyworm and other noctuid outbreaks during 1946–47. *Queensland Agriculture Journal*, **65**, 396–401.

Symmons, P.M. (1986). Locust displacing winds in eastern Australia. *International Journal of Biometeorology*, **30**, 53–64.

Tothill, J.C. & Hacker, J.B. (1973). *The Grasses of Southeast Queensland*. St Lucia: University of Queensland Press.

Walker, T.J. (1980). Migrating Lepidoptera: are butterflies better than moths? *Florida Entomologist*, **63**, 79–98.

# 8

## Insect migration in an arid continent.
## II. *Helicoverpa* spp. in eastern Australia

### P. C. GREGG, G. P. FITT, M. P. ZALUCKI
### AND D. A. H. MURRAY

### *Helicoverpa* spp. in Australia

Of the three species of *Helicoverpa* in Australia, two, the endemic *H. punctigera* and the cosmopolitan *H. armigera*, are major pests. In eastern Australia the two species frequently occur together, and can cause severe damage in the cropping regions of the southeast of the continent (Fig. 8.1). Summer crops affected include cotton, sunflowers, sorghum, soybeans, maize, and many vegetables and horticultural crops. Winter/spring crops include chickpeas, field peas and faba beans. It is often difficult to explain changes in the numbers of *Helicoverpa* spp. in the cropping areas, and this has led to speculation that immigration from non-cropping areas further inland could account for some of the discrepancies (Zalucki *et al.*, 1986).

The pest status of *Helicoverpa* spp. arises from four characteristics exhibited to varying degrees by both species: high mobility, polyphagy, high fecundity and facultative diapause (Fitt, 1989). The evidence for migration in *Helicoverpa* spp. has been reviewed by Farrow & Daly (1987). They rated *H. punctigera* as the most migratory and *H. armigera* as the least, with the two major American pest species *H. zea* and *Heliothis virescens* intermediate.

There is compelling evidence that *H. punctigera* is highly migratory, and is probably an obligate migrant. Specimens of presumed Australian origin have been collected in New Zealand (Fox, 1978) and Norfolk Island (Holloway, 1977), indicating migrations of at least 2200 and 1600 km, respectively. Radar observations, combined with light trapping and nets suspended from kites, suggest that *H. punctigera* is one of the most common large insects moving above the boundary layer at night (Drake *et al.*, 1981; Drake & Farrow, 1985). The species frequently appears in large numbers in early spring, arriving almost synchronously

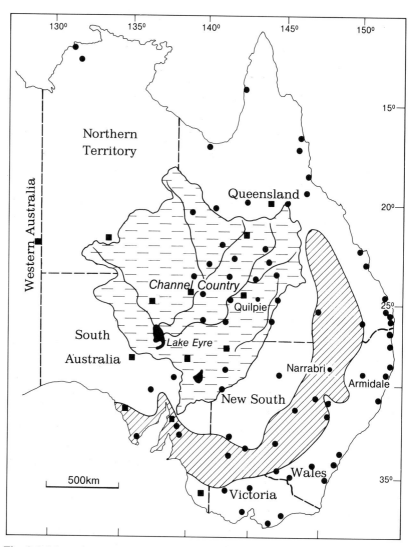

Fig. 8.1. Map of eastern Australia showing the States and locations of towns and regions mentioned in the text. Horizontal hatching, internal drainage into Lake Eyre basin; diagonal hatching, major cropping belt of southeastern Australia; there are also smaller, localised cropping areas along the east coast of Queensland and New South Wales. Circles, sites of pheromone traps only; squares, sites of light and pheromone traps.

over a wide area of eastern Australia (Gregg *et al.*, 1993), including regions where local overwintering is uncommon (Fitt & Daly, 1990).

*H. armigera* is also a migrant. It has been recorded arriving on islands distant from the Australian coast (Holloway, 1977; Farrow, 1984). Studies of genetic variation with the use of electrophoretic techniques (Daly & Gregg, 1985) suggest high levels of gene flow between geographically distant populations, similar to those of *H. punctigera*. This conclusion is supported by the pattern of temporal changes in the frequency of genes for insecticide resistance in agricultural regions (Daly & Fitt, 1990) and regions remote from spraying (Gunning & Easton, 1989). However, in many regions *H. armigera* is less dependent on migration than is *H. punctigera*. It commonly persists in larger numbers during the late summer and autumn, and overwinters in some areas of northern New South Wales and southern Queensland. In cropping regions it is possible to explain at least some population trends without invoking migration (Wardhaugh, Room & Greenup, 1980).

### The far inland of Australia as a source for *Helicoverpa* spp.

From a worldwide viewpoint, arid and semi-arid regions are not noted as sources for *Helicoverpa* spp. However, the far inland of Australia (i.e. the Lake Eyre basin and adjacent regions (Fig. 8.1)) is unusual among such regions, because of its extremely unpredictable rainfall and the heterogeneity of its edaphic and biological components (Stafford Smith & Morton, 1990; Chapter 7, this volume). This variation virtually ensures that at any season of any year, there will be suitable habitat for *Helicoverpa* spp. somewhere within the broad region extending from western Queensland across the Northern Territory and South Australia to Western Australia (Fig. 8.1). Even major droughts (e.g. 1981–83) affect only parts of this huge area.

The far inland is least suitable for *Helicoverpa* spp. in summer. It is very hot (maximum temperatures frequently exceed 40 °C) and, because evaporation is so high, frequently dry, even though much of the region has a summer rainfall maximum. Incursions of moist tropical air, especially during the passage of tropical depressions, occasionally (and unpredictably) produce heavy summer rainfalls. The frequency with which these rain-bearing depressions pass over the area depends on the intensity of the northern Australian monsoon, and this is in turn linked to the El Niño / Southern Oscillation phenomenon. Interannual variation in the number of depressions contributes substantially to the variability of rainfall in tropical and subtropical inland Australia (McBride, 1987). In some seasons, rain-bearing weather systems pass through in autumn, and heavy falls at this time play a key role in making areas of the far inland suitable for winter breeding.

Winter temperatures in the far inland, in contrast to those in the cropping areas, are mild enough to permit the development of *Helicoverpa* larvae if autumn and winter rain produces growth of suitable hosts. The availability of hosts can also be affected by flooding following summer and autumn rain. In the internal drainage systems of far inland Australia (the large Lake Eyre basin and 'Channel Country' shown in Fig. 8.1, and a series of smaller catchments further to the west), heavy rainfall leads to inundation of a network of floodplains (which may be up to 50 km wide in places) and broad depressions. Lush plant growth develops as the floods recede, often some months after any rain-fed vegetation on higher ground nearby has senesced.

The landforms and vegetation associations of far inland Australia are varied and frequently distributed in a fine-grained matrix. Beard (1981) recognises 13 primary vegetation types, of which four widespread types are particularly important in the population dynamics of *Helicoverpa* spp:

1 Low woodland or shrubland, dominated by mulga (*Acacia aneura*) or gidgee (*Acacia cambagei*) and occurring mainly in the higher-rainfall regions;
2 Bunch grasslands dominated by Mitchell grass (*Astrebla* spp.);
3 Hummock grasslands dominated by spinifex (*Triodia* spp.), occurring on sandy soils, including desert sand-dune systems;
4 The grasses and forbs of the floodplains of the Channel Country, on grey clay soils.

In all four of these vegetation associations, the plants that are important *Helicoverpa* hosts are native annuals that germinate and grow in response to rain or flooding in autumn and winter. Some of the major *Helicoverpa* hosts are listed in Table 8.1. Some species, particularly the daisies *Helipterum floribundum* and *Myriocephalus stuartii*, can form dense stands which cover many hundreds of square kilometers. Extensive breeding of *Helicoverpa* spp. can occur in these stands. In most years there will be some regions which receive sufficient rain to produce these conditions, though their location, extent and contiguity will vary considerably from year to year (Nicholls, 1991).

The species that form extensive stands are frequently short-lived

Table 8.1. *Some of the major hosts of* Helicoverpa *spp. in far inland Australia, and the vegetation associations in which they occur*

| Species | Vegetation associations |
|---|---|
| *Asteraceae* | |
| *Helipterum floribundum* (white everlasting) | 1, 3, 4 |
| *Helipterum strictum* | 1, 2, 3 |
| *Craspedia* spp. (billy buttons) | 1, 2, 3, 4 |
| *Myriocephalus stuartii* (poached egg daisy) | 3 |
| *Senecio gregorii* (fleshy groundsel) | 3 |
| *Senecio lautus* (variable groundsel) | 4 |
| *Calotis cuneifolia* (blue burr daisy) | 1 |
| *Calotis multicaulis* (small burr daisy) | 4 |
| | |
| *Goodeniaceae* | |
| *Velleia glabrata* (smooth velleia) | 1, 3, 4 |
| | |
| *Malvaceae* | |
| *Abutilon* spp. | 1, 3 |
| *Sida* spp. | 1, 2, 3, 4 |
| | |
| *Solanaceae* | |
| *Nicotiana* spp. (wild tobacco) | 1, 3 |
| | |
| *Fabaceae* | |
| *Psoralea* spp. (verbines) | 2, 4 |
| *Trigonella suavissima* (Cooper clover) | 4 |

1, *Acacia* woodland and shrubland; 2, tussock grassland; 3, hummock grassland; 4, floodplain.

ephemerals. Other hosts, notably those in the Fabaceae, are more persistent but more restricted in their distribution, germinating in response to flooding in the channel systems. They can remain favourable for breeding well after the surrounding regions have dried out, forming a refuge in which remnant *Helicoverpa* populations can persist in dry times.

In early spring (September), temperatures increase rapidly in the far inland and daily maxima often exceed 30 °C by this time. This period is dry in most regions. The ephemeral plants set seed and senesce quickly and synchronously in these conditions, so that habitat quality (for *Helicoverpa* spp.) deteriorates rapidly over wide areas. Even if a heavy summer rainfall occurs, most growth at this season consists of perennials and grasses, such as *Astrebla* spp. and *Triodia* spp., and these are unsuitable as hosts for *Helicoverpa*.

Between the far inland and the cropping zone is a belt of grazing country. This sometimes provides hosts for immigrants from the far inland, and in the north may be a source of migrants back into the inland. However, its role generally appears to be a secondary one, and it is not considered further here.

### Evidence supporting an origin in the far inland for spring migrants

Gregg *et al.* (1993) have reviewed the evidence for migration from the far inland as the source of most of the *H. punctigera*, and some of the *H. armigera*, appearing at eastern sites during the springs of 1989–91. During this period, an extensive network of pheromone and light traps (Gregg & Wilson, 1991) was operated throughout eastern Australia (Fig. 8.1). The traps were cleared weekly by local collaborators and the catch posted to a central sorting laboratory. *Helicoverpa* moths were identified to species, and the mated status of females caught in light traps determined by dissection. This information was used to identify regions where potential source populations existed. The suitability of such regions for *Helicoverpa* breeding was assessed using rainfall records, with particular attention to rain in the period April–June. Variations in normalised difference vegetation indices (NDVI), recorded in successive months from data from the advanced very high resolution radiometer (AVHRR) instrument on National Oceanographic and Atmospheric Administration (NOAA) polar-orbiting satellites (Bryceson, Hunter & Hamilton, 1993), were used to identify areas where the vegetation had responded to rainfall. Surveys to assess the extent of breeding were conducted where these areas coincided with the four vegetation associations known to support many *Helicoverpa* hosts. Larval populations were measured on various hosts using a quantitative sweep-netting technique. Larvae were reared using artificial diet so that identification could be confirmed at the adult stage, since there are other Heliothine species that cannot be reliably separated from *H. punctigera* and *H. armigera* in the immature stages.

Likely migration trajectories were inferred from synoptic-scale airflows. A number of types of weather system, most notably the strong northwesterly winds that precede cold fronts, have been associated with major insect migrations in Australia (Drake & Farrow, 1985; Farrow & McDonald, 1987). Such systems cross southern Australia at least once weekly in early spring. When the passage of one of these systems was associated with sudden increases in trap catches in the southeast of the continent, trajectories were calculated by backtracking, using winds measured at ~600 m at 22.00 and 04.00 LST at stations of the Australian Bureau of Meteorology's operational upper-air observing network. The winds measured at the closest station were normally used, but data from more appropriate nearby stations were substituted when a frontal discontinuity in the windfield (identified by inspecting the surface synoptic chart) was located between the closest station and the trajectory point. Flight durations of 10 h per night were assumed, and no allowance was made for downwind flight or the possible effects of low-level wind jets (Drake, 1985). Despite these conservative assumptions, backtracking has indicated numerous occasions when moths could have reached the eastern cropping areas from the far inland after covering 1000–1500 km in 2–3 nights of flight (Gregg *et al.*, 1993).

Scanning electron microscopy of moth-borne pollen (Gregg, 1993) was used in conjunction with backtracking to establish likely areas of origin for immigrant moths. Moths feed on a wide range of plants prior to migration, and probably also during pauses in migration, perhaps during the day (Coombs, 1992). Differences in morphology and surface sculpture allow identification of pollen retained on the proboscis, to the level of family and sometimes species. If the distribution and flowering phenology of the plants are known, the regions where the moths fed can be deduced (cf. Hendrix *et al.*, 1987).

These studies have enabled winter breeding areas in far inland Australia during 1989, 1990 and 1991 to be identified (Fig. 8.2). Breeding followed above-average rainfall in the April–June period in each case. The larvae found were predominantly *H. punctigera*; the proportion of *H. armigera* ranged from 20–30% in some samples from central Queensland to less than 1% in South Australia and Western Australia during late winter. Deterioration of the habitat in September (when moths were emerging) led to emigration, and the timing of this coincided with the influx of large numbers of *H. punctigera* across much of southeastern Australia (Gregg *et al.*, 1993; Fig. 8.2).

### A case study: 1989

In early April 1989, tropical cyclone 'Aivu' crossed the coast of northern Queensland in a southwestward direction and degenerated into a rain depression. This system combined with a trough and cold front moving in from the southwest, bringing heavy rain to central and southwestern Queensland. A similar tropical depression associated with tropical cyclone 'Meena' occurred in early May, bringing further heavy falls. Most of Queensland recorded above-average rainfall for the April–June period, and an area of about 150 000 km$^2$ in the southwest corner of the

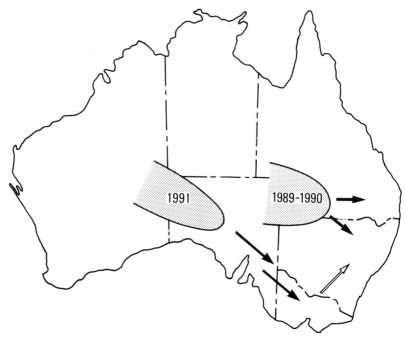

Fig. 8.2. Major winter breeding areas 1989–91, and migration pathways in early spring. The breeding areas reflected the different distribution of April–June rainfall in each year. Moths moved on pre-frontal winds to invade the cropping areas of northern New South Wales and southern Queensland in 1989 and 1990, and western Victoria in 1991. In 1991, subsequent migration on post-frontal southwesterlies led to the invasion of cropping regions in northern New South Wales.

state received the highest rainfall on record (Fig. 8.3). This region was centred on the town of Quilpie, which received more than its average annual rainfall of 319 mm from these two systems. Major flooding followed in the Channel Country.

Extensive germination of *Helicoverpa* hosts occurred in May and June, leading to dramatic changes in NDVI on the NOAA images (Fig. 8.4). Initial surveys indicated the presence of *Helicoverpa* larvae over a wide area, but particularly in a region of approximately 200 000 km² in southwestern Queensland, which was given closer attention in subsequent surveys (Fig. 8.5). The hosts that responded to the rain, as indicated by roadside vegetation samples, and the numbers of *Helicoverpa* larvae found during surveys in this intensively sampled region in July and early September are shown in Table 8.2.

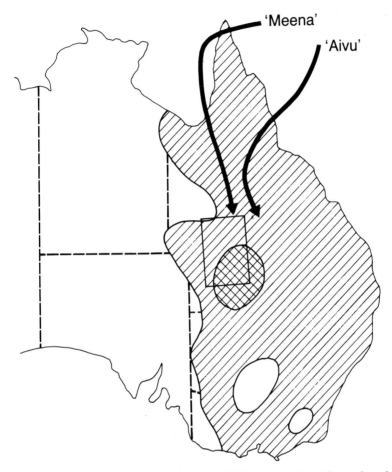

Fig. 8.3. Rainfall for the period April–June 1989 in eastern Australia, and tracks of two tropical cyclones ('Aivu', 2–6 April 1989 and 'Meena', 7–12 May), which degenerated into rain depressions after crossing the coast. Hatching, rain in the 9th decile (wettest 20% of years); cross-hatching, highest rain on record. The rectangle indicates the intensively surveyed area where very high numbers of winter-breeding *Helicoverpa* spp. were found.

Host plants were present over 81% of the intensively sampled area in July, and 90% in early September. Temporal differences in the extent of some plants reflect different germination and flowering times, and the inaccessibility of some areas, particularly floodplains, during the first of the two surveys. The vegetation was generally green in July, but by September it had dried off considerably, except in the floodplains (Fig. 8.4).

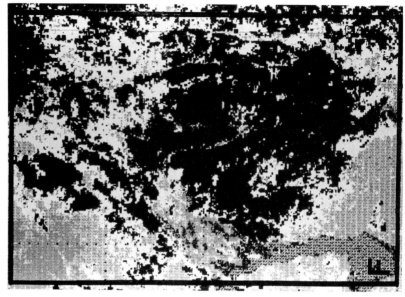

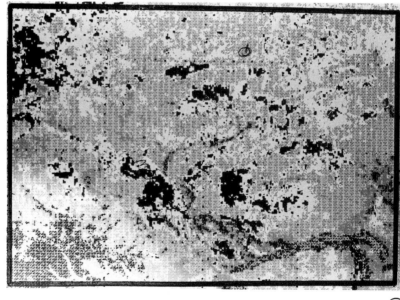

(b)

(a)

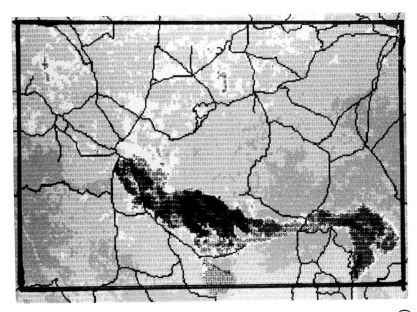

(c)

Fig. 8.4. Changes in normalised difference vegetation index (NDVI) in the intensively surveyed area of western Queensland in 1989. Dark areas have high NDVI values, indicating extensive green vegetation, mostly ephemeral hosts for *Helicoverpa* spp. Lighter area (F) indicates flooding in Cooper Creek, one of the major floodplains of the Channel Country. The rectangle that almost fills each image marks the border of the intensively surveyed area (see Fig. 8.3). (*a*) early May, soon after heavy rainfall associated with tropical cyclone 'Aivu', but before the vegetation had responded; (*b*) early June, showing large areas of ephemeral growth in response to the rains associated with both tropical cyclones; (*c*) early September, showing drying off of ephemeral host plants, except in the floodplain of Cooper Creek, where hosts remained green because of soil moisture resulting from the May–June floods.

Table 8.2. *Host plants and* Helicoverpa *numbers in the intensively sampled region of southwestern Queensland (see Fig. 8.3)*

| Species | July | | | | September | | | |
|---|---|---|---|---|---|---|---|---|
| | Frequency | Sweeps | H.p. | H.a. | Frequency | Sweeps | H.p. | H.a. |
| *Helipterum floribundum* | 35.7 | 9 | 17.1 | 0.0 | 74.4 | 23 | 8.3 | 0.1 |
| *Helipterum strictum* | 8.3 | 2 | 3.2 | 0.8 | 2.2 | 2 | 2.0 | 0.0 |
| *Craspedia* spp. | 2.0 | 1 | 0.0 | 0.0 | 11.1 | 4 | 7.9 | 0.0 |
| *Myriocephalus stuartii* | 0.8 | – | – | – | 0.8 | 1 | 11.5 | 0.0 |
| *Senecio lautus* | 1.6 | 2 | 0.0 | 0.0 | 5.6 | 1 | 10.5 | 0.0 |
| *Calotis cuneifolia* | 18.3 | 6 | 35.4 | 0.0 | 7.8 | 2 | 15.6 | 0.0 |
| *Calotis multicaulis* | 20.2 | 1 | 5.0 | 0.0 | 6.7 | 4 | 6.5 | 0.3 |
| Other *Asteraceae*[a] | 2.8 | 6 | 11.9 | 0.9 | 7.7 | 2 | 0.3 | 0.0 |
| *Velleia glabrata* | 21.8 | 13 | 2.7 | 0.1 | 13.3 | 8 | 0.4 | 0.0 |
| *Nicotiana* spp. | 1.6 | 4 | 13.0 | 0.8 | 2.2 | 1 | 12.3 | 0.0 |
| *Psoralea* spp. | 3.2 | 1 | 0.0 | 0.0 | 3.3 | 1 | 0.0 | 0.0 |
| *Trigonella suavissima* | – | – | – | – | 2.2 | 1 | 12.0 | 0.0 |
| Other hosts[b] | 9.5 | – | – | – | 2.2 | – | – | – |
| All hosts | 81.0 | 45 | 11.9 | 0.2 | 90.0 | 50 | 6.7 | 0.1 |

Frequency: the percentage of roadside samples, made at intervals of about 10 km, where the host was conspicuous. These frequencies were based on 252 such samples in July, and 90 in September. Sweeps: the number of times the host was sampled with a sweep net, each sample usually consisting of 200 sweeps. H.p., H.a.: the numbers of *H. punctigera* and *H. armigera*, respectively, that were recovered per 100 sweeps.
[a]*Brachycome* spp., *Calotis* spp., *Verbesina encelioides*, *Helipterum moschatum*, *Ixiolaena brevicompta*.
[b]*Erodium crinitum* (Geraniaceae), *Swainsona* spp. (Fabaceae).

Larvae were found on approximately 20 host species, with some (such as *Calotis cuneifolia* and *Helipterum floribundum*) supporting more than others (such as *Helipterum strictum* and *Velleia glabrata*) (Zalucki *et al.*, 1994). The larvae were overwhelmingly of *H. punctigera*, with only about 1.6% of the July sample and 1.4% of the September sample of *H. armigera*. Higher numbers of larvae were found in July than in September. Approximate calculations indicated that about $2 \times 10^{11}$ larvae would have been present in the intensively sampled area in July (P.C. Gregg, unpublished data), some 3–5 orders of magnitude more than were

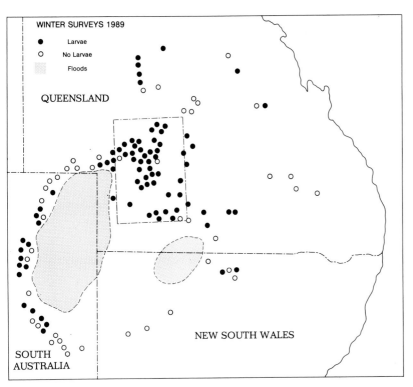

Fig. 8.5. Sites where *Helicoverpa* spp. larvae were found (●) or not found (○) during July–August 1989. Shaded areas were flooded, preventing access for surveys.

overwintering (as pupae) in cropping regions of comparable size in northern New South Wales (Fitt & Daly, 1990).

Simulation modelling by means of the HEAPS model (Fitt, Dillon & Hamilton, 1995) predicted that the July larvae should have emerged in September, with a peak in mid-September. Catches from light and pheromone traps in the breeding region confirmed that many moths were present at this time. Simultaneously, large numbers of moths were being caught in pheromone and light traps at eastern sites that varied widely in latitude and local temperature (Gregg *et al.*, 1993). In some cases, phenological modelling and/or emergence-cage studies showed that local emergence was unlikely to be the source of moths trapped in these eastern locations (Fig. 8.6).

Backtracking indicated that pre-frontal northwesterly winds, which might have brought moths from the Quilpie region to the southeastern

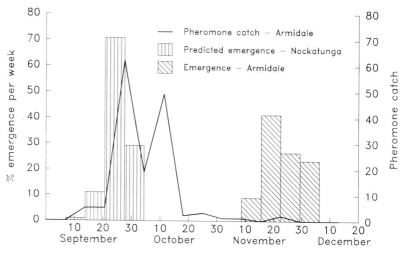

Fig. 8.6. Emergence of *Helicoverpa* spp. moths at Nockatunga (within the intensively surveyed area shown in Fig. 8.3), as predicted by simulation modelling with the HEAPS model (Fitt *et al.*, 1995), compared to pheromone trap catches and emergence at Armidale (in northeastern New South Wales). Emergence results for Armidale were from pupae buried in the previous autumn (M. Coombs, unpublished data); no natural overwintering populations could be found in this region. The phenologies of local and far inland populations suggest that the spring moths had migrated into the area rather than having emerged locally.

cropping areas, occurred on five occasions in August and September 1989. Backtracks over three nights for the strongest of these systems, that of 10–13 September, are shown in Fig. 8.7. The track for moths arriving at Narrabri on 12 September passed over the putative source area. That for 13 September passed further to the southwest, in a region that was not surveyed but is likely to have contained many *Helicoverpa* hosts. These dates coincided with a sudden increase in catches at Narrabri.

Scanning electron microscopy revealed the presence of pollen from plants common in the far inland but absent, or not flowering, in the southeast (Gregg, 1993). These included *Ptilotus* spp. (Amaranthaceae), *Velleia* spp. (Goodeniaceae) and various Asteraceae. Pollen from these plants was found on 30% of *H. punctigera* and 35% of *H. armigera* trapped at Narrabri and Armidale.

## Implications for forecasting

A forecasting system for spring outbreaks of *Helicoverpa* spp. in the cropping areas requires information on the distribution of adults in the far inland during the winter, the availability of host plants in a suitable condition for larval survival, the phenology of any developing populations, and the likelihood of suitable winds for migration to the cropping areas. Some of this information can be obtained from trapping networks, rainfall data and satellite images, but given current knowledge these techniques are useful primarily to guide surveys to determine the extent and timing of winter breeding. The synoptic weather systems associated with migration, particularly cold fronts, are common in early spring and migration to the east or southeast is virtually assured, but these systems are transient and the speed and precise direction of winds about them are difficult to predict. Thus, at present, we can only provide forecasts in general terms.

For crops sown later in the season, such as cotton, sunflowers and sorghum, further complications arise. Such crops are grown in areas where some local overwintering occurs, particularly of *H. armigera*. These moths usually emerge a few weeks after the arrival of the spring immigrants. Progeny of both immigrants and locally overwintering moths can contribute significantly to problems during late spring and summer. Forecasts for these periods must take account of the extent of local overwintering, the timing of local emergence and the availability and quality of local breeding habitats (including both crop and non-crop hosts). We also need information about the patterns of migration within the southeastern region during late spring and summer, a time when the far inland produces very few moths.

## Some unanswered questions

### *Return migrants or Pied Piper victims?*

A major unanswered question is whether migration out of the far inland has adaptive value, or whether it represents a genetic 'dead end' for the inland populations. The latter situation might arise because of a combination of frequent unsuccessful migrations and lack of a return migration pathway (the 'Pied Piper' effect; Rabb & Stinner, 1979).

There is no doubt that many spring migrations have catastrophic consequences for a large proportion of the migrants. The detection of migrants on remote islands suggests that some migrations overfly suitable destinations, and many moths must be lost at sea (Fox, 1978; Farrow, 1984). Similarly, early spring migrants to cooler regions in the eastern highlands of Australia often do not breed successfully (P.C. Gregg, unpublished data). Oviposition can be suppressed by low night-time

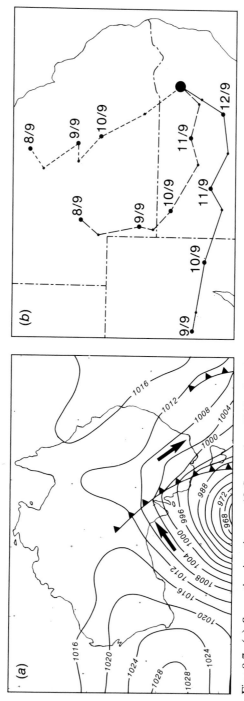

Fig. 8.7. (a) Synoptic situation on 10 September 1989, showing pre-frontal winds that brought *Helicoverpa* spp. from western Queensland to cropping areas of northern New South Wales. (b) Backtracks over three nights for trajectories arriving at Narrabri in northern New South Wales on 11 September 1989 (– – –), 12 September 1989 (— —) and 13 September 1989 (———). Methods for backtracking are described in the text. The track for each night is divided into two legs, before and after midnight. Large dots indicate the end of the track for each night; associated numbers indicate dates.

temperatures (Coombs, 1992), and those eggs that are laid often die, presumably because of cold and late frosts. In drought years, the lower-elevation parts of northeastern New South Wales and southeastern Queensland, though warm enough for successful breeding, can be too dry at the time the moths arrive. On the other hand, conditions further south in Victoria, South Australia and Tasmania (regions where winter rainfall is relatively reliable) consistently permit some breeding. Many crops and weeds, including species that are less common further north, provide good hosts at this time (G. McDonald, personal communication).

We have little information about when and how often (if at all) *Helicoverpa* spp. migrate from the southeast back to source areas in the far inland. The numbers of *H. punctigera* frequently decline in crops in the southeast during the late summer and autumn (Wardhaugh *et al.*, 1980). At this time, easterly and southeasterly winds are common throughout inland Australia (Symmons, 1986). However, these winds are usually much lighter, especially at night, than those from the opposite direction that occur in spring as cold fronts pass. If return migration to the far inland occurs in summer and autumn, it is likely to be a slower and more uncertain process than spring migration. It seems very likely that westward and northwestward movements of up to a few hundred kilometers from eastern coastal regions are possible in summer and autumn (Farrow & McDonald, 1987), because the southeasterly winds are often stronger near the coast. Whether these potential return migrants can reach the Lake Eyre basin remains to be determined.

The genetic losses associated with unsuccessful migrations, together with the lack of a clear return pathway, may appear to constitute an evolutionary dilemma. However, the dilemma may be more apparent than real, because the same migration strategy that occasionally produces catastrophes may be of great survival benefit in the ephemeral environment of the far inland (Farrow & McDonald, 1987). The ability to locate regions (often widely spaced) where sufficient rain has fallen at the right time for good host growth might enable the maintenance of a permanent, if nomadic, population of *H. punctigera* in the continental interior.

### Why so few **H. armigera**?

*H. punctigera* is the dominant *Helicoverpa* species in the far inland. In contrast, in cropping regions, particularly near the coast and in late summer and autumn, *H. armigera* often predominates. This may be due partly to the greater propensity for winter pupal diapause exhibited by *H. armigera*, which would lead to progressively smaller proportions of larvae of this species in the far inland during the autumn and winter. However, even during the autumn, and again in the spring when the moths emerge, the proportion of *H. armigera* is usually small.

The lower numbers of *H. armigera* in the far inland could also be due to poor adaptation to some native hosts. However, it seems unlikely that this is the major cause, since *H. armigera* has a known host range almost as wide as that of *H. punctigera* (Zalucki *et al.,* 1986), and it has been detected in the far inland on all *H. punctigera* hosts that have been sampled with sufficient frequency. Rather, it appears that the migratory strategy of *H. armigera* is less well adapted to the patchy and ephemeral habitats of this region than is that of *H. punctigera*. The patterns of electrophoretic variation between widely separated populations of *H. armigera* suggest that migration occurs with sufficient frequency to maintain genetic similarity and a widespread distribution of rare alleles (Daly & Gregg, 1985). However, migration in *H. armigera* may be too infrequent, or many migrations may not cover sufficient distances, to prevent the repeated extinction of the species from substantial regions of the far inland following immigration. Work in Australia (Wardhaugh *et al.*, 1980), Africa (Roome, 1975) and India (Riley *et al.*, 1992) suggests that *H. armigera* is a facultative migrant, which undertakes short movements if suitable hosts are present, and longer ones only when there is a general deterioration in the habitat. This may represent a strategy for exploiting more predictable habitats than are found in the far inland of Australia, where the more regular migrations of the endemic *H. punctigera* perhaps provide a better balance between the risks of staying and those of migration.

### What are the origins of the winter populations in the far inland?

The absence of obvious return migrations poses problems in understanding the origins of winter populations in the far inland. Summer surveys of this region have yielded very few larvae. At this time of the year, even potential refuges such as the floodplains are likely to be dry, or covered with unsuitable host plants, such as grasses. A simple calculation, taking account of both the smaller areas of potential hosts and the lower larval densities found on those hosts that were present, suggests that summer populations in the far inland are $<10^{-4}$ times the size of winter populations (P.C. Gregg, unpublished data). Light trapping has shown that some moths may be present in summer. However, the males are rarely caught in pheromone traps and the females are almost exclusively unmated. It is unlikely that summer breeding in refuges is extensive enough to produce the enormous numbers of larvae that can be found by late autumn.

Another possibility is that the complex dormancy of *Helicoverpa* spp., especially *H. punctigera*, allows them to cope with highly variable conditions (Cullen & Browning, 1978; Wilson, Lewis & Cunningham, 1979). True overwintering diapause, spring diapause and summer dormancy are strategies for surviving harsh conditions. Overwintering pupae form in autumn (March–May) in response to decreasing photoperiod and low temperatures (<22 °C). Moths emerge from these overwintering pupae in spring (September–October), although a proportion (<5%) may not emerge until early summer. Spring diapause, in *H. punctigera* only, occurs in pupae formed in late October to early November, when photoperiod corresponds to that which is most effective in inducing autumn diapause. Spring diapausing pupae produce moths in summer (December–January). Neither autumn nor spring diapause is likely to carry pupae through to the following autumn in the far inland, since all moths would emerge by mid-summer.

Pupal diapause in response to high temperatures (>38 °C) has been recorded in *H. punctigera* and *H. armigera* (Murray & Zalucki, 1990), although the factors responsible for induction and termination of this dormancy are not well understood. When placed at lower temperatures (22 °C), dormant pupae resumed development within 2–4 days. Pupae entering dormancy in far inland areas when temperatures are high during late spring and early summer could survive the harsh summer conditions and remain reproductively viable, as shown for *H. virescens* (Henneberry & Butler, 1986). The lowering of soil temperatures following rainfall may act as a trigger to resume development. Moths would emerge from these pupae 2–4 weeks afterwards, depending on soil temperature. Where sufficient rain fell, this timing would coincide with the early growth of host plants. This hypothesis requires testing through field studies of pupal development in the far inland.

Further research is needed to clarify the relative contributions of summer dormancy, summer breeding in refuges, and return migration to the increase of *Helicoverpa* populations in the far inland during autumn and winter.

## References

Beard, J.S. (1981). Vegetation of central Australia. In *Flora of Central Australia*, ed. J. Jessop, pp. xxi-xxvi. Sydney: Australian Systematic Botany Society.

Bryceson, K.P., Hunter, D.M. & Hamilton, J.G. (1993). Use of remotely sensed data in the Australian Plague Locust Commission. In *Pest Control and Sustainable Agriculture*, ed. S.A. Corey, D.J. Dall & W.M. Milne, pp. 435–9. Melbourne: CSIRO.

Coombs, M. (1992). Diel feeding behaviour of *Helicoverpa punctigera* (Wallengren) adults (Lepidoptera: Noctuidae). *Journal of the Australian Entomological Society*, **31**, 193–7.

Cullen, J.M. & Browning, T.O. (1978). The influence of photoperiod and temperature on the induction of diapause in pupae of *Heliothis punctigera* (Lepidoptera: Noctuidae). *Journal of Insect Physiology*, **24**, 595–601.

Daly, J.C. & Fitt, G.P. (1990). Resistance frequencies in overwintering pupae and the first spring generation of *Helicoverpa armigera* (Lepidoptera: Noctuidae): selective mortality and migration. *Journal of Economic Entomology*, **83**, 1682–8.

Daly, J.C. & Gregg, P.C. (1985). Genetic variation in *Heliothis* in Australia: species identification and gene flow in the two pest species *H. armigera* (Hübner) and *H. punctigera* Wallengren (Lepidoptera: Noctuidae). *Bulletin of Entomological Research*, **75**, 169–84.

Drake, V.A. (1985). Radar observations of moths migrating in a nocturnal low-level jet. *Ecological Entomology*, **10**, 259–65.

Drake, V.A. & Farrow, R.A. (1985). A radar and aerial-trapping study of an early spring migration of moths (Lepidoptera) in inland New South Wales. *Australian Journal of Ecology*, **10**, 223–35.

Drake, V.A., Helm, K.F., Readshaw, J.L. & Reid, D.G. (1981). Insect migration across Bass Strait during spring: a radar study. *Bulletin of Entomological Research*, **71**, 449–66.

Farrow, R.A. (1984). Detection of transoceanic migration of insects to a remote island in the Coral Sea, Willis Island. *Australian Journal of Ecology*, **9**, 253–72.

Farrow, R.A. & Daly, J.C. (1987). Long-range movements as an adaptive strategy in the genus *Heliothis* (Lepidoptera: Noctuidae): a review of its occurrence and detection in four pest species. *Australian Journal of Zoology*, **35**, 1–24.

Farrow, R.A. & McDonald, G. (1987). Migration strategies and outbreaks of noctuid pests in Australia. In *Recent Advances in Research on Tropical Entomology*, ed. M.F.B. Chaudhury, pp. 531–42. *Insect Science and its Application*, vol. 8, Special Issue. Nairobi: ICIPE Science Press.

Fitt, G.P. (1989). The ecology of *Heliothis* species in relation to agroecosystems. *Annual Review of Entomology*, **34**, 17–52.

Fitt, G.P. & Daly, J.C. (1990). Abundance of overwintering pupae and the spring generation of *Helicoverpa* spp. (Lepidoptera: Noctuidae) in New South Wales, Australia: implications for pest management. *Journal of Economic Entomology*, **83**, 1827–36.

Fitt, G.P., Dillon, M.L. & Hamilton, J.G. (1995). Spatial dynamics of *Helicoverpa* populations in Australia: simulation modelling and empirical studies of adult movement. *Computers and Electronics in Agriculture* (in press).

Fox, K.J. (1978). The transoceanic migration of Lepidoptera to New Zealand – a history and a hypothesis on colonisation. *The New Zealand Entomologist*, **6**, 368–80.

Gregg, P.C. (1993). Pollen as a marker for migration of *Helicoverpa armigera* and *H. punctigera* (Lepidoptera: Noctuidae) from western Queensland. *Australian Journal of Ecology*, **18**, 209–19.

Gregg, P.C., Fitt, G.P., Zalucki, M.P., Murray, D.A.H. & McDonald, G. (1993). Spring migrations of *Helicoverpa* spp. from inland Australia 1989–1991: implications for forecasting. In *Pest Control and Sustainable Agriculture*, ed. S.A. Corey, D.J. Dall & W.M. Milne, pp. 460–3. Melbourne: CSIRO.

Gregg, P.C. & Wilson, A.G.L. (1991). Trapping methods for adults. In *Heliothis: Research Methods and Prospects*, ed. M.P. Zalucki, pp. 30–48. New York: Springer-Verlag.

Gunning, R.V. & Easton, C.S. (1989). Pyrethroid resistance in *Heliothis armigera* (Hübner) collected from unsprayed maize crops in New South Wales 1983–1987. *Journal of the Australian Entomological Society*, **28**, 57–62.

Hendrix, W.H., Mueller, T.F., Phillips, J.R. & Davis, O.K. (1987). Pollen as an indicator of long-distance movement of *Heliothis zea* (Lepidoptera: Noctuidae). *Environmental Entomology*, **9**, 483–5.

Henneberry, T.J. & Butler, G.D. (1986). Effects of high temperature on tobacco budworm (Lepidoptera: Noctuidae) reproduction, diapause, and spermatocyst development. *Journal of Economic Entomology*, **79**, 410–13.

Holloway, J.D. (1977). *The Lepidoptera of Norfolk Island: Their Biogeography and Ecology*, pp. 141–61. The Hague: Junk.

McBride, J.L. (1987). The Australian summer monsoon. In *Monsoon Meteorology*, ed. C.P. Chang & T. Krishnamurti, pp. 203–31. Oxford: Oxford University Press.

Murray, D.A.H. & Zalucki, M.P. (1990). Effect of soil moisture and simulated rainfall on pupal survival and moth emergence of *Helicoverpa punctigera* (Wallengren) and *H. armigera* (Hübner) (Lepidoptera: Noctuidae). *Journal of the Australian Entomological Society*, **29**, 193–6.

Nicholls, N. (1991). The El Niño / Southern Oscillation and Australian vegetation. *Vegetatio*, **91**, 23–36.

Rabb, R.L. & Stinner, R.E. (1979). The role of insect dispersal and migration in population processes. In *Radar, Insect Population Ecology, and Pest Management*. NASA Conference Publication 2070, ed. C.R. Vaughn, W.W. Wolf & W. Klassen, pp. 3–16. Wallops Island, Virginia: NASA Wallops Flight Centre.

Riley, J.R., Armes, N.J., Reynolds, D.R. & Smith, A.D. (1992). Nocturnal observations on the emergence and flight behaviour of *Helicoverpa armigera* (Lepidoptera: Noctuidae) in the post-rainy season in central India. *Bulletin of Entomological Research*, **82**, 243–56.

Roome, R.E. (1975). Activity of adult *Heliothis armigera* (Hb) (Lepidoptera: Noctuidae) with reference to the flowering of sorghum and maize in Botswana. *Bulletin of Entomological Research*, **69**, 149–60.

Stafford Smith, D.M. & Morton, S.R. (1990). A framework for the ecology of arid Australia. *Journal of Arid Environments*, **18**, 255–78.

Symmons, P.M. (1986). Locust displacing winds in eastern Australia. *International Journal of Biometeorology*, **30**, 53–64.

Wardhaugh, K.G., Room, P.M. & Greenup, L.R. (1980). The incidence of *Heliothis armigera* (Hübner) and *H. punctigera* Wallengren (Lepidoptera: Noctuidae) on cotton and other host plants in the Namoi Valley, New South Wales. *Bulletin of Entomological Research*, **70**, 113–31.

Wilson, A.G.L., Lewis, T. & Cunningham, R.B. (1979). Overwintering and spring emergence of *Heliothis armigera* (Hübner) (Lepidoptera: Noctuidae) in the Namoi valley, New South Wales. *Bulletin of Entomological Research*, **69**, 97–109.

Zalucki, M.P., Daglish, G., Firempong, S. & Twine, P. (1986). The biology and ecology of *Heliothis armigera* (Hübner) and *H. punctigera* Wallengren (Lepidoptera: Noctuidae) in Australia: what do we know? *Australian Journal of Zoology*, **34**, 779–814.

Zalucki, M.P., Murray, D.A.H., Gregg, P.C., Fitt, G.P., Twine, P.H. & Jones, C. (1994). Ecology of *Helicoverpa armigera* (Hübner) and *H. punctigera* (Wallengren) in the inland of Australia: larval sampling and host plant relationships during winter and spring. *Australian Journal of Zoology* **42**, 329–46.

# 9

# Insect migration in an arid continent. III. The Australian Plague Locust *Chortoicetes terminifera* and the Native Budworm *Helicoverpa punctigera* in Western Australia

K. J. WALDEN

## Western Australia: an environment for migration?

The state of Western Australia (WA) occupies approximately one-third of the Australian continent. It consists principally of a gently undulating plateau 300–600 m above sea level; there are no outstanding mountain ranges. Three broad climatic zones are recognised (Fig. 9.1): a tropical region in the north, with 90% of annual rainfall (500–1200 mm) falling between November and March (summer); a Mediterranean region in the southwest with 80% of annual rainfall (400–1200 mm) falling between April and September (autumn, winter and early spring); and an arid (desert and semidesert) zone, occupying the great majority of the state's land area, in which rainfall averages less than 300 mm per year. Cropping is almost completely confined to the Mediterranean region, and occurs principally in a broad band towards its boundary with the arid zone.

Within the vast arid zone, rainfall occurs mainly during summer in the north, during both summer and winter in the centre, and during winter in the southwest. It is non-seasonal in the southeast (Fig. 9.1*b*). Rain usually falls on only a few days each year and drought conditions can persist for months or even several years. Temperatures are generally high, with mean monthly maxima exceeding 37 °C in the hottest 3–5 months. Average annual evaporation is more than ten times the rainfall.

Over 60% of the arid zone has a vegetation of tree and shrub steppe: arid or semi-desert grasslands dominated by perennial hummock grasses (*Triodia* spp., 'spinifex') with scattered trees (*Eucalyptus* spp.) in the north, and scattered tall shrubs (*Acacia* spp.) in the south (Beard, 1974–76). In the remaining 40%, located in a 250-km band running east–west through the centre and in areas to the west and southwest, there is a Mulga (*Acacia aneura*) shrubland which becomes a low woodland at its southwestern edge. Throughout the arid zone, forbs and both perennial

173

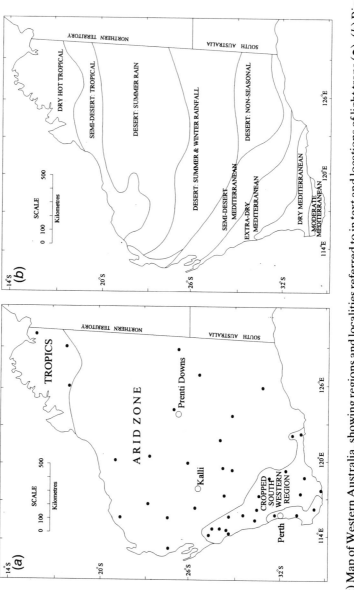

Fig. 9.1. (a) Map of Western Australia, showing regions and localities referred to in text and locations of light traps (●). (b) Bioclimatic regions. Classification based on seasonal rainfall and length of dry period (after Beard (1990)). Months when precipitation does not support plant growth are defined as 'dry': tropical, 7–8 months dry (over winter); semi-desert, 9–11 months dry; desert, 12 months dry (i.e. no assured growing season); Mediterranean, 3–8 months dry (over summer).

and annual grasses respond to erratic drought-breaking rains, and often complete their life cycles after a single rainfall event. Grasses include *Aristida* spp. and *Eragrostis* spp. while forbs are from many families with Amaranthaceae (mulla mullas) and Asteraceae (daisies) common, the latter especially so in the Mulga regions. The forbs often form continuous carpets (Fig. 9.2*a*) extending over wide areas, but these typically senesce after no more than a few weeks and eventually disappear (Fig. 9.2*b*).

The rain-induced vegetation flushes produce habitats that are temporarily highly favourable for suitably adapted herbivores. Insect species that combine short development times with a highly mobile adult stage appear to be particularly effective at exploiting these rapidly changing conditions. Such species can locate newly favourable habitats quickly, then develop rapidly through the almost immobile larval stages before the hosts senesce. In WA, this strategy appears to be employed by the Australian Plague Locust *Chortoicetes terminifera* and the Native Budworm *Helicoverpa punctigera*. Both species breed in the arid zone and are also major agricultural pests in the adjacent croplands to the southwest; their major outbreaks appear to arise from sequences of breeding in, and migration between, these two regions.

The population dynamics and spatial distribution of *C. terminifera* and *H. punctigera* in WA have been the subject of the author's research since 1983 and 1987, respectively. Both species have been studied extensively in eastern Australia (Farrow, 1990; Chapter 8, this volume). This chapter summarises the author's observations (mostly still unpublished elsewhere) and relates them to the regional weather and climate; it also discusses how the species' life strategies are adapted to the WA environment.

### *Chortoicetes terminifera* in Western Australia

*Chortoicetes terminifera* is always present in both the arid zone and the southwestern croplands, usually at very low densities; it has also been collected from the subtropics. Its abundance depends primarily on rainfall, which determines the condition of the species' grass hosts. Following heavy and widespread rains, plagues sometimes develop and cause extensive damage if they form within, or enter, the cropping region.

*C. terminifera* populations in WA have been monitored by ground surveys since 1983, and by a network of light traps since 1987 (Fig. 9.1*a*). Female locusts collected during surveying and trapping are dissected to assess their store of lipid, as this provides an indication of whether they

(a)

(b)

Fig. 9.2. (*a*) Everlasting daisy *Cephalipterum drummondii* flowering in open Mulga woodland at Kalli (see Fig. 9.1*a*) during winter (August 1989). This widespread annual is an important host of *Helicoverpa punctigera*. (*b*) Open Mulga woodland at Kalli during summer (February 1990).

are likely to breed locally, migrate and breed elsewhere, or perish before breeding (Symmons & McCulloch, 1980; Hunter, McCulloch & Wright, 1981; Hunter 1982). Development rates, maturation periods, migration and egg-laying, all strongly dependent on weather factors (Farrow & Daly, 1987; Wright, 1987; Farrow, 1990), are inferred from measurements made routinely at stations of the Australian weather and climate observing network. If conditions are sufficiently favourable for reproductive rates $R_0$ in excess of 10 to be sustained for two or three successive generations, a plague will develop (Farrow 1979); usually this requires each adult generation to migrate to new temporary habitats before breeding. Flight trajectories are estimated by backtracking, using winds from the upper-air observing stations of the weather network and taking account of the species' flight behaviour (Clark, 1971; Drake & Farrow, 1983). A simulation model of locust population dynamics is used to assist with these analyses, which will now be illustrated with an example.

### The 1989–90 locust plague

A major *C. terminifera* plague occurred in WA in the spring and early summer of 1990. Adult locusts swarmed over most of the southwestern croplands, causing extensive damage to agricultural and horticultural crops. Ground surveys a year earlier had shown that very few locusts were present in this region, and development simulations indicate that a local origin was extremely unlikely. It is believed the plague developed in the arid zone.

The locusts probably migrated into the southwestern croplands (area B2 in Fig. 9.3*a*) on two occasions: in late November 1989, when large numbers of adults were collected in light traps in the northern part of the cropping zone, and again in late January 1990. On the nights the locusts were trapped in November, strong northeasterly winds, produced by a high-pressure cell to the southeast and a low-pressure trough to the northwest (Fig. 9.4*a*), prevailed. Such systems often occur in late spring and draw warm tropical air over the arid zone. In January, southwestward flights could have occurred on the eastern side of a tropical depression (Fig. 9.4*b*). Backtracks from the light-trap sites in the southwestern croplands indicate that the migrants probably originated in the arid regions to the northeast. As the November synoptic weather pattern persisted for several days, and the tropical depression produced strong winds throughout the two days of its passage down the coast, it is possible that both migrations covered very large distances.

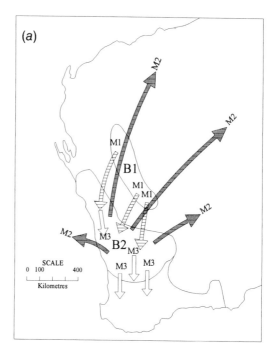

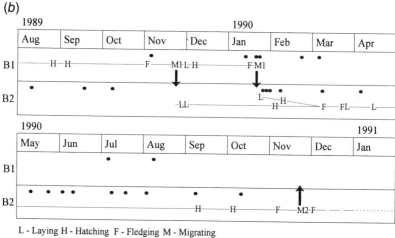

L - Laying  H - Hatching  F - Fledging  M - Migrating
• >20 mm rain   •• >40 mm rain   ••• >100 mm rain

Fig. 9.3. (*a*) Breeding grounds and flights of *Chortoicetes terminifera*, 1989–1990. B1 – spring breeding, 1989; B2 – autumn breeding, overwintering (in eggbeds), and spring breeding, 1990; M1 – night migration, November 1989 and January 1990; M2 – night migration, November 1990; M3 – daytime flights (altitude <10 m), November–December 1990. (*b*) Flow chart showing development of the 1989–91 plague.

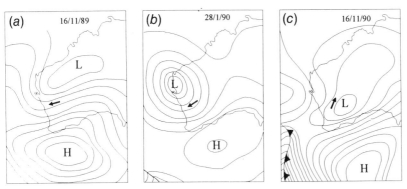

Fig. 9.4. Synoptic charts showing air circulation when *Chortoicetes terminifera* migrated into the southwestern croplands in November 1989 (*a*) and January 1990 (*b*) and into the arid zone in November 1990 (*c*). Arrows indicate the wind direction over areas where *C. terminifera* was fledging.

The most likely source region for these migrants is an area 300–500 km to the northeast (area B1 in Fig. 9.3*a*) which received effective rainfalls during six of the seven months December 1988–June 1989 (i.e. through summer, autumn and early winter). Such rains would have led to successful breeding (i.e. high $R_0$) by any low-density population present. Overwintering would probably have occurred in the egg stage with hatching in early spring (September) 1989. The sequence of breeding and migration that followed is illustrated in Fig. 9.3*b*. Very little rain fell in the area during spring 1989 until mid-November, when there were isolated thunderstorms. The locusts that survived to reach the adult stage would have been able to increase their lipid stores and would have been ready to migrate at the time the large catches were made in the croplands. Some adults may have persisted in area B1, producing a further genera-tion; this seems the most likely origin of the migrants that reached the croplands at the end of January 1990.

Following their arrival in the cropping zone, the November immigrants would either have reproduced immediately, laying eggs that eventually entered a state of quiescence, or they would have persisted and delayed reproduction until rain fell. In January 1990 the tropical depression arrived, bringing not only additional immigrant locusts, but also heavy rainfall that broke the usual summer drought. This allowed the locusts to complete a generation by March (early autumn), and to increase their numbers considerably. The usual mid- to late-autumn rains allowed these adults to persist and breed locally, the eggs overwintering in diapause.

Early the following spring (September–October 1990), locusts hatched from these eggbeds in plague proportions. Despite successful control at many sites, numerous swarms formed by the beginning of summer, and low-altitude daytime flights gradually extended the infested area southwards by over 200 km (Fig. 9.3a). Major displacements, probably at night and at high altitudes, also dispersed the adults over large areas. Locusts were washed up along the coastline and were seen floating on the water some 50 km offshore. Swarms were also found in the arid zone, where there had been no spring hatching, at locations including Meekatharra (26°36′ S, 118°29′ E), Carnegie (25°42′ S, 123°5′ E), and Newman (23°22′ S, 119°44′ E) over 800 km to the northwest of the croplands (Fig. 9.3a). The flights into the arid interior occurred to the west of large low-pressure cells (e.g. Fig. 9.4c); these are common at this time of year and persist for 2–4 days.

The summer of 1990–91 in the cropping zone was a typical one (i.e. very dry and hot), and most adult locusts that did not disperse in November perished before breeding. By the end of summer, the plague was over. The following spring, *C. terminifera* was again at extremely low densities throughout the croplands.

### *Helicoverpa punctigera* in Western Australia

*Helicoverpa punctigera* occurs throughout WA. It is a major pest in the southwestern cropping region where it infests lupins *Lupinus angustifolius*, the state's second largest grain crop (over 800 000 ha sown annually), in spring. Infestation and damage levels vary greatly from year to year.

Information about *H. punctigera* populations in WA has been obtained primarily from a network of 40 light traps. The traps are cleared throughout the year, either daily or twice weekly. Relative abundance of moths is estimated from the catch by correcting for the influences of temperature, wind and nocturnal behaviour (Morton, Tuart & Wardhaugh, 1981). Female moths are dissected to determine their mating status, ovarian development and lipid content. In regions where rainfall and the condition of female moths suggest that breeding may be occurring, ground surveys are conducted to determine the extent of infestation.

The distribution of *H. punctigera* in WA has been found to change continually with the seasons, with extensive breeding in the arid zone in winter, followed by extensive breeding in the southwestern croplands in spring. It appears that major infestations in the croplands are initiated by moths dispersing from the arid zone, and that moths collected in the arid zone in late spring and summer have migrated there from the croplands.

### Development in the arid zone and migration to the southwest

During the winters of 1986, 1989 and 1991, *H. punctigera* populations developed over extensive areas of the central and southern arid zone (Fig. 9.5*a*) on annual wildflowers that germinated after autumn and early-winter rains (Fig. 9.2*a*). Moths emerged at the end of winter, having completed one generation. In each of these years, further local breeding was impossible, as the hosts had senesced by early spring. Female *H. punctigera* caught at these inland emergence sites remained unmated, indicating that there was no local egg-laying. By early spring, the moths had departed.

Detailed observations of moth emergence at one of these inland breeding sites was made in August–September 1991 at Prenti Downs, a remote grazing property in the far centre of the arid zone (Fig. 9.1*a*). For over two weeks, considerable flight activity was seen each night after dusk, with moths rapidly ascending to altitudes of over 20 m. This behaviour is typical of moths initiating migratory flights (Farrow and Daly, 1987).

Approximate migration trajectories for these moths have been estimated from the known flight behaviour of the species (Drake & Farrow, 1985; Farrow and Daly, 1987; Chapter 8, this volume) and the meteorological conditions. Moths have been assumed to remain airborne for 10 h following dusk, and to move with the geostrophic wind. Forward trajectories calculated from Prenti Downs suggest that moths leaving this site could have been displaced over great distances and dispersed over a wide area. Displacement directions were very variable, because the large-scale airflow changed from night to night as a succession of high-pressure cells and cold fronts moved slowly eastwards over the region (Fig. 9.6*a*). Northward movements towards the tropics and subtropics occurred ahead of high-pressure cells, whereas movements into the southwestern cropping zone took place during or following the passage of these cells to the east. The strong north-to-northwesterly winds ahead of cold fronts would have carried moths into the southern and southeastern parts of the arid zone, and possibly beyond the eastern border of WA. Some of these moths may have reached favourable habitats in southeastern Australia after travelling over 2000 km in 2–3 nights (Gregg *et al.*, 1993).

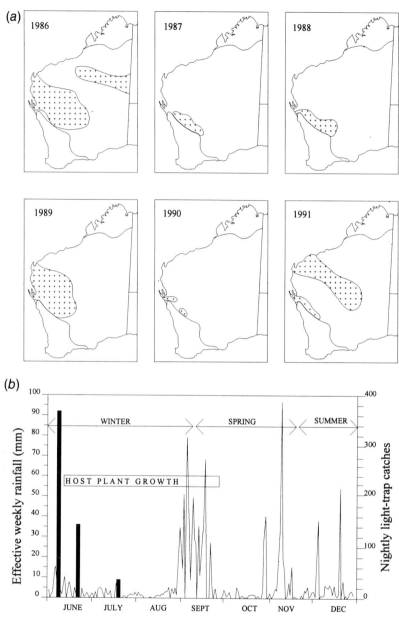

Fig. 9.5. (*a*) Regions of the arid zone within which forbs germinated during winters of 1986–91, and which would have been able to support a generation of *Helicoverpa punctigera*. Winter surveys for larval populations provided support for these broad delineations of potential breeding areas. (*b*) Nightly trap catches, weekly rainfall and period of host-plant growth at Prenti Downs in 1991.

*Helicoverpa punctigera* moths appear synchronously over much of the southwestern cropping zone at the beginning of spring. The initial large catches are invariably associated with warm northeasterly airflows and the female moths are unmated. The number of moths invading the croplands depends not on local rainfall but rather on late-autumn and winter rainfall in the arid zone and the associated extent of winter breeding there. The amount and distribution of rainfall in the arid zone is highly variable. Over the six years 1986–1991, the estimated area within which *H. punctigera* bred during winter varied from 20 000 to over 500 000 km$^2$ (Fig. 9.5a). Major spring outbreaks in the croplands occurred in those years (1986, 1989 and 1991) with the most extensive area of winter breeding in the arid zone. These observations suggest that the moths appearing in the southwest in early spring are immigrants from elsewhere rather than the result of local emergence.

### *Development in the Mediterranean zone and migration to the north and east*

During spring (September–November), extensive breeding of *H. punctigera* is usually confined to the southwestern croplands, where reliable late-autumn and winter rains germinate large areas of cultivated and wild host plants. During summer this region usually experiences drought, and *H. punctigera* numbers diminish rapidly as first the cultivated hosts and later also the wild hosts senesce. Although further generations occur over summer and autumn, numbers decrease so dramatically that breeding populations are very difficult to find.

In October and November 1991, *H. punctigera* moths were observed taking off from lupin crops that had senesced or been harvested. Their flight behaviour was typical of moths initiating a long-distance migration (Farrow & Daly, 1987). These moths had completed only one spring generation in the southwest. By late autumn (April–May) of 1992, very few *H. punctigera* remained in the croplands and significant population movements into the arid zone were no longer possible.

Forward trajectories estimated for the October–November emigrants suggest that they were dispersing over a wide area. Moths were collected from light traps along the south coast, over 200 km from the major spring breeding grounds, and from boats 40 km off the west coast. Large collections were also made in traps 200–1000 km to the northeast in the arid zone. For example, at Prenti Downs there were three large catch peaks in October and November (late spring) and two during December

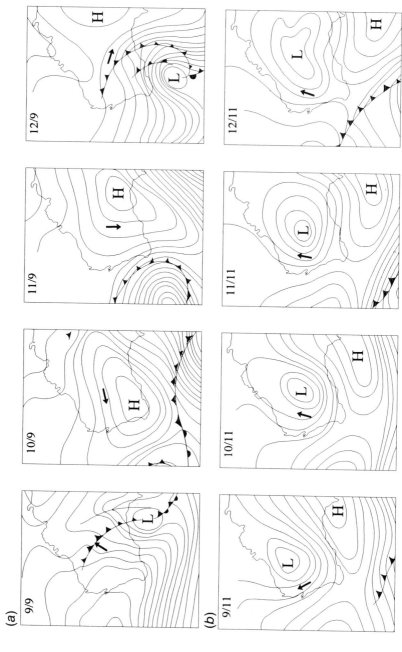

Fig. 9.6. Daily synoptic charts (10.00 LST) for the periods 9–12 September 1991 (*a*) and 9–12 November 1991 (*b*). Arrows indicate the wind direction over emergence sites.

(early summer), with each peak lasting 1–4 nights (Fig. 9.5*b*). Forbs that had hosted a winter generation (which produced the peaks in late August and September in Fig. 9.5*b*) following drought-breaking rains in June had senesced by the end of September, and there were no further significant rainfalls before the end of the year. The October–December catches are therefore thought to be due to immigrants, presumably from the south-western croplands, rather than to local emergence. Large numbers of *H. punctigera* were also collected from other areas where drought conditions had persisted for over 12 months. No local refuges, where spring breeding could have occurred, were discovered during extensive surveys of the region, and late-instar caterpillars and pupae collected at the end of winter showed no signs of delayed development. Delayed emergence of the winter generation as a result of either diapause (Cullen & Browning, 1978) or a high-temperature-induced summer dormancy (Murray, 1991) both seem unlikely, because the conditions at the end of winter (August) were not appropriate.

Flights into the arid zone could have occurred on southerly winds on the western side of low-pressure areas. Low-pressure cells and tropical troughs often dominate the region by late spring and summer, as the high-pressure areas which prevailed in early spring have moved south (Fig. 9.6*b*). During the late spring and early summer of 1991, peak catches of *H. punctigera* moths in arid-zone light traps (e.g. Fig. 9.5*b*) occurred on nights when strong southerly to westerly winds blew.

### Migration as an adaptive strategy for *C. terminifera* and *H. punctigera* in Western Australia

#### *C. terminifera*

These observations suggest that in WA, as in the eastern half of the Australian continent (Farrow, 1990), *C. terminifera* survives by migrating between ephemeral breeding grounds that are widely separated. This strategy balances high mortality rates associated with dispersal and the inhospitable climate of the arid zone with an increased net reproductive rate associated with drought-breaking rains.

In 1988–90, these migrations carried locusts to and fro between lower-latitude summer-rainfall areas in the arid zone and higher-latitude winter-rainfall areas in the southwestern croplands. However, the sequence of breeding in these zones that resulted in a plague appears to have arisen more from chance (the atypical occurrence of summer rainfall in the croplands in January 1990) than from a strategy of seasonal movements between summer and winter rainfall areas. Although attempts to reproduce locally eventually failed, the migration to the southwest cannot be regarded as ultimately unsuccessful: following a dramatic increase in numbers, a redistribution of populations occurred by trivial and migratory flights, and while most of these locusts would have perished either offshore or in areas that were experiencing extended droughts, a small proportion were apparently able to reach summer-rainfall areas north of 25° S.

### *H. punctigera*

As with *C. terminifera*, the migrations of *H. punctigera* observed during this study can be interpreted as part of a strategy of random long-distance dispersal from senescing breeding grounds. However, unlike with *C. terminifera*, the success of migrations into the southwest is not dependent on aseasonal summer rainfalls: *H. punctigera* arrives at the beginning of spring, when favourable habitats are abundant. Subsequent breeding produces numerous progeny, which disperse before the onset of the usual summer drought. During late spring (October–November) of 1991, some probably reached the south coast (a flight of over 200 km) and others may have reached parts of the northern summer-rainfall areas of the arid zone (a flight of ~1000 km) where favourable conditions for further breeding would have been encountered.

This seasonality of movement contrasts with the situation in North America (Chapter 2, this volume), where moths make return migrations to lower latitudes in autumn ahead of a lethally cold winter.

### *Random dispersal: an appropriate strategy for Western Australia?*

If migration is to evolve as a strategy for avoiding unfavourable conditions, it must enhance the survival of the participants and their offspring, and at some stage the progeny of insects that emigrated from a region can be expected to return to the same general area to breed. There have been cases reported where poleward spring migrations during disturbed weather appear to direct insects beyond their permanent ranges (Farrow, 1975; Drake *et al.*, 1981; Hartstack *et al.*, 1982; Domino *et al.*, 1983; Wolf *et al.*, 1987; and most contributions in Part one of this volume) and, in the absence of a subsequent return movement, the initial poleward migrations are interpreted as non-adaptive. Rabb & Stinner (1978) used the term 'Pied Piper effect' to describe such movements, which may be an incidental outcome of an otherwise successful general dispersal strategy: e.g. one in which individuals move in all directions to enable some to colonise scattered ephemeral habitats (Farrow & McDonald, 1987). However, poleward migration in spring will be successful in itself if it leads to successful breeding at higher latitudes and if the offspring are able to return to suitable breeding habitats or overwintering areas at lower latitudes.

On the basis of the limited evidence presented here, migration in *C. terminifera* and *H. punctigera* may be interpreted as part of a strategy for finding and exploiting ephemeral habitats opportunistically. Suitable habitats are apparently found by random dispersal: an effective strategy in much of WA where rainfall is predominantly unreliable. Migrations take place during the passage of all types of weather systems, so that dispersal occurs in all directions, and often to areas unsuitable for breeding. Major movements often occur during settled weather when the region is dominated by either high- or low-pressure cells which sometimes persist for a number of days and often do not produce rain.

Some degree of random dispersal apparently also occurs in eastern Australia. *C. terminifera* have been observed moving northwards in November (late spring) (Drake & Farrow, 1983) and *H. punctigera* eastwards and northwards in September (early spring) (Drake & Farrow, 1985), although movements towards higher latitudes are more prominent in both species at this season (Farrow, 1979; Drake & Farrow, 1983; Farrow & McDonald, 1987; Chapter 8, this volume). The spring migrations ahead of cold fronts that carry both species to favourable habitats in the agricultural regions of the southeast of the continent may well be incidental 'Pied Piper' movements, as populations of *C. terminifera* in these areas appear to be unable to return to their permanent range in the interior (Farrow, 1975, 1979). The same may be true for *H. punctigera* (Chapter 8, this volume). On the other hand, a looping return path has been tentatively identified for another noctuid that sometimes exploits the ephemeral habitats of inland eastern Australia, the Common Armyworm *Mythimna convecta* (Chapter 7, this volume). In both eastern and western Australia, movements over large distances often occur ahead of cold fronts. However, in WA they result in the migrants being taken further into the desert and away from favourable breeding grounds, and are therefore apparently movements into ecological dead-ends. Only if the migrants reach breeding grounds in the eastern half of the continent, and their progeny are eventually able to return, would such flights be adaptive.

Mortality rates associated with these wide-area random dispersals will often be very high, because of the scarcity of favourable habitats; however, both species have high fecundities which will to some extent compensate for these losses (Farrow & McDonald, 1987). The costs associated with migrating long distances before breeding must be weighed against the alternative: to stay put and delay breeding until conditions become favourable again. Diapause or dormancy enables insects to avoid seasonal adversities, like the cold winters of the southwestern croplands, without having to migrate. Both *C. terminifera* and *H. punctigera* exhibit a number of forms of diapause or quiescence (Cullen & Browning, 1978; Wardhaugh, 1988; Murray, 1991), but observations in the southwest of WA indicate that very few individuals of either species survive through summer to experience the conditions that might cause them to enter any form of dormancy.

For both species, return flights resulting in successful breeding in the arid zone are essential if movements to the southwest and subsequent breeding there are to form part of an overall survival strategy rather than a 'Pied Piper' dead-end. Dispersal from the southwest occurs in late spring or early summer, before the onset of unfavourable conditions. The nearer regions of the arid zone, where winter breeding often occurs, are also usually experiencing drought at this time and the emigrants have to travel further afield. There is now evidence that some of these migrants do reach summer-rainfall areas north of ~25° S, but whether they breed there, and perhaps produce progeny that infest the winter-rainfall areas further south in autumn, remains unknown.

## Acknowledgements

The work on *C. terminifera* was supported by the Agriculture Protection Board of Western Australia, while that on *H. punctigera* is funded by the Grain Research and Development Corporation and the Grain Research Committee of Western Australia. Alistair Drake, Peter Gregg and Don Reynolds made helpful comments on a draft.

## References

Beard, J.S. (1974–76). *Vegetation Survey of Western Australia.* [*1 : 1 000 000 Series* maps 2–6, and accompanying *Explanatory Notes.*] Nedlands: University of Western Australia Press.

Beard, J.S. (1990). *Plant Life of Western Australia.* Kenthurst, New South Wales: Kangaroo Press.

Clark, D.P. (1971). Flights after sunset by the Australian plague locust, *Chortoicetes terminifera* (Walk.), and their significance in dispersal and migration. *Australian Journal of Zoology*, **19**, 159–76.

Cullen, J.M. & Browning, T.O. (1978). The influence of photoperiod and temperature on the induction of diapause in pupae of *Heliothis punctigera*. *Journal of Insect Physiology*, **24**, 595–601.

Domino, R.P., Showers, W.B., Taylor, E. & Shaw, R.H. (1983). Spring weather patterns associated with suspected black cutworm moth (Lepidoptera: Noctuidae) introduction to Iowa. *Environmental Entomology*, **12**, 1863–72.

Drake, V.A. & Farrow, R.A. (1983). The nocturnal migration of the Australian plague locust, *Chortoicetes terminifera* (Walker) (Orthoptera: Acrididae): quantitative radar observations of a series of northward flights. *Bulletin of Entomological Research*, **73**, 567–85.

Drake, V.A. & Farrow, R.A. (1985). A radar and aerial-trapping study of an early spring migration of moths (Lepidoptera) in inland New South Wales. *Australian Journal of Ecology*, **10**, 223–35.

Drake, V.A., Helm, K.F., Readshaw, J.L. & Reid, D.G. (1981). Insect migration across Bass Strait during spring: a radar study. *Bulletin of Entomological Research*, **71**, 449–66.

Farrow, R.A. (1975). Offshore migration and the collapse of outbreaks of the Australian plague locust (*Chortoicetes terminifera* Walk.) in south-east Australia. *Australian Journal of Zoology*, **23**, 569–95.

Farrow, R.A. (1979). Population dynamics of the Australian plague locust, *Chortoicetes terminifera* (Walker), in central western New South Wales I. Reproduction and migration in relation to weather. *Australian Journal of Zoology*, **27**, 717–45.

Farrow, R.A. (1990). Flight and migration in acridoids. In *Biology of Grasshoppers*, ed. R.F. Chapman & A. Joern, pp. 227–314. New York: John Wiley & Sons.

Farrow, R.A. & Daly, J.C. (1987). Long-range movements as an adaptive strategy in the genus *Heliothis* (Lepidoptera: Noctuidae): a review of its occurrence and detection in four pest species. *Australian Journal of Zoology*, **35**, 1–24.

Farrow, R.A. & McDonald, G. (1987). Migration strategies and outbreaks of noctuid pests in Australia. In *Recent Advances in Research on Tropical Entomology*, ed. M.F.B. Chaudhury, pp. 531–42. *Insect Science and its Application*, vol. 8, Special Issue. Nairobi: ICIPE Science Press.

Gregg, P.C., Fitt, G.P., Zalucki, M.P., Murray, D.A.H. & McDonald, G. (1993). Spring migrations of *Helicoverpa* spp. from inland Australia 1989–1991: implications for forecasting. In *Pest Control and Sustainable Agriculture*, ed. S.A. Corey, D.J. Dall & W.M. Milne, pp. 460–3. Melbourne: CSIRO.

Hartstack, A.W., Lopez, J.D., Muller, R.A., Sterling, W.L., King, E.G., Witz J.A. & Eversull, A.C. (1982). Evidence of long range migration of *Heliothis zea* (Boddie) into Texas and Arkansas. *Southwestern Entomologist*, **7**, 188–201.

Hunter, D.M. (1982). Adult development in the Australian plague locust, *Chortoicetes terminifera* (Walker) (Orthoptera: Acrididae). *Bulletin of Entomological Research*, **72**, 589–98.

Hunter, D.M., McCulloch, L. & Wright, D.E. (1981). Lipid accumulation and migratory flights in the Australian plague locust, *Chortoicetes terminifera*

(Walker) (Orthoptera: Acrididae). *Bulletin of Entomological Research*, **71**, 543–6.

Morton, R., Tuart, L.D. & Wardhaugh, K.G. (1981). The analysis and standardisation of light-trap catches of *Heliothis armigera* (Hübner) and *H. punctigera* (Wallengren) (Lepidoptera: Noctuidae). *Bulletin of Entomological Research*, **71**, 207–25.

Murray, D.A.H. (1991). Investigations into the development and survival of *Heliothis* spp. pupae in south-east Queensland. PhD thesis, University of Queensland, Brisbane.

Rabb R.L. & Stinner, R.E. (1978). The role of insect dispersal in population processes. In *Radar, Insect Population Ecology and Pest Management*. NASA Conference Publication 2070, ed. C.R. Vaughn, W.W. Wolf & W. Klassen, pp. 3–16. Wallops Island, Virginia: NASA Wallops Flight Centre.

Symmons, P.M. & McCulloch, L. (1980). Persistence and migration of *Chortoicetes terminifera* (Walker) (Orthoptera: Acrididae) in Australia. *Bulletin of Entomological Research*, **70**, 197–201.

Wardhaugh, K.G. (1988). Diapause strategies in the Australian Plague Locust (*Chortoicetes terminifera* Walker). In *The Evolution of Insect Life Cycles*, ed. F. Taylor & R. Karban, pp. 89–104. New York: Springer-Verlag.

Wolf, R.A., Pedigo, L.P., Shaw, R.H. & Newsom, L.D. (1987). Migration/transport of the green cloverworm, *Plathypena scabra* (F.) (Lepidoptera: Noctuidae), into Iowa as determined by synoptic-scale weather patterns. *Environmental Entomology*, **16**, 1169–74.

Wright, D.E. (1987). Analysis of the development of major plagues of the Australian plague locust *Chortoicetes terminifera* (Walker) using a simulation model. *Australian Journal of Ecology*, **12**, 423–37.

# Part two

## Adaptations for migration

# 10

# Migratory potential in insects: variation in an uncertain environment

## A.G. GATEHOUSE AND X.-X. ZHANG

### Introduction

When and where a female insect lays her eggs or deposits her larvae have a determining influence on her fitness, and the evolution of flight has extended the spatial range over which that choice can be expressed. It has also enabled insects faced with deteriorating environmental conditions to reach more distant sites, where their chances of survival are improved. Thus, the flight capacities and strategies of insects are shaped by the distribution, within and between habitats, of resources on which their own and their offspring's survival and reproduction depend. Some species, including those with which this book is concerned, have evolved the capacity to make flights that have specific behavioural characteristics, usually taking them beyond their current habitat – flights which are defined as migratory on these behavioural criteria (Kennedy, 1985; Gatehouse, 1987*a*; see also Preface, this volume).

Migration can evolve only when the fitness achieved as a consequence of migration exceeds the fitness that would have been achieved by remaining in the current habitat (Southwood, 1977; Solbreck, 1978). Any such fitness differential must depend on the quality of the current habitat relative to the quality of habitats that can be reached by migration, in terms of their capacity to support reproduction by the migrant and the growth, development and survival of its offspring. It is a consequence, then, of either deteriorating quality of the current habitat, or the occurrence elsewhere of more favourable habitats or, more usually, both (Baker, 1978; Southwood, 1981). Migration may be obligatory, in species exploiting ephemeral habitats that can support only a single generation, but most insects are facultative migrants, emigrating in response to environmental cues of changes in habitat availability and quality. Furthermore, as Southwood (1977) points out, the ranges of values for reproductive success (or fitness) associated with reproducing in the current habitat on the one hand, or reproducing in a new habitat on the other, frequently overlap. When this happens, he argues, a poly-morphism for migration and/or a mixed strategy is likely to evolve, so that only some individuals migrate to survive and reproduce elsewhere, or individuals partition their reproductive output between the current and new habitats.

The flight strategies that have evolved to distribute insects and their offspring in space, within and between habitats, are extremely varied. To understand the evolution of migration by flight, it is necessary to identify the component traits that contribute to an insect's *migratory potential*. These are its capacity for persistent, migratory flight and the length of the interval, during its life, over which this capacity can be expressed. When realised, therefore, an insect's migratory potential determines the extent of the displacements it can achieve by migration.

In this chapter, we discuss the components of migratory potential in species of the orders Orthoptera, Hemiptera and Lepidoptera (with emphasis on moths), which include the majority of migratory insects of agricultural importance. We then examine the nature of the selection imposed on these traits by the spatiotemporal distribution of habitats, and by the environment in which migration takes place (see also Chapter 11, this volume).

### Flight behaviour

In insects, most migrations are by flight, the migrants either travelling under their own power, within their flight boundary layer (FBL; Taylor, 1958, 1974), or climbing rapidly to altitudes generally well above it, to be carried on the wind (Drake & Farrow, 1988; Zhai & Zhang, 1993; Part one, this volume).

Migrants travelling within their FBL retain control over their direction of travel. However, progress is necessarily slow and energetically expen-sive, so that, over long distances, lipids fuelling flight must be replenished by feeding (see e.g. Brower, 1985). Some of these insects may, like the Monarch butterfly *Danaus plexippus*, leave their FBL temporarily to utilise favourable winds, by soaring within thermals and then gliding, saving on energy expended, although at the cost of some relaxation of control over their direction of travel (Gibo, 1986).

Other insect migrants fly on the wind, well above their FBL, which allows them to achieve speeds and distances of travel generally orders of magnitude greater than would be possible under their own power. During the day, the planetary boundary layer (PBL) is disturbed by convective upflows, limiting the potential for rapid horizontal transport by the wind (Drake & Farrow, 1988). Many wind-borne, diurnal migrants are relatively small, weak-flying species whose climb to altitude may be assisted by these convective updraughts. Among them are the aphids, in which these migrations last from less than a minute to a few hours and cover a few metres to tens of kilometres, enough to enable them to track the distribution of their host plants in temperate regions (Loxdale *et al.*, 1993).

At night, however, the PBL stabilises and strong, steady winds (low-level jet streams) frequently become established at altitudes of a few hundred metres, immediately above the temperature inversion (Drake & Farrow, 1988). These winds provide a vehicle that can transport flying insects at speeds typically of 10–20 m s$^{-1}$ but sometimes considerably higher, so that displacements of several hundreds of km can be achieved in a single night's flight. Such flights enable migratory species to track and exploit habitats whose availability changes on a spatiotemporal scale of a different order, for example to follow seasonal changes on a geographic scale over hundreds or thousands of kilometres (see e.g. Part one of this volume).

Although some long-distance migrations, particularly over the sea, are known to involve flight through night and day (Johnson, 1969; Pedgley, 1982; Wolf *et al.*, 1986), direct and indirect evidence indicates that others occur, largely or exclusively, either nocturnally or diurnally. For example, entomological radar studies suggest a general descent of night-flying insects, particularly the larger moths and acridoids, at dawn (Schaefer,1976; Riley, Reynolds & Farmery, 1981, 1983; Reynolds & Riley, 1988; Chen *et al.*, 1989); 92% of *Cnaphalocrocis medinalis* moths caught in a mountain-top net and 91% caught at sea over three years were caught at night (National Cooperative Research Group on Rice Leaf Roller, 1981); direct observations indicate that migratory flights by non-swarming *Melanoplus sanguinipes* grasshoppers are predominantly diurnal (McAnelly & Rankin, 1986*a*). Indirect evidence includes back-tracking of migrations of Lepidoptera (see e.g. Mikkola, 1971, 1986; Tucker, Mwandoto & Pedgley, 1982; Tucker, 1994) and studies of the diurnal distribution of tethered-flight activity in *Oncopeltus fasciatus* (Caldwell & Rankin, 1974) and *Nilaparvata lugens* (Padgham, Perfect & Cook, 1987). However, radar observations show *N. lugens* to have a second peak of emigration at dawn, leading to flights of shorter duration and involving smaller numbers than migrate by night (Riley *et al.*, 1991). Shorter range displacements are also achieved by the essentially crepuscular migration shown by other microinsects observed by radar in the Philippines (Riley, Reynolds & Farrow, 1987).

Some species (e.g. *Locusta migratoria migratorioides* – Farrow, 1975a; *N. lugens* – Padgham *et al.*, 1987; aphids – Loxdale *et al.*, 1993) appear usually to make a single, long-distance migratory flight (although this may be followed by significant movements – see below), while others achieve their displacements by a sequence of migratory flights, often over a number of nights or days (e.g. *O. fasciatus* – Caldwell, 1974; *Patanga (Nomadacris) guttulosa* – Baker & Casimir, 1986; many Lepidoptera including *Mythimna separata* – Li, Wong & Woo, 1964, Chapter 4, this volume; *C. medinalis* – Zhang *et al.*, 1980; *D. plexippus* – Brower, 1985; *Agrotis ipsilon* – Showers *et al.*, 1989a,b, 1993). In some species, e.g. *Chortoicetes terminifera*, whether migration is restricted to a single night or takes place over several nights may depend on the persistence of weather conditions favourable for emigration, (Symmons & Wright, 1981).

For all species that migrate above their FBL, variations in wind velocity and direction impose a strong stochastic influence on the distance covered by, and direction of, a particular migratory flight, and so on its destination. For short-duration, diurnal flights by, for example, many aphids (Loxdale *et al.*, 1993), where most migrations start and end within regions in which suitable habitat patches are more or less evenly distributed, these variations in windspeed and direction are probably immaterial. However, the prevalence of this mode of transport in species migrating over very much longer distances, between habitat patches that are much more discontinuously distributed, shows that the probabilities of reaching favourable habitats remain high enough for the behaviour to have evolved and become established, in spite of this stochastic influence of the wind on how far and in what direction migrants are carried.

## Migration and reproduction

Displacements achieved as a result of migratory behaviour, *sensu* Kennedy, may end in a habitat offering resources and conditions for immediate reproduction (Johnson's Type I). Alternatively, the initial migration may terminate in a habitat providing conditions that allow the migrant to survive seasons or periods unfavourable for growth, development and reproduction, in a reproductive diapause of greater or lesser intensity (Type III migration – Johnson, 1969). Following the return of favourable conditions, these insects may migrate to breeding habitats (e.g. *P. guttulosa* – Baker & Casimir, 1986; *Diabolocatantops axillaris* – Reynolds & Riley, 1988; several Noctuidae – Oku, 1982; *D. plexippus* – Malcolm, Cockrell & Brower, 1993). Less is known of the behavioural characteristics of Johnson's Type II category of movements (between breeding and feeding sites) and they are not considered here. Also largely excluded from the discussion in this chapter are movements by gregaria-phase locusts in swarms. While some of these appear to meet Kennedy's (1985) criteria for migratory behaviour, others are probably better considered as extended foraging behaviour (Farrow, 1990).

In this section, we review some evidence on the occurrence of migration in relation to reproduction, with the aim of identifying when, in the life cycle, migratory flight occurs. It is the behaviour of the female insect that determines the timing and location of deposition of eggs or larvae. Fitness in males, on the other hand, depends on maximising the number of females they mate with, and male physiology and behaviour may evolve to increase the fitness of the females they fertilise. Reproductive and migratory strategies, as well as the pattern of integration of migration with reproduction are likely, therefore, to differ between the sexes (Johnson, 1969) and it is appropriate to consider females and males separately.

### Females

As earlier workers observed, migration by females of many species of insects takes place post-tenerally and pre-reproductively, often while oocyte development is arrested or at least incomplete (Kennedy, 1961; Johnson, 1969; Dingle, 1972). This timing is obviously appropriate where migration occurs in anticipation of, or in response to, rapid deterioration of conditions in the current habitat. Speculation on the role of juvenile hormone in their control (Kennedy, 1961; Johnson, 1969) led to the proposal that migration and reproductive development were physio-logically incompatible processes and that an 'oogenesis-flight syndrome' was characteristic of insect migration (Johnson, 1969).

Clearly, an oogenesis-flight syndrome is necessarily evident in species that undergo wing-muscle histolysis (and/or dealation) when they attain reproductive maturity and mate, whether this is irreversible, as in some Hemiptera and Hymenoptera, or reversible, as in some Coleoptera (Johnson, 1969, 1976; Dingle 1985). However, the assumption, in much of the insect migration literature, that the the oogenesis-flight syndrome is a general phenomenon has been challenged by several authors. For example, Baker (1978) criticised it on the grounds that it does not cater for males, although, as pointed out above, the pattern of integration of migration and reproduction can be expected to differ between the sexes. More recently, doubt has been cast on the general applicability of the oogenesis-flight syndrome (Sappington & Showers, 1992).

Thus, while there is evidence from a wide range of taxa that migratory flight commences while females are immature, less is known of the reproductive status of migrants when migratory flights end, or of possible causal interactions between migratory behaviour and reproductive development. The occurrence of inter-reproductive migratory behaviour (except in one or two species), and the extent to which it can be distinguished from the extensive flights associated with oviposition made by some species, is even less well documented.

### Reproductive status during migration

Direct and indirect evidence of the reproductive status of migrating female insects is available for a range of species from several orders. Some evidence for Orthoptera, Hemiptera and Lepidoptera, with emphasis on species migrating outside their FBL, is summarised in Tables 10.1–10.4 (for earlier work, see Johnson, 1969).

In the field, emigrations and immigrations have been observed and the insects sampled immediately before or after migration, with reproductive status assessed by dissection (Table 10.1). Migrants flying within their FBL have been sampled during migration and, in spite of their relative inaccessibility, a few species flying above it have been captured in traps on tall towers or mountain tops, or carried by balloons, kites or aircraft; migrants have also been trapped on ships at sea (Table 10.2). Indirect evidence on the reproductive status of migrating females is provided by the demonstration, in the laboratory, of delayed and/or arrested reproductive development in individuals captured in the field from emigrant generations, or reared in the laboratory under environmental conditions in which migration is known to occur (Table 10.3). Also included in Table 10.3 are examples where extensive variation in pre-reproductive period has been demonstrated in migratory species, which may be indicative of pre-reproductive migration (McNeil, 1986). Finally, laboratory studies on insects in free or tethered flight have investigated the incidence of prolonged flights, assumed to reflect migratory behaviour, in relation to adult age and reproductive status (Table 10.4).

Table 10.1. *Evidence for pre- and inter-reproductive occurrence of migration in Orthoptera, Hemiptera and Lepidoptera from assessment of the reproductive status of migrants sampled before or during emigration, or immediately after immigration. Many movements of gregaria-phase locusts in swarms can be considered to represent extended foraging rather than migratory behaviour sensu Kennedy (Farrow, 1990) and these are not included. Evidence relating to other movements that may not be migratory on Kennedy's (1985) behavioural criteria is preceeded by a question mark. d, migration by day; n, migration by night.*

| Species | d/n | Pre-reproductive migration | Inter-reproductive migration | References |
|---|---|---|---|---|
| ORTHOPTERA (Acrididae) | | | | |
| *Chortoicetes terminifera* – solitaria | n | Yes. Emigration of immature fledglings[1,2] | Yes. Emigration of mature females that have already laid, when weather conditions prevent pre-reproductive flight[3] | [1]Symmons & McCullogh (1980) [2]Clark (1969, 1972) [3]Farrow (1977b) |
| gregaria | n | Yes. Night flights from swarms by individuals in post-teneral, pre-reproductive state[1,2,3] | | |
| *Diabolocatantops axillaris* | n | Yes. Light-trap catches associated with migration observed on radar – all females with undeveloped ovaries and no vitellogenesis[1,2] | | [1]Reynolds & Riley (1988) [2]Farrow (1990) |
| *Gastrimargus africanus, G. nigericus* | n | Yes. Disappearance of immature young adults from breeding areas[1] | | [1]Descamps (1962, 1965) |
| *Locusta migratoria migratorioides* – solitaria | n | Yes. Disappearance of immature male and female fledglings from fledgling areas[1] Long-distance migration by immature locusts[2] | Yes. Immigration of mature, mated females that have already laid, with males[1] | [1]Farrow (1975a) [2]Farrow (1977a) |

Table 10.1. (cont.)

| Species | d/n | Pre-reproductive migration | Inter-reproductive migration | References |
|---|---|---|---|---|
| transiens/gregaria | n | Yes. Individual night flights from first-generation aggregations soon after fledging and before swarm flights[1] | | [1]McAnelly & Rankin (1986b) |
| Melanoplus sanguinipes – solitaria | d | Yes. Flight testing and dissection of field-caught insects[1] | | [2]Parker et al (1955) |
| gregaria | d | Yes. Emigration of immature adults and evidence that most mature females stop prolonged flight to lay all their eggs in one location[2] | Yes. Emigration of a minority of mature females that have already laid[2] | |
| Oedaleus senegalensis | n | Yes. Light-trap catches associated with emigration/immigration – high proportion of females immature[1,2,3] Light-trap catches indicate that adults likely to be making prolonged flights are > 5 d old and sexually immature[4] | Yes. Light-trap catches associated with radar studies – females having laid/resorbed first egg batch with second undeveloped[1] | [1]Riley & Reynolds (1983) [2]Farrow (1990) [3]Jago (1979) [4]Cheke, Fishpool & Forrest (1980) |
| Patanga (Nomadacris) guttulosa – solitaria | n | Yes. Long-distance, nocturnal migrations by immature locusts in early summer[1]. Immigration of young females that had never laid[2] | ? Probably not; no evidence of emigration of females after oviposition[2] | [1]Baker & Casimir (1986) [2]Farrow (1977c) |

| | | | |
|---|---|---|---|
| gregaria | d | ? Diurnal migrations of swarms of immature locusts in winter and early summer[2] | [1]Tétefort & Wintrebert (1967)<br>[2]Chapman (1959)<br>[3]Uvarov (1977) |
| *Patanga (Nomadacris) septemfasciata* – solitaria | n | Yes. Emigration of immature males and females and immigration of newly matured females with males[1] | |
| transiens/gregaria | d | ? Local movements by individuals in the dry season to aggregate in favourable aestivation sites[2]<br>? Immature swarms at start of dry season in southern Africa[3] | |
| *Schistocerca gregaria* – solitaria | n | Yes. Emigration of mixed populations of *S. gregaria* and *L. migratoria*[1,2]<br>Emigration of newly fledged adults and/or immigration of immature adults[3,4,5,6] | Yes. Immigration of mature females that had already laid[5] | [1]Farrow (1990)<br>[2]Farrow (1975a)<br>[3]Roffey & Popov (1968)<br>[4]Waloff (1963)<br>[5]Rao (1960)<br>[6]Roffey (1963)<br>[7]Rao (1942) |
| transiens/gregaria | n | Yes. Individual night flights from first-generation aggregations soon after fledging and before swarm flights[4] | |
| | d | Yes. Immigration of immatures from remote breeding areas[6,7]<br>? Behaviour leading to towering swarms usually confined to immature swarms where it may persist for long periods before reproduction[1] | |

Table 10.1. (cont.)

| Species | d/n | Pre-reproductive migration | Inter-reproductive migration | References |
|---------|-----|----------------------------|------------------------------|------------|
| **HEMIPTERA** | | | | |
| *Cicadulina* spp. (Cicadellidae) | d | Yes. Hand-net catches of insects in flight – high proportion of females non-gravid[1,2] | Yes. Hand-net catches of insects in flight – non-gravid females that had oviposited[1,2] | [1]Rose (1972) [2]Rose (1973) |
| *Empoasca biguttula* (Cicadellidae) | d? | Yes. Trap catches following emergence of emigratory generation – females immature[1] | | [1]B.-Z. Zhong (personal communication) |
| *Nilaparvata lugens* (Delphacidae) | n | Yes. Light and net-trap catches before and/or after migration – females immature[1,2,3,4,5] | Probably not[5] | [1]Ohkubo & Kisimoto (1971) [2]Kisimoto (1976) [3]Chen et al. (1979) [4]Perfect et al. (1985) [5]Padgham et al. (1987) |
| *Sogatella furcifera* (Delphacidae) | n&d | Yes. Trap catches following emergence of emigratory generation – females immature[1] | | [1]Zhang & Zhang (1991) |
| *Nysius vinitor* (Lygaeidae) | n&d | Yes. Suction and light-trap catches of emigrating bugs – high proportion of immature females[1] | ? Probably not. Very small proportions of mature females among suction and light-trap catches of emigrating bugs[1] | [1]Kehat & Wyndham (1973a) |
| **LEPIDOPTERA** | | | | |
| *Choristoneura fumiferana* (Tortricidae) | n | No. Net catches of emigrating moths immediately after takeoff include no virgin females – all females had oviposited[1] | Yes. Net catches of emigrating moths immediately after takeoff – females mature with 50–60% of their eggs expended[1] | [1]Greenbank et al. (1980) |

| Species | | Description | References |
|---|---|---|---|
| *Cnaphalocrocis medinalis* (Pyralidae) | n | Yes. Trap catches following emergence of emigratory generation – large majority of males and females immature[1,2,3,4,5] | [1]Zhang *et al.* (1979) [2]Zhang *et al.* (1980) [3]Zhang (1991) [4]Wada & Kobayashi (1980, 1982, 1985) [5]Wada *et al.* (1988) |
| *Danaus plexippus* (Nymphalidae) | d | Yes. Catches before and after autumn migration – large majority of females immature in reproductive diapause. In eastern population, substantial proportion of spring migrants leave diapause sites immature and unmated[1] | [1]Brower (1985) |
| *Parnara guttata* (Hesperiidae) | n? | Yes. Trap catches following emergence of the emigratory generation – all females immature and unmated[1,2] | [1]Nakasuji & Isii (1988) [2]Li *et al.* (1985) |
| *Agrotis ipsilon* (Noctuidae) | n | Yes. Trapping following emergence of emigratory generation in China – large majority of females immature and unmated[1]. Light-trap catches following emergence of autumn emigratory generation in USA and Europe – large majority of females unmated[2,3,4,6]. Pheromone-trap catches before autumn emigratory generation in USA – near zero catches of males[2,4,5] | [1]Zhang *et al.* (1979) [2]Kaster & Showers (1982) [3]Spitzer (1972) [4]Clement *et al.* (1985) [5]Swier *et al.* (1976) [6]Mulder *et al.* (1989) [7]Sappington & Showers (1992) |

Possibly. Western population – >95% females mated before spring migration. Eastern population – majority of females mated before spring migration; maturation of ovaries as butterflies fly north[1]

Possibly (see text). Light-trap catches following emergence of spring emigratory generation in USA – majority of females mature and mated[7]. Pheromone-trap catches following emergence of spring emigratory generation in USA – substantial catches of males[7]

Table 10.1. (cont.)

| Species | d/n | Pre-reproductive migration | Inter-reproductive migration | References |
|---------|-----|----------------------------|------------------------------|------------|
| *Helicoverpa* spp. (Noctuidae) | n | Yes. Light-trap catches of immature moths associated with immigration of known migrants[1,2,3] | | [1]Roome (1972) [2]Bowden & Gibbs (1973) [3]Walden, Chapter 9, this volume |
| *Mythimna separata* (Noctuidae) | n | Yes. Lure-trap catches following emergence of emigratory generation – 79–97% females immature, 98% unmated[1,2]. Pheromone-trap catches following emergence of emigratory generation – near zero catches of males[1]. Lure-trap catches of immature females following immigration[3] | | [1]Han *et al.* (1990) [2]Zhang *et al.* (1979) [3]Hirai (1988) |
| *Mythimna convecta* (Noctuidae) | n | Yes. Lure- and light-trap catches during period of emigration (summer) – large proportions immature, unmated females[1,2] | | [1]Smith & McDonald (1986) [2]McDonald & Cole (1991) |
| *Pseudaletia unipuncta* (Noctuidae) | n | Yes. Light-trap catches before autumn emigration – large majority of females immature and unmated[1]. Pheromone-trap catches before autumn emigration – near zero catches of males[1] | | [1]McNeil (1987) |

| Species | | | |
|---|---|---|---|
| *Spodoptera exempta* (Noctuidae) | n | Yes. Catches following emergence – females immature and unmated[1]. Pheromone-trap catches following emergence – near zero catches of males[1] | No. Evidence of immigrants staying in same area and reproducing for 7 d[1]. | [1]Rose *et al.* (1987) |

Table 10.2. *The reproductive status of female insects sampled during flight within or outside their flight boundary layer, or trapped on mountain tops or at sea, and assumed to be captured during migration (but see text). For details of stages of ovarian development represented in these samples, see references.*

| Species | Sampling method | Location | No. caught | % immature | % mated | Reference |
|---|---|---|---|---|---|---|
| **HEMIPTERA** | | | | | | |
| *Athysanus argentarius* (Cicadellidae) | Suction trap 2 and 9 m | Ohio, USA | 132 | 96 | – | Teraguchi (1986) |
| *Doratura stylata* (Cicadellidae) | Suction trap 2 and 9 m | Ohio, USA | 23 | 100 | – | Teraguchi (1986) |
| *Laodelphax striatellus* (Delphacidae) | At sea | East China Sea during seasonal migration to Japan | 14[a] | 100 | 0 | Noda (1986) Kisimoto & Rosenberg (1993) |
| *Nilaparvata lugens* (Delphacidae) | Aircraft 500–2000 m | China 1978–79 | 128 | 99 | 0 | Dung (1981) |
| | Alpine net 1840 m | China 1980 | 36 | 97.2 | 0 | Shexian Forestry Station (unpublished data) |
| | | China July 1977 | ? | 98.7 | 0 | Chen *et al.* (1978) |
| | | China July 1977 | ? | 100 | 0 | |
| | | China Oct 1977 | ? | 96.6 | 0 | |
| | | China ? | 114 | 99.1 | 0 | Chen *et al.* (1979) |
| | At sea | East China Sea during seasonal migration to Japan | 26[a] | 100 | 0 | Noda (1986) Kisimoto & Rosenberg (1993) |
| *Sogatella furcifera* (Delphacidae) | Alpine net 1840 m | China 1980 | 20 | 100 | 0 | Shexian Forestry Station (unpublished data) |

| | Method | Location | n | % | | Reference |
|---|---|---|---|---|---|---|
| | At sea | East China Sea during seasonal migration to Japan | 53[a] | 100 | 0 | Noda (1986) Kisimoto & Rosenberg (1993) |
| *Nysius vinitor* (Lygaeidae) | Kite net 100–300 m | Australia | 2935 | 94.3–100 | – | McDonald & Farrow (1988) |
| **LEPIDOPTERA** | | | | | | |
| *Cnaphalocrocis medinalis* (Pyralidae) | Alpine net | China | 836 (3 yrs) | 100 | 0 | Zhang *et al.* (1980) |
| | At sea | China | 960 (3 yrs) | 92.7 | 6 | National Cooperative Research Group on Rice Leaf Roller (1981) |
| | Aerial (kytoon) net 380 m | China | 7 | 100 | 0 | D.R. Reynolds (personal communication) |
| *Danaus plexippus* (Nymphalidae) | Netted on autumn, southward migration | USA Massachusetts, New Jersey, Kansas Texas | ? | 100 | 0 | Brower (1985) |
| | | | ? | ? | 5–9 | Brower (1985) |
| *Parnara guttata* (Hesperiidae) | Mountain-top trap catches 940 m | Japan | ? | >95 | 4.7 | Nakasuji & Isii (1988) |
| *Agrotis ipsilon* (Noctuidae) | Alpine net 4300–4500 m | China | 2 | 100 | 0 | P.-H. Jia (personal communication) |
| *Helicoverpa armigera* (Noctuidae) | Tower light trap | Eastern Australia 1985–89 | 132 | 88.6 | 10.6 | Coombs *et al.* (1993) |
| *Helicoverpa punctigera* (Noctuidae) | Tower light trap | Eastern Australia 1985–89 | 410 | 91.0 | 8.5 | Coombs *et al.* (1993) |

Table 10.2. (*cont.*)

| Species | Sampling method | Location | No. caught | % immature | % mated | Reference |
|---|---|---|---|---|---|---|
| *Helicoverpa zea* (Noctuidae) | Tower light trap >150 m | USA 1967 | ? | – | 22.7 | Callahan *et al.* (1972) |
| | | 10/1967–6/1968 | 12 | – | 8.3 | |
| | | 1968 | ? | – | c. 35 | |
| | Tower light trap 320 m | USA 9/1970 | 9 | – | 0 | |
| | Unmanned oil rig | Gulf of Mexico | | | | Sparks *et al.* (1975) |
| | 20 miles from shore | | 15 | – | 33.3 | |
| | 46 miles from shore | | 9 | – | 66.6 | |
| | 66 miles from shore | | 2 | – | 50 | |
| | 110 miles from shore | | 2 | – | 50 | |
| *Mythimna convecta* (Noctuidae) | Tower light trap | Eastern Australia 1988–89 | 783 | – | 3 | Coombs *et al.* (1993) |
| *Mythimna separata* (Noctuidae) | Kite trap 200–300 m | China | 7 | 71.4 | 28.6 | G. McDonald & R.A. Farrow (personal communication) |

[a]No. dissected.

Table 10.3. *Evidence of delayed or arrested reproductive development, in insects captured in the field from emigrant generations, or reared in the laboratory under environmental conditions known to be associated with migration. Species showing substantial variation in pre-reproductive period, that may be associated with the occurrence of pre-reproductive migration, are also included.*

| Species | Evidence | References |
| --- | --- | --- |
| ORTHOPTERA (Acrididae) *Melanoplus sanguinipes –* solitaria | Intra-population variation in rate of maturation of the first batch of oocytes | McAnelly & Rankin (1986b) |
| HEMIPTERA *Nilaparvata lugens* (Delphacidae) | Ovarian development delayed in macropters | Kisimoto (1965); Ohkubo (1967); Mochida (1970) |
| | Duration of ovarian grades I–III in macropters extended at high (28–30 °C) and low (18–21 °C) temperatures and when feeding on mature rice plants | Chen et al. (1979) |
| *Sogatella furcifera* (Delphacidae) | Ovarian development delayed in macropters | Ichikawa (1979) |
| | Duration of ovarian grades I–II delayed at high (30 °C) and low (18 °C) temperatures | Zhang & Zhang (1991) |
| *Dysdercus* spp. (Lygaeidae) | Wing-muscle histolysis delayed in starved females | Dingle & Arora (1973); Fuseini & Kumar (1975) |
| *Oncopeltus fasciatus* (Lygaeidae) | Pre-reproductive period in females extended in short photoperiod/low temperature conditions and by poor food quality and crowding. High degree of variation in pre-reproductive period under these conditions | Dingle (1968b); Caldwell & Rankin (1972); Dingle (1974); Rankin & Riddiford (1977); Dingle et al. (1980) |
| LEPIDOPTERA *Danaus plexippus* (Nymphalidae) | Reproductive diapause demonstrated in insects exposed to late-summer/autumn conditions in the field | Herman (1981) |
| | Reproductive diapause induced by short photoperiods and high or low temperatures in the laboratory | Barker & Herman (1976) |

Table 10.3. (*cont.*)

| Species | Evidence | References |
|---|---|---|
| *Autographa gamma* (Noctuidae) | Pre-reproductive periods in females extended under short photoperiod/low temperature conditions | Hill & Gatehouse (1992) |
| *Helicoverpa armigera* (Noctuidae) | Intra-population variation in pre-reproductive periods in females (1–10 d) and males (1–8 d). Pre-reproductive period in females prolonged in insects deprived of sucrose | Colvin & Gatehouse (1993*a*,*b*,*c*) |
| *Mythimna convecta* (Noctuidae) | Intra-population variation in ovarian development to maturity (3–10 d). Crowding and long photoperiod increased the proportions of females with arrested oocyte development | McDonald & Cole (1991) |
| *Mythimna separata* (Noctuidae) | Intra-population variation in pre-reproductive periods 3–21 d | Han & Gatehouse (1991*a*) |
| | Intra-population variation in pre-reproductive periods 2–11 d | Wang (1991) |
| | Pre-reproductive period in females extended under short photoperiod/low temperature conditions | Han & Gatehouse (1991*b*) |
| *Pseudaletia unipuncta* (Noctuidae) | Pre-reproductive period in females extended under short photoperiod/low temperature conditions | Turgeon & McNeil (1983); Delisle & McNeil (1986); McNeil (1986) |
| *Spodoptera exempta* (Noctuidae) | Variation in pre-reproductive periods in field-collected insects 0–13 d in females and 0–5 d in males | Wilson & Gatehouse (1993) |

Table 10.4. *Free- or tethered-flight performance of migratory insects in the laboratory, in relation to reproductive status or to environmental conditions associated with migration. For details of the designation of grades/stages of ovarian development, see references.*

| Species | Evidence | References |
|---|---|---|
| **ORTHOPTERA** (Acrididae) *Melanoplus sanguinipes* – solitaria | Prolonged tethered flight occurs pre-reproductively, and inter-reproductively in some females. Clear negative relationship between performance of prolonged flights and ratio of reproductive tract to body weight in females, but not in males | McAnelly & Rankin (1986b) |
| *Hemiptera* *Cicadulina mbila* (Cicadellidae) | Tethered gravid females fly significantly less than males or non-gravid females | Rose (1972) |
| *Nilaparvata lugens* (Delphacidae) | Tethered-flight durations maximal in ovarian development grades late I–II. Little flight in grade III and none in grade IV | Chen (1983) |
| *Nysius vinitor* (Lygaeidae) | Decline in free-flight activity in a wind tunnel with maturation in females | Kehat & Wyndham (1973b) |
| *Oncopeltus fasciatus* (Lygaeidae) | Tethered flight suppressed after maturation and mating in females. Males show a second peak of prolonged flight after mating | Dingle (1965, 1966); Caldwell (1974) |
| | Starvation and poor-quality food may induce prolonged inter-reproductive flight | Dingle (1968a, 1972); Rankin & Riddiford (1977) |
| *Nezara viridula* (Pentatomidae) | Decline in tethered-flight duration with mating in females | Gu & Walter (1989) |

Table 10.4. (cont.)

| Species | Evidence | References |
|---|---|---|
| *Lepidoptera* (Noctuidae) | | |
| *Agrotis ipsilon* | No evidence of a decline in tethered-flight duration with reproductive maturation in females or males | Sappington & Showers (1992) |
| *Chorizagrotis auxiliaris* | Equivocal evidence of some decline in tethered-flight activity during reproduction in females | Koerwitz & Pruess (1964) |
| *Helicoverpa armigera* | Decline in duration of tethered flight with reproductive maturation in females | Colvin & Gatehouse (1993*a*) |
| | Decline in duration of tethered flight with mating in females | Armes & Cooter (1991) |
| *Mythimna separata* | Decline in duration of tethered flight with reproductive maturation in females | Hwang & How (1966); Han & Gatehouse (1993) |
| *Spodoptera exempta* | Decline in duration of tethered flight with reproductive maturation in females | Parker (1983); R.E. Wood & A.G. Gatehouse (unpublished data) |

The evidence summarised in these tables shows that most species that have been investigated commence migration pre-reproductively, when the ovaries are immature. Good evidence that migration is *exclusively* pre-reproductive is available for many fewer species, including those that histolyse flight muscles as they mature and armyworm moths, like the African Armyworm *Spodoptera exempta,* that retain their capacity for flight but lay eggs in large batches over a relatively small area (Rose *et al.,* 1987). There is evidence for inter-reproductive migration with un-developed ovaries in a number of species but evidence for embarkation on unequivocally migratory flights by females with mature ovaries, whether or not they have already started reproduction, is scarce. The large proportions of mature and mated Black Cutworm moths *Agrotis ipsilon* in light-trap catches in Louisiana, USA, over the period in which emigration to the north was thought to occur, has been interpreted as evidence for migration by reproductive individuals in this species (Sappington & Showers, 1992). However, more extensive data on the timing of emigration and on the relative numbers emigrating or remain-ing to reproduce locally are required to substantiate this interpretation.

### Reproductive status and the termination of migration

Migrations culminating in a period of diapause must end with the insect in reproductive arrest. However, the relationships and interdependence between migratory behaviour and reproduction in migrations that termi-nate in breeding habitats are less clearcut.

There is some evidence that migration may cease before reproductive maturity is attained and/or mating occurs. For example, *Helicoverpa punctigera* females migrating into the southwestern cropping zone in Western Australia are unmated (Chapter 9, this volume). Also, *Mythimna separata* immigrants to Japan had slender abdomens, imma-ture ovaries and, if they had mated, contained no more than one spermatophore (Hirai, 1988); these moths fed on nectar before complet-ing maturation, mating and starting oviposition. More usually in species that start these migrations as unmated immatures, first captures after the immigrants have arrived are found to comprise entirely or over-whelmingly mature (i.e. with oocyte development at an advanced stage or complete), usually mated females (see e.g. Têtefort & Wintrebert, 1967; Spitzer, 1972; Myers & Pedigo, 1977; Chen *et al.*, 1979; Kaster & Showers, 1982; Li, Li & Liu, 1985; McNeil, 1987; Wada, Ogawa & Nakasuga, 1988; Han, Zhen & Song, 1990). Whether or not this indicates that migratory behaviour persists after maturation and mating, as has been claimed for *A. ipsilon* and, by implication, other species (Sappington & Showers, 1992), is uncertain. If migration ends as maturity is achieved and if there is selection to minimise delay in starting to reproduce (Roff, 1992; Stearns, 1992), immigrants might be expected to mate immediately they arrive, before they become available to traps. Female acridoids with immature ovaries may mate before emigration (Farrow, 1990), and those that reach habitats suitable for reproduction and development can also be expected to commence reproduction without delay.

If migration persists for significant periods after maturation and mating, this should be reflected in a significant proportion of mature, mated females among insects trapped actually during migratory flight. However, for most species for which these data are available, the large majority of migrants are immature and unmated (Table 10.2). The fact that some mature and/or mated individuals are consistently recorded in these samples shows that migration is not terminated *immediately* with maturation or mating. However, their small numbers suggest that it does not persist for long and that these insects may be in the final stages of their flights.

An exception seems to be *Helicoverpa zea* trapped on a television tower in Georgia, USA (Callahan *et al.*, 1972) when relatively large proportions (22%–c. 35%) of mated females were caught at heights between 150 and 320 m. This result contrasts with tower-trap data for the closely related *Helicoverpa armigera* and *H. punctigera* in eastern Australia (Coombs *et al.*, 1993) which showed 90% of females to be immature and unmated. A possible explanation for this apparent disagreement may lie in the fact that the television tower in Georgia was sited in an agricultural area, planted with mixed crops supporting breeding populations of *Helicoverpa zea*. There was, on the other hand, no evidence of local breeding by the species trapped on towers in eastern Australia (Gregg *et al.*, 1993). The proportions of mated females caught high on the Georgia tower during the summer months appear, therefore, to confirm that reproducing *Helicoverpa* (and *Heliothis*) spp. females undertake extensive flights associated with oviposition (Farrow & Daly, 1987), and indicate that they ascend to greater heights than these authors supposed. That these flights by mature, mated females of these genera may not cover long distances is indicated by the results of Coombs *et al.* (1993) and by the fact that only one mated female was trapped above 150 m on the Georgia tower from October to June, when there was no local breeding (Callahan *et al.*, 1972). Similarly high proportions of mated females among small numbers of *H. zea* caught on oil rigs in the Gulf of Mexico (Sparks, Jackson & Allen, 1975) are not inconsistent with this conclusion, as they may have been constrained to continue flight over the sea. The same was probably true of the mature, mated *Mythimna separata* captured in a kite net at 200–300 m altitude after they had crossed the Bohai Sea in northeastern China (G. McDonald & R.A. Farrow, personal communication; Table 10.2).

There is other evidence that migration continues as ovarian development is progressing, in species in which it is effectively pre-reproductive. For example, *Danaus plexippus* and *Hippodamia convergens* leave their hibernacula with rapidly developing ovaries (Rankin, McAnelly & Bodenhamer, 1986) and *Melanoplus sanguinipes* females continue to give long flights while oogenesis proceeds, stopping only at maturation (Parker *et al.*, 1955; McAnelly & Rankin, 1986*b*).

A decline in tethered-flight performance with the attainment of maturity and mating was even evident in *Helicoverpa amigera* (Armes & Cooter, 1991; Colvin & Gatehouse, 1993*a*; Table 10.4) which flies extensively during oviposition, laying its eggs singly over a large area (Farrow & Daly, 1987). However, tethered-flight data for *A. ipsilon* (Sappington & Showers, 1992) provided no evidence of any such decrease with maturation, although mated females initiated flight later in the night and showed some reduction in flight duration compared with unmated ones. An earlier tethered-flight study of another cutworm species *Chorizagrotis auxiliaris* gave equivocal results (Koerwitz & Pruess, 1964). Although these data indicate pre-reproductive migration in several species, but not in the cutworms, the results of all laboratory studies of migratory flight must be interpreted with caution. This is particularly true of tethered-flight data, because of the highly intrusive character of the techniques, and because of the failure to distinguish behaviourally migratory from non-migratory flight (Chapter 12, this volume).

In a few of these species, there is evidence that oviposition occurs in the period of the diurnal cycle occupied by migratory flight in pre-reproductive females, the implication being that reproduction supersedes migration (e.g. in *Oncopeltus fasciatus* – Caldwell & Rankin, 1974; *S. exempta* – C.F. Dewhurst, unpublished report). Further evidence of a causal relationship between reproductive maturation and the termination of migratory flight, in at least six species from three orders (Hemiptera, Coleoptera, Lepidoptera), is provided by physiological studies of the influence of juvenile hormone on flight and reproductive development

(Chapter 13, this volume and references therein). However, this work also relies on tethered flight as an index of migratory performance and so the strictures on cautious interpretation apply.

There is also some evidence of a reciprocal effect of flight on reproduction. Expression of flight capacity in the laboratory accelerated maturation in aphids (Johnson, 1958), *O. fasciatus* (Slansky, 1980), and *M. sanguinipes* (McAnelly & Rankin, 1986*b*), resulting in significantly earlier reproduction than in unflown insects. In *Hippodamia convergens* (Williams, 1958) and in migratory alates of *Aphis fabae* (Johnson, 1958; Kennedy & Booth, 1963), a period of flight may be necessary before sexual maturation or reproduction can be achieved.

### *Timing of migration and reproduction in females – conclusions*

The pattern of migration in relation to reproduction observed in particular species is influenced by the permanence of their habitats, by the characteristics of the resources they exploit and by the food available to the immigrants in the new habitat. Pre-reproductive migration can be expected when, at emergence, there is a high probability of achieving greater fitness in new habitats, following migration, than in the current habitat. In extreme cases, it occurs in response to, or in close anticipation of, rapid deterioration of the current habitat, so that it cannot support a further generation. When the new habitat is favourable for breeding and provides abundant resources for the offspring, within the immigrants' walking or non-migratory flight range and/or within the range over which the larvae themselves can disperse, further migration is likely to be unusual. In some species, it may be precluded by wing-muscle histolysis and/or dealation. In others, immigrants retain the capacity for flight but lay eggs in batches over a relatively small area, relying on larval dispersal to reach their host plants. Migration may also be exclusively pre-reproductive in species in which the female locates the larval resource and lays her eggs singly, when the distribution of those resources is clumped.

Current evidence (Tables 10.1–10.4) suggests that, in several species, long-distance displacement by migration is achieved before reproduction commences, although it may continue until maturation of the ovaries is complete. It may involve a single flight, with little or no associated delay in reproductive maturation, or flights over several successive days or nights, when it is often associated with substantially delayed or arrested maturation (reproductive diapause) which, in diapause habitats, may persist after the migration. Migration to diapause habitats may or may not be followed by further migration to another diapause habitat and, ultimately, to a breeding area.

Good evidence of migratory behaviour after reproductive maturation is available for many fewer insects. It might be expected in species for which conditions for larval survival and development, in the current habitat and patches accessible by migration, are to a greater or lesser extent unpredictable. A 'bet-hedging' strategy may then be advantageous. In some species that migrate pre-reproductively, further migration is known to occur between bouts of reproduction when oocytes are immature, in a second or subsequent cycle of egg development (Table 10.1). In others, there is unequivocal evidence that, at least under some circumstances, the first migration is by reproductive females, carrying embryos or mature oocytes, that may or may not have already commenced ovi- or larviposition in their current habitat. These include *A. fabae* (Shaw, 1970a,b) and *Choristoneura fumiferana* (Greenbank, Schaefer & Rainey, 1980; see also Gatehouse, 1989), but migratory flight by reproductively mature cutworm moths (Sappington & Showers, 1992) must be regarded as unproven on present evidence (see Tables 10.1 and 10.4). Migration by reproductively mature *Chortoicetes terminifera* has also been observed when earlier emigration was inhibited by unfavourable weather conditions (Farrow, 1977b).

However, migratory behaviour by reproductive females may be very difficult to distinguish, on the basis of Kennedy's (1985) behavioural criteria, from extensive oviposition flights which, in some species, distribute eggs, sometimes over very large areas (e.g. *Helicoverpa* and *Heliothis* spp. – Farrow & Daly, 1987; Fitt, 1989; locusts – Farrow, 1990; *Agrotis ipsilon* – Sappington & Showers, 1992; see also Kennedy, 1985). In locusts, these movements occur over reproductive lives of several weeks.

## Costs of migration in females

Implicit in discussions of the evolution of migration is the assumption that migratory behaviour is associated with some cost to fitness. Costs of migration have been thoroughly reviewed recently by Rankin & Burchsted (1992; see also Dixon, Horth & Kindlmann, 1993; Denno, 1993) and only a few general comments are necessary here. They can be subsumed under two headings: costs to survival and costs to reproduction.

Departure from the current habitat is generally perceived to entail a risk of mortality during migration or as a result of failing to find a suitable habitat afterwards. The latter risk must vary greatly between species, and with location and time, according to the pattern of distribution of habitat patches and the winds. In species dependent on generally distributed resources (e.g. phytophagous insects exploiting conifers in northern temperate forests or grasses in savannah regions), they must often be minimal and there is some evidence that they may be small even in species with more patchily distributed habitats (Farrow, 1975*a*). In all cases, they are difficult, if not impossible, to quantify.

Possible costs to reproduction include effects of migration on fecundity, due to delayed onset of reproduction, effects on the length of reproductive life, and depletion of reserves by the metabolic demands of flight. Because of the sensitivity of fitness to delays in the onset of reproduction (Roff, 1992; Stearns, 1992), insects whose migrations entail reproductive delay can be expected to suffer some cost to fitness compared with their non-migratory conspecifics. However, the magnitude of the effect may be minimised by physiological mechanisms by which flying itself accelerates reproductive maturation (see above and Rankin & Burchsted, 1992). Genetic correlations that link migratory potential to traits determining the lifetime schedule of fecundity within a 'migration syndrome' (Palmer & Dingle, 1986; Gunn & Gatehouse, 1993) may have a similar effect in eliminating or even reversing the fecundity differential between migratory and non-migratory phenotypes.

The real impact of metabolic costs of flight on fitness is even more difficult to assess, especially on the basis of laboratory experiments. Many migrants are well able to replenish expended reserves by feeding during (*D. plexippus* – Brower, 1985) or after flight (e.g. *A. fabae* – Cockbain, 1961; *O. fasciatus* – Slansky, 1980; *M. sanguinipes* – McAnelly & Rankin, 1986*b*; *S. exempta* – Gunn, Gatehouse & Woodrow, 1989; Gunn & Gatehouse, 1993). However, food may sometimes be unavailable or scarce in the field.

Therefore, the whole question of the actual costs of migration to reproduction in the field is problematic. One can only conclude that selection for adaptations to minimise or eliminate them can be expected to be strong.

### *Males*

Our understanding of the relationship between migration and reproduction in male insects has advanced very little since Johnson's (1969) book, in which he discusses differing mating strategies in relation to migration.

Field data for a number of species indicate that migration by males starts before the onset of reproductive behaviour, very low catches being obtained in pheromone traps before emigration (Table 10.1). In other insects, there is evidence that at least some individuals mate before migration. For example, *D. plexippus* males terminate diapause approximately one month before females and substantial proportions of females are mated before they leave diapause sites on the northward migration in spring (Herman, 1985). However, it is still not clear whether or not these mated females reproduce more or less locally, leaving the unmated individuals to migrate northwards. Males certainly take part in the northward migration, at least in the eastern population (Brower, 1985). There is also evidence of some mating before migration in *Nysius vinitor* (McDonald & Farrow, 1988) but their substantial catches of males migrating at altitudes of 100–300 m show that Kehat & Wyndham's (1974) conclusion, that only females are capable of long-distance migrations, requires revision. Male acridoids also generally fledge first and often mate with newly fledged females at the end of the teneral period, before emigration (Farrow, 1990).

In *O. fasciatus* (Caldwell & Rankin, 1972) and *Pseudaletia unipuncta* (Turgeon, McNeil & Roelofs, 1983; Dumont & McNeil, 1992), the male (like the female) pre-reproductive period has been shown to be prolonged in response to short photoperiods and low temperatures. In other species, genetic variation in male pre-reproductive period, similar to that in females, has been demonstrated, e.g. in *Helicoverpa armigera* (Chapter 12, this volume) and *Autographa gamma* (Hill & Gatehouse, 1993).

For some species or individuals in which migration commences before mating, there is circumstantial evidence that males may continue migration after they attain reproductive maturity (e.g *O. fasciatus* – Dingle, 1965; *Locusta migratoria migratorioides* – Farrow, 1975a; *Neacoryphus bicrucis* – Solbreck, 1978; *Leucoma salicis* and *Pontia daplidice* – Mikkola, 1986; *D. plexippus* – Brower, 1985; *S. exempta* – Wilson & Gatehouse, 1993).

It is important to reiterate that there is no reason to expect the pattern of integration of migration with reproduction in males to be the same as that in females. Thus, to achieve maximum fitness, males may need to continue migration throughout their reproductive lives, even in species in which migration is exclusively pre- or inter-reproductive in females. This appears to happen in *S. exempta* (Wilson & Gatehouse, 1993), but males have been neglected in most studies of insect migration.

## Migratory potential

The migratory potential of an individual insect determines the extent of the displacement it can achieve by migration. In wing polymorphic species, macroptery is clearly a prerequisite for migration but we are concerned here primarily with the physiological and behavioural components of migratory potential.

Migratory flight, defined behaviourally (Kennedy, 1985; Gatehouse, 1987*a*), involves some suppression of responsiveness to stimuli associated with resources necessary to the vegetative phases of the life cycle. The migratory phase (*sensu stricto*) therefore lasts for a finite length of time and flights are not terminated by encounters with such stimuli, until it is over. Therefore, most migratory flights, like those of aphids (Hardie, 1993), can be assumed to comprise two elements; an initial phase of truly migratory behaviour, followed by a phase in which the migrant regains responsiveness to stimuli associated with favourable habitats. When migration is above the FBL, descent presumably commences with the transition to this second phase. Where migration occurs on several successive days or nights, this applies to each flight, as the migrant descends and locates a hiding/roosting site for the night or day. It is less clear whether or not it is true for insects that migrate to specific locations, for example particular diapause or breeding sites, as nothing is known of the factors terminating the migrations of these species.

In insects that accomplish their migrations in a single flight, migratory potential is a function of flight capacity, expressed as flight duration. However, in species whose migrations involve a sequence of nocturnal or diurnal flights, the displacements achieved are a function of both flight capacity – defined as the time spent flying each day (24-h period; see below) – and the number of days on which this flight capacity is expressed.

For relatively short-lived species, migrating pre-reproductively to breeding habitats and flying on successive days/nights, this latter component is related to, but not equal to (migration is preceded by a teneral period immediately following emergence), the pre-reproductive period. For similar inter-reproductive migrations, the length of the inter-reproductive period is presumably of equivalent significance (Johnson, 1969). For other insects, including some locusts and grasshoppers, that may migrate once or several times during a prolonged diapause, the pre-reproductive period does not have this significance. The number of days/ nights on which these migratory flights occur may be determined by the incidence of weather conditions favourable for migration (see above).

The remainder of this chapter is concerned primarily with species for which the pre-reproductive period is a component of migratory potential. However, much of the discussion is obviously relevant to species in which migratory potential is a function simply of flight capacity.

### Selection by the habitat template

Because migratory behaviour, *sensu* Kennedy, is not by definition terminated in response to environmental stimuli, it is to be expected that both the major traits contributing to migratory potential, flight capacity and pre-reproductive period, are determined genetically, though they may be and often are modulated by environmental cues or stimuli. In all species in which the question has been addressed, both have indeed been found to show significant narrow sense heritabilities, ranging from 0.14 to 0.88, and so can be expected to respond rapidly to selection (Chapter 11, this volume). In these studies, flight capacity has been estimated using tethered-flight duration.

However, a major deficiency of migration research has been the failure to investigate, except in aphids (see below), the duration of the truly migratory phase of these flights. For now, it is necessary to assume that tethered-flight duration provides an index of the duration of this phase, which is probably not unreasonable, at least for species that migrate outside their FBL, at altitude.

To understand how flight capacity and pre-productive period contribute to migratory potential and the outcome of migration in the field, it is helpful to consider how they may be subjected to selection by the habitat 'template' (Southwood, 1977).

As Wilson (Chapter 11, this volume) discusses, the habitats of migratory species vary greatly in their heterogeneity and predictability, in both spatial and temporal dimensions. Even apparently homogeneous habitats, such as the extensive northern-temperate coniferous forests, have been shown to contain considerable heterogeneity, in terms of the quality of different individual trees or stands of trees, as host plants for phytophagous insects exploiting them (Denno & McClure, 1984).

Consider a cohort of a hypothetical, migratory insect emigrating over a few days from a single location in such a forest. The destination of each migrant will depend on its migratory potential and, if it migrates above its FBL, on the strength and direction of the winds that carry it. Variation in windspeed and direction over the period of a single flight introduces a significant stochastic element into the course and extent of the displacement achieved and this effect will be amplified for migrants making sequential flights over two or more nights. Thus, the vagaries of the wind may result in an individual flying for a short time over several nights reaching a favourable habitat patch, while one flying for a similar total number of hours on one night fails to do so – or *vice versa*. In the field, therefore, the influence of the variability of the winds distributing migrants over a mosaic of varying habitat quality will ensure the selection of a wide range of phenotypes of flight capacity and pre-reproductive period within and between migrations.

An extreme case may serve to reinforce this important point. Several migratory species, including *Spodoptera exempta*, depend on the flush of growth of vegetation following tropical rainstorms, at least at the start of the rainy season (Rose, 1979). The location and timing of such storms is notoriously unpredictable; from the point of view of an emerging migrant, a storm may occur nearby or at any distance, soon or at any time during the period available to it for migration. Subject to the effects of variation in the winds that carry them, emigrants with low flight capacity and short pre-reproductive periods may reach new habitats and prosper if storms occur soon and nearby, and those with low flight capacity and longer pre-reproductive periods may do so if storms occur nearby but later. Similarly, emigrants with higher flight capacities and shorter or longer pre-reproductive periods will succeed if storms occur further away and sooner or later, respectively.

The fitness of these migrants, and so the frequencies of the various phenotypes of flight capacity and pre-reproductive period in the subsequent generations, depends on the quality of the habitat patch they reach. Habitat quality and its distribution in space and time is, therefore, the selective agent that drives the changes in these frequencies. The *intensity* of that selection is related to the degree of heterogeneity of a species' habitat, being most intense for those species exploiting the most highly heterogeneous habitats.

Two important predictions emerge from this model:

1 That high levels of variation in flight capacity (in most migrants) and in pre-reproductive period (in sequential migrants) will be maintained by selection;

2 In sequential migrants, that flight capacity and pre-reproductive period will be inherited independently.

Both these predictions appear to be borne out by the available data described in the following two sections.

## Variation in migratory potential

A high level of intra-specific and intra-population variation in flight capacity, as indexed by tethered flight in the laboratory, has been demonstrated in several migratory species (Johnson, 1976), including *O. fasciatus* (Dingle, 1966, 1968*a*), *Nilaparvata lugens* (Padgham *et al.*, 1987), *Melanoplus sanguinipes* (McAnelly, 1985; McAnelly & Rankin, 1986*a*), *S. exempta* (Parker & Gatehouse, 1985*a,b*; Gatehouse, 1989) and other noctuids (e.g. *Mythimna separata* – Han & Gatehouse, 1993; *Helicoverpa armigera* – Colvin & Gatehouse, 1993*a*). Other examples of this variation, and evidence for its association with the spatiotemporal distribution of habitats, are discussed by Wilson (Chapter 11, this volume).

The high level of variation in flight duration between individuals within clones of the aphids, *Aphis fabae* and *Rhopalosiphum padi,* is of particular interest. Studies of the migratory behaviour of these species in an automated update of Kennedy's vertical wind tunnel (Kennedy & Booth, 1963), have used the return of responsiveness to a host-plant-related stimulus (an illuminated green target) as the criterion for the termination of the migratory phase of these flights (David & Hardie, 1988; Nottingham & Hardie, 1989; Nottingham, Hardie & Tatchell, 1991). The duration of this phase in spring, summer and autumn migrants ranges from <1 min to several hours. Whether or not this variation has a genetic basis (e.g. stochastic polyphenism – Walker, 1986) or is the result of extreme sensitivity to environmental stimuli (e.g. differential crowding – J. Hardie, personal communication), there seems little doubt that it is adaptive.

Evidence for the range of variation in pre-reproductive period is available for many fewer species, but, where it has been measured, a high level of variation has been found. Substantial variation in pre-reproductive period is apparent in *O. fasciatus* (Dingle *et al.*, 1980; Groeters & Dingle, 1987) and *Autographa gamma* (Hill & Gatehouse, 1993) under both long and short photoperiods. In *M. separata*, pre-reproductive periods ranging from 3 to 21 d were demonstrated by Han & Gatehouse (1991*a*) in first-generation laboratory insects reared from eggs collected in northern China.

There is no evidence of variation in the length of inter-reproductive periods in insects known to migrate at that time.

*Independent inheritance of flight capacity and pre-reproductive period*

The only comprehensive studies of the genetic correlations among life history traits and those associated with migration (wing length and flight capacity) have been by Dingle and his co-workers on *O. fasciatus* (Hegmann & Dingle, 1982; Dingle, 1984; Palmer & Dingle, 1986, 1989). They showed that time to first reproduction (= pre-reproductive period) was free of genetic correlations with the other traits. Circumstantial evidence suggests that this is true also of *S. exempta*; there was no evidence that the capacity for prolonged flight was associated with delayed onset of reproductive behaviour (Parker, 1983). This allows both flight capacity and pre-reproductive period to respond rapidly and independently to selection (see also Dingle, 1984).

### Expression of flight capacity within the pre-reproductive period

There is little evidence on how consistently sequential migrants fly through their pre-reproductive period. Laboratory studies indicate that some individuals of some species do appear to fly for comparable durations on two or more nights (e.g. *S. exempta* – Parker, 1983; *Melanoplus sanguinipes* – McAnelly & Rankin, 1986*a,b*; *H. armigera* – Colvin & Gatehouse, 1993*a*; *Mythimna separata* – Han & Gatehouse, 1993). However, in species with much longer pre-reproductive periods, such as *O. fasciatus*, tethered-flight studies suggest that, although a female giving a long flight on one night is likely to give long flights on several nights during the pre-reproductive period, individuals do not appear to express their potential for prolonged flight every night (Dingle, 1965; Caldwell & Rankin, 1972). As implied above, tethered-flight studies of the grasshopper, *Melanoplus sanguinipes*, suggest that individuals differ in this respect; some give long flights only once or twice, while others do so repeatedly (McAnelly, 1985).

Therefore, the precise relationship between the duration of the pre-reproductive period and migratory potential depends on the expression of flight capacity through that period. Where survival is likely to depend on rapid displacement, for example southward in autumn to escape the approaching winter, migrants might be expected to realise their maximum flight capacity every night or day.

**Variation in responses to environmental cues for migration**

The overwhelming majority of, but not all (Gatehouse, 1989), migratory insects are facultative migrants, migration occurring in response to environmental cues coinciding with or anticipating changes in habitat availability and quality. Such changes occur within and between seasons, on a wide range of spatial and temporal scales, and vary in their predictability. The environmental cues anticipating or associated with the changes, and the migration, diapause or other responses to these cues, have been extensively reviewed (see e.g. Johnson, 1969; Dingle, 1972, 1980, 1985; Tauber, Tauber & Masaki, 1984; Danks, 1987; Gatehouse, 1989).

Within favourable seasons, deterioration in habitat condition must generally be monitored as it occurs, by responses to proximal biotic (e.g. changes in host-plant quality, availability of prey or food) and abiotic cues (e.g. moisture, temperature) that induce maternal responses to increase the frequency of migratory phenotypes produced, and/or that initiate migratory behaviour. A response to crowding may allow some anticipation of actual deterioration in habitat favourability and it is an important biotic cue, especially when maternal effects are involved and there is a delay before emigration can occur. However, its reliability may not always be high (for discussion of the reliability of environmental cues, see Danks, 1987).

Seasonal changes may often be anticipated by responses to abiotic, 'token' cues, as well as monitored as they occur by the perception of proximal cues. Except at very low latitudes, photoperiod and temperature (the latter usually modulating the response to the former) predict changes with a relatively high degree of reliability, which accounts for their ubiquity as cues inducing diapause and migration responses (Danks, 1987).

Responses to environmental cues generally determine the occurrence or otherwise of migration. Pre-reproductive period and flight capacity, and consequently the extent of the displacements achieved, are less strongly influenced by these cues. However, where migration to escape seasonal disappearance of habitats requires migratory potential of a different order from that needed to track habitat distribution within seasons, responses of pre-reproductive period to photoperiod and temperature are found (e.g. *O. fasciatus* – Dingle 1974; *Pseudaletia unipuncta*- Delisle & McNeil, 1987; *Mythimna separata* – Han & Gatehouse, 1991b; *Autographa gamma* – Hill & Gatehouse, 1993). Modulation, by larval density, of the expression of genetically determined flight capacity has also been shown in *Spodoptera exempta* (Parker & Gatehouse, 1985*a*; Woodrow, Gatehouse & Davies, 1987; Gatehouse, 1987*b*) and there is some evidence for it in *Melanoplus sanguinipes* (McAnelly, 1985). Finally, variation in flight capacity in aphids has been attributed to a high degree of sensitivity in their response to density (J.Hardie, personal communication; but see above).

### Reliability of environmental cues and variation in response

A response to photoperiod, modulated by temperature, can provide a precise indicator of contemporary seasonal change (except at equatorial latitudes). However, biotic and abiotic variables introduce a stochastic element to the outcome of decisions to migrate, made in response to any environmental cue anticipating habitat deterioration, especially if, as in maternal effects, there is a significant delay before emigration can occur. For example, unseasonally low temperatures over the period between perception of the cue and implementation of the response may lower rates of development and maturation and, at the same time, accelerate habitat deterioration, so that the optimal time for emigration is missed. Alternatively, a sudden improvement in conditions after the response is induced may slow the rate of habitat deterioration, so as to change the consequences for fitness of migration at that time; for example, the condition of the habitat may permit the completion of a further generation. Aphids have evolved mechanisms to accommodate this latter eventuality (Fig. 10.1; Shaw, 1970*a,b*), allowing the actual and potential favourability of the current habitat to be monitored long after the initial induction of alate production, so that subsequent change in conditions can contribute to the ultimate decision on the timing of migration.

This continuing sensitivity to crowding in alate nymphs of *Aphis fabae* ensures that the condition of the current habitat is monitored up to the time of emigration. However, intra-population variation in responses to this cue is also evident, as a given level of crowding has been shown to induce different proportions of alate offspring in different clones derived from the same field population of *Acyrthosiphon pisum* (Lamb & MacKay, 1979). Similar variation in responsiveness to environmental cues is known in other species (Chapter 11, this volume; and see below) and presumably reflects variation in the reliability with which the environmental cue predicts the actual progress of subsequent changes in habitat quality or availability.

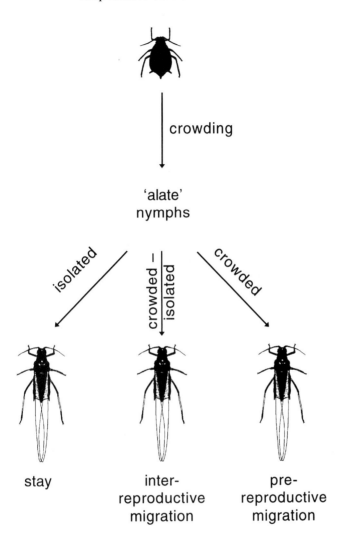

Fig. 10.1. Crowding and the incidence and timing of migration in *Aphis fabae* (Shaw, 1970*a*). Apterous virginoparae respond to crowding by producing 'alate' nymphs. These alate offspring maintain their sensitivity to crowding during development, its intensity determining the characteristics of flight behaviour shown by the resulting adults. If the nymphs are isolated from birth or early in life, the resulting alates do not fly but reproduce on the current host plant. If nymphs are crowded initially and isolated later in life, migration in the majority of individuals takes place inter-reproductively, the alates partitioning their reproductive investment between the current and a new host plant. Only if nymphs are subjected to intense crowding throughout their development does pre-reproductive migration occur.

Migration by several species involves significant displacements across latitude. For example, several subtropical or tropical species regularly penetrate to high latitudes in the temperate-zone summer by migration over several generations (Part one, this volume). The stochastic influence of the winds on the direction and extent of these migrations has a determining influence on the latitude at which individuals of any particular line of descent are located at the end of successive seasons, following these migrations. As the optimal response to photoperiodic and temperature cues must vary with latitude, any single genetic line can be expected to experience intense but varying selection over successive seasons. This must maintain variation within the population in the critical photoperiod that induces extension of the pre-reproductive period and so allows the insects to escape the approaching winter (Dingle, Brown & Hegmann, 1977; Gatehouse, 1989).

A response to a critical rate of change of photoperiod, rather than to a critical absolute photoperiod, might be expected to provide a better cue for these insects (Solbreck, 1979; Han & Gatehouse, 1991*b*). Because the amplitude of the seasonal change increases with increasing latitude, a critical rate of change of photoperiod will be reached earlier at higher than at lower latitudes. Migration is thus likely to be cued at a time more closely related to habitat deterioration, irrespective of latitude. There is evidence for both types of response (see e.g. Dingle *et al.*, 1977; Han & Gatehouse, 1991*b*; Hill & Gatehouse, 1993).

Studies of the response of wing morph and pre-reproductive period to photoperiodic cues have, without exception, revealed substantial variation between individuals and the genetic basis of the variation has been confirmed in some species. For example, variation between clones of *Myzus persicae* (Blackman, 1971) in the proportion of sexuals produced in response to short photoperiods reveals intra-population variation in responses to this cue. Similarly, populations of *Oncopeltus fasciatus* collected in Iowa, Georgia, California and Florida vary in the mean pre-reproductive periods recorded in response to a short photoperiod. More striking, however, are the large differences in the extent of variation in individual pre-reproductive periods between the populations (Dingle *et al.*, 1980). In this species, which responds to critical rather than changing photoperiods, these differences were interpreted as reflecting the varying reliability with which photoperiod predicts seasonal changes in the spatial and temporal distribution of favourable habitats available to these populations (Dingle *et al.*, 1977). Studies of genotype–environment interactions for the response of pre-reproductive period to photoperiod in *O. fasciatus* have confirmed the maintenance of genetic variance for phenotypic plasticity in this trait (Groeters & Dingle, 1987) and similar evidence of variation in phenotypic plasticity has been obtained for the migratory noctuid moth, *Autographa gamma* (Hill & Gatehouse, 1993).

This variation is an inevitable consequence of the stochastic effect of climate on the reliability of environmental cues of changing habitat availability and quality, as well as that of weather, particularly winds, on the direction and extent of migrations.

## Conclusions

This review has drawn attention to the shortcomings in current understanding of the integration of migration with reproduction in insect life cycles, of the persistence of true migratory behaviour (*sensu* Kennedy) during migratory flights, and of the factors involved in the termination of migration, by insects that migrate within or outside their FBL. This is as true for insects migrating to specific diapause or breeding areas as it is for those whose migrations end in habitat patches that are unpredictably distributed in space and time. We are especially ignorant of these aspects of the migration of males. Furthermore, the evidence that is available is very unevenly distributed across the diversity of migratory species, relating almost exclusively to particular families of the Orthoptera, Hemiptera and Lepidoptera.

However, in spite of these limitations, for females of at least some of those species, there is accumulating evidence for extensive variation in traits associated with both the initiation of facultative migration and with migratory potential itself. This variation reflects the uncertain reliability of environmental cues of changing habitat quality on the one hand, and the unpredictability of the spatiotemporal distribution of habitats, together with the stochastic influence of the winds on the direction and extent of migratory displacements, on the other.

## Acknowledgements

We thank Roger Farrow, Don Reynolds, Ken Wilson and Alistair Drake for useful comments on an earlier draft of this chapter.

## References

Armes, N.J. & Cooter, R.J. (1991). Effects of age and mated status on flight potential of *Helicoverpa armigera* (Lepidoptera: Noctuidae). *Physiological Entomology*, **16**, 131–44.

Baker, R.R. (1978). *The Evolutionary Ecology of Animal Migration*. London: Hodder & Stoughton.

Baker, G.L. & Casimir, M. (1986). Current locust and grasshopper control technology in the Australasian region. *Proceedings of the Pan American Acridological Society, 4th Triennial Meeting*, 29 July–2 August, 1985, Saskatoon, Canada, ed. D.A. Nickle, pp. 191–220. Ann Arbor: Michigan University.

Barker, J.F. & Herman, W.S. (1976). Effect of photoperiod and temperature on reproduction of the Monarch butterfly, *Danaus plexippus*. *Journal of Insect Physiology*, **22**, 1565–8.

Blackman, R.L. (1971). Variation in the photoperiodic response within natural populations of *Myzus persicae* (Sulz.). *Bulletin of Entomological Research*, **60**, 533–46.

Bowden, J. & Gibbs, D.G. (1973). Light trap and suction trap catches of insects in the northern Gezira, Sudan in the season of the southward movement of the inter-tropical front. *Bulletin of Entomological Research*, **62**, 571–96.

Brower, L.P. (1985). New perspectives on the migration biology of the Monarch butterfly, *Danaus plexippus*. In *Migration: Mechanisms and Adaptive Significance*, ed. M.A. Rankin, pp. 749–85. *Contributions in Marine Science*, **27**, Suppl. Port Aransas, Texas: Marine Science Institute, The University of Texas at Austin.

Caldwell, R.L. (1974). A comparison of the migratory strategies of two milkweed bugs, *Oncopeltus fasciatus* and *Lygaeus kalmii*. In *Experimental Analysis of Insect Behaviour*, ed. L. Barton Browne, pp. 304–16. Berlin: Springer-Verlag.

Caldwell, R.L. & Rankin, M.A. (1972). Effects of juvenile hormone mimic on flight in the milkweed bug, *Oncopeltus fasciatus*. *General and Comparative Endocrinology*, **19**, 601–5.

Caldwell, R.L. & Rankin, M.A. (1974). The separation of migratory from feeding and reproductive behaviour in *Oncopeltus fasciatus*. *Journal of Comparative Physiology*, **88**, 383–94.

Callahan, P.S., Sparks, A.N., Snow, J.W. & Copeland, W.W. (1972). Corn earworm moth: vertical distribution in nocturnal flight. *Environmental Entomology*, **1**, 497–503.

Chapman, R.F. (1959). Observations on the flight activity of the red locust *Nomadacris septemfasciata* (Serville). *Behaviour*, **14**, 300–44.

Cheke, R., Fishpool, L.D.C. & Forrest, G.A. (1980). *Oedaleus senegalensis* (Krauss) (Orthoptera: Acrididae: Oedipodinae): an account of the 1977 outbreak in West Africa and notes on eclosion under laboratory conditions. *Acrida*, **9**, 107–32.

Chen, R.-C. (1983). Studies on lipids, the energy source for flight in the brown planthopper. *Acta Entomologica Sinica*, **26**, 42–8. In Chinese, English summary.

Chen, R.-C., Cheng, X.-N., Yang, L.-M. & Yin, Z.-D. (1979). The ovarial development of the brown planthopper (*Nilaparvata lugens* Stål) and its relation to migration. *Acta Entomologica Sinica*, **22**, 280–8. In Chinese, English summary.

Chen, R.-L., Bao, X.-Z., Drake, V.A., Farrow, R.A., Wang, S.-Y., Sun, Y.-J. & Zhai, B.-P. (1989). Radar observations of the spring migration into northeastern China of the oriental armyworm moth, *Mythimna separata*, and other insects. *Ecological Entomology*, **14**, 149–62.

Chen, Z.-W., Wang, Z.-G. & Cheu, Y.-D. (1978). [Observations on migration of the brown planthopper.] *Kunchong Zhishi [Entomological Knowledge]*, **6**, 165–70. In Chinese.

Clark, D.P. (1969). Night flights of the Australian plague locust *Chortoicetes terminifera* (Walk.) in relation to storms. *Australian Journal of Zoology*, **17**, 329–52.

Clark, D.P. (1972). The plague dynamics of the Australian plague locust *Chortoicetes terminifera* (Walk.). In *Proceedings of the International Study Conference on the Current and Future Problems in Acridology*, ed. C.F. Hemming & T.H.C. Taylor, pp. 275–87. London: Centre for Overseas Pest Research.

Clement, S.L., Kaster, L.V., Showers, W.B. & Schmidt, R.S. (1985). Seasonal changes in the reproductive condition of female black cutworm moths (Lepidoptera: Noctuidae). *Journal of the Kansas Entomological Society*, **58**, 62–8.

Cockbain, A.J. (1961). Viability and fecundity of alate alienicolae of *Aphis fabae* Scop. after flights to exhaustion. *Journal of Experimental Biology*, **38**, 181–7.

Colvin, J. & Gatehouse, A.G. (1993a). The reproduction-flight syndrome and the inheritance of tethered-flight activity in the cotton-bollworm moth, *Heliothis armigera*. *Physiological Entomology*, **18**, 16–22.

Colvin, J. & Gatehouse, A.G. (1993b). Migration and genetic regulation of the pre-reproductive period in the cotton-bollworm moth, *Helicoverpa armigera*. *Heredity*, **70**, 407–12.

Colvin, J. & Gatehouse, A.G. (1993c). Migration and the effect of three environmental factors on the pre-reproductive period of the cotton-bollworm moth, *Helicoverpa armigera*. *Physiological Entomology*, **18**, 109–13.

Coombs, M., Del Socorro, A.P., Fitt, G.P. & Gregg, P.C. (1993). The reproductive maturity and mating status of *Helicoverpa armigera*, *H. punctigera* and *Mythimna convecta* (Lepidoptera: Noctuidae) collected in tower-mounted light traps in northern New South Wales, Australia. *Bulletin of Entomological Research*, **83**, 529–34.

Danks, H.V. (1987). *Insect Dormancy: an Ecological Perspective*. Ottawa: Biological Surveys of Canada Monograph 1.

David, C.T. & Hardie, J. (1988). The visual responses of free-flying summer and autumn forms of the black bean aphid, *Aphis fabae*, in an automated flight chamber. *Physiological Entomology*, **13**, 277–84.

Delisle, J. & McNeil, J. N. (1986). The effect of photoperiod on the calling behaviour of virgin females of the true armyworm, *Pseudaletia unipuncta* (Haw.) (Lepidoptera: Noctuidae). *Journal of Insect Physiology*, **32**, 199–206.

Delisle, J. & McNeil, J. N. (1987). The combined effect of photoperiod and temperature on the calling behaviour of the true armyworm, *Pseudaletia unipuncta*. *Physiological Entomology*, **12**, 157–68.

Denno, R.F. (1993). Life history variation in planthoppers. In *Planthoppers:*

232 *Migratory potential in insects*

*their Ecology and Management*, ed. R.F. Denno & T.J. Perfect, pp. 163–215. New York: Chapman & Hall.

Denno, R.F. & McClure, M.S. (1984). *Variable Plants and Herbivores in Natural and Managed Systems*. New York: Academic Press.

Descamps, M. (1962). Le cycle biologique de *Gastrimargus nigericus* (Orth. Acrididae) dans la vallée du Bani (Mali). *Revue de Pathologie Végétale et d'Entomologie Agricole de France*, **40**, 187–99.

Descamps, M. (1965) Acridoides du Mali. Régions de San et Sikasso (Zone soudanaise). *Bulletin de l'Institut Fondamental d'Afrique Noir Série A*, **27**, 922–62, 1259–314.

Dingle, H. (1965). The relation between age and flight activity in the milkweed bug, *Oncopeltus*. *Journal of Experimental Biology*, **42**, 269–83.

Dingle, H. (1966). Some factors affecting flight activity in individual milkweed bugs (*Oncopeltus*). *Journal of Experimental Biology*, **44**, 335–43.

Dingle, H. (1968*a*). The influence of environment and heredity on flight activity in the milkweed bug, *Oncopeltus*. *Journal of Experimental Biology*, **48**, 175–84.

Dingle, H. (1968*b*). Life history and population consequences of density, photoperiod and temperature in a migrant insect, the milkweed bug, *Oncopeltus*. *American Naturalist*, **102**, 149–63.

Dingle, H. (1972). Migration strategies of insects. *Science*, **175**, 1327–35.

Dingle, H. (1974). Diapause in a migrant insect, the milkweed bug *Oncopeltus fasciatus* (Dallas) (Hemiptera: Lygaeidae). *Oecologia (Berlin)*, **17**, 1–10.

Dingle, H. (1980). Ecology and evolution of migration. In *Animal Migration, Orientation and Navigation*, ed. S.A. Gauthreaux, pp. 1-101. New York: Academic Press.

Dingle, H. (1984). Behaviour, genes and life histories: complex adaptations in uncertain environments. In *A New Ecology: Novel Approaches to Interactive Systems*, ed. P.W. Price, C.N. Slobodchikoff & W.S. Gaud, pp.169–94. New York: John Wiley.

Dingle, H. (1985). Migration. In *Comprehensive Insect Physiology, Biochemistry and Pharmacology*, vol. 9, ed. G.A. Kerkut & L.I. Gilbert, pp. 375–415. Oxford: Pergamon Press.

Dingle, H., Alden, B.A., Blakely, N.R., Kopec, D. & Miller, E.R. (1980). Variation in photoperiodic response within and among species of milkweed bug (*Oncopeltus*). *Evolution*, **34**, 356–70.

Dingle, H. & Arora, G. (1973). Experimental studies of migration in bugs of the genus *Dysdercus*. *Oecologia (Berlin)*, **12**, 119–40.

Dingle, H., Brown, C.K. & Hegmann, J.P. (1977). The nature of genetic variance influencing photoperiodic diapause in a migrant insect, *Oncopeltus fasciatus*. *American Naturalist*, **111**, 1047–59.

Dixon, A.F.G., Horth, S. & Kindlmann, P. (1993). Migration in insects: costs and strategies. *Journal of Animal Ecology*, **62**, 182–90.

Drake, V.A. & Farrow, R.A. (1988). The influence of atmospheric structure and motions on insect migration. *Annual Review of Entomology*, **33**, 183–210.

Dumont, S. & McNeil, J. N. (1992). Responsiveness of *Pseudaletia unipuncta* (Lepidoptera: Noctuidae) males, maintained as adults under different temperature and photoperiodic conditions, to female sex pheromone. *Journal of Chemical Ecology*, **18**, 1797–807.

Dung, W.-Z. (1981). A general survey on seasonal migrations of *Nilaparvata lugens* (Stål) and *Sogatella furcifera* (Horváth) (Homoptera:

Delphacidae) by means of airplane collections. *Acta Phytophylacica Sinica*, **8**, 73–82. In Chinese, English summary.

Farrow, R.A. (1975*a*). The African migratory locust in its main outbreak area of the middle Niger: quantitative studies of solitary populations in relation to environmental factors. *Locusta*, 11.

Farrow, R.A. (1975*b*). Offshore migration and the collapse of outbreaks of the Australian plague locust (*Chortoicetes terminifera* Walk.) in southeast Australia. *Australian Journal of Zoology*, **23**, 569–95.

Farrow, R.A. (1977*a*). First captures of the migratory locust, *Locusta migratoria* L., at light traps and their ecological significance. *Journal of the Australian Entomological Society*, **18**, 59–61.

Farrow, R.A. (1977*b*). Origin and decline of the 1973 plague locust outbreak in central western New South Wales. *Australian Journal of Zoology*, **25**, 455–89.

Farrow, R.A. (1977*c*). Maturation and fecundity of the spur-throated locust, *Austracris guttulosa* (Walker), in New South Wales during the 1974/75 plague. *Journal of the Australian Entomological Society*, **16**, 27–39.

Farrow, R.A. (1990). Flight and migration in Acridoids. In *Biology of Grasshoppers*, ed. R.F. Chapman & A. Joern, pp. 227–314. New York: John Wiley.

Farrow, R.A. & Daly, J.C. (1987). Long-range movements as an adaptive strategy in the genus *Heliothis* (Lepidoptera: Noctuidae): a review of its occurrence and detection in four pest species. *Australian Journal of Zoology*, **35**, 1–24.

Fitt, G.P. (1989). The ecology of *Heliothis* species in relation to agroecosystems. *Annual Review of Entomology*, **34**, 17–52.

Fuseini, B.A. & Kumar, R. (1975). Ecology of cotton stainers (Heteroptera: Pyrrhocoridae) in southern Ghana. *Biological Journal of the Linnean Society*, **7**, 113–46.

Gatehouse, A.G. (1987*a*). Migration: a behavioural process with ecological consequences? *Antenna*, **11**, 10–12.

Gatehouse, A.G. (1987*b*). Migration and low population density in armyworm (Lepidoptera: Noctuidae) life histories. In *Recent Advances in Research on Tropical Entomology*, ed. M.F.B. Chaudhury, pp. 573–80. *Insect Science and its Application*, vol. 8, Special Issue. Nairobi: ICIPE Science Press.

Gatehouse, A.G. (1989). Genes, environment and insect flight. In *Insect Flight*, ed. G.J. Goldsworthy & C.H. Wheeler, pp. 115–38. Boca Raton, Florida: CRC Press.

Gibo, D.L. (1986). Flight strategies of migrating monarch butterflies (*Danaus plexippus* L.) in southern Ontario. In *Insect Flight: Dispersal and Migration*, ed. W. Danthanarayana, pp. 172–84. Berlin: Springer-Verlag.

Greenbank, D. O., Schaefer, G. W. & Rainey, R. C. (1980). *Spruce Budworm (Lepidoptera: Tortricidae) Moth Flight and Dispersal: New Understanding from Canopy Observations, Radar, and Aircraft*. Memoirs of the Entomological Society of Canada 110. Ottawa: Entomological Society of Canada.

Gregg, P.C., Fitt, G.P., Coombs, M. & Henderson, G.S. (1993). Migrating moths (Lepidoptera) collected in tower-mounted light traps in northern New South Wales, Australia: species composition and seasonal abundance. *Bulletin of Entomological Research*, **83**, 563–78.

Groeters, F.R. & Dingle, H. (1987). Genetic and maternal influences on life history plasticity in response to photoperiod by milkweed bugs (*Oncopeltus fasciatus*). *American Naturalist*, **129**, 332–46.

Gu, H.-N. & Walter, G.H. (1989). Flight of green vegetable bugs *Nezara viridula* (L.) in relation to environmental factors. *Journal of Applied Entomology*, **108**, 347–54.

Gunn, A. & Gatehouse, A.G. (1993). The migration syndrome in the African armyworm moth, *Spodoptera exempta*: allocation of resources to flight and reproduction. *Physiological Entomology*, **18**, 149–59.

Gunn, A., Gatehouse, A.G. & Woodrow, K.P. (1989). Trade-off between flight and reproduction in the African armyworm moth, *Spodoptera exempta*. *Physiological Entomology*, **14**, 419–27.

Han, E.-N. & Gatehouse, A.G. (1991*a*). Genetics of precalling period in the oriental armyworm, *Mythimna separata* (Walker) (Lepidoptera: Noctuidae) and implications for migration. *Evolution*, **45**, 1502–10.

Han, E.-N. & Gatehouse, A.G. (1991*b*). Effect of temperature and photoperiod on the calling behaviour of a migratory insect, the oriental armyworm *Mythimna separata*. *Physiological Entomology*, **16**, 419–27.

Han, E.-N. & Gatehouse, A.G. (1993). Flight capacity: genetic determination and physiological constraints in a migratory moth *Mythimna separata*. *Physiological Entomology*, **18**, 183–8.

Han, E.-N., Zhen, Z.-Q. & Song, Z.-S. (1990). Differences in the responses of male oriental armyworm, *Mythimna separata*, to sex pheromone traps. *Journal of Nanjing Agricultural University*, **13**, 54–6. In Chinese, English summary.

Hardie, J. (1993). Flight behaviour in migrating insects. *Journal of Agricultural Entomology*, **10**, 239–45.

Hegmann, J.P. & Dingle, H. (1982). Phenotypic and genetic covariance structure in a milkweed bug life history. In *Evolution and Genetics of Life Histories*, ed. H. Dingle & J.P. Hegmann, pp. 177–84. New York: Springer-Verlag.

Herman, W.S. (1981). Studies on the reproductive diapause of the monarch butterfly, *Danaus plexippus*. *Biological Bulletin*, **160**, 89–106.

Herman, W.S. (1985). Hormonally mediated events in adult monarch butterflies. In *Migration: Mechanisms and Adaptive Significance*, ed. M.A. Rankin, pp. 799–815. *Contributions in Marine Science*, **27**, Suppl. Port Aransas, Texas: Marine Science Institute, The University of Texas at Austin.

Hill, J. K. & Gatehouse, A. G. (1992). Effects of temperature and photoperiod on development and pre-reproductive period of the silver Y moth *Autographa gamma* (Lepidoptera: Noctuidae). *Bulletin of Entomological Research*, **82**, 335–41.

Hill, J. K. & Gatehouse, A. G. (1993). Phenotypic plasticity and geographic variation in the pre-reproductive period of *Autographa gamma* (Lepidoptera: Noctuidae) and its implications for migration in this species. *Ecological Entomology*, **18**, 39–46.

Hirai, K. (1988). Sudden outbreaks of the armyworm, *Pseudaletia separata* Walker and its monitoring systems in Japan. *Japan Agricultural Research Quarterly*, **22**, 166–74.

Hwang G.-H. & How, W.-W. (1966). Studies on the flight of the armyworm moth (*Leucania separata* Walker). I. Flight duration and wing-beat frequency. *Acta Entomologica Sinica*, **15**, 96–104. In Chinese, English summary.

Ichikawa, T. (1979). [Studies on the mating behaviour of four species of auchenorrhynchous Homoptera which attack the rice plant.] *Memoirs of the Faculty of Agriculture Kagawa University*, **34**, 1–60. In Japanese.

Jago, N. D. (1979). Light trap sampling of the grasshopper *Oedaleus sengalensis* (Krauss, 1877) (Acrididae, Oedipodinae) and other species in West Africa, a critique. In *Proceedings of the Pan American Acridological Society, 2nd Triennial Meeting*, 21–25 July 1979, Montana, USA, ed. M. Tyrkus, I.J. Cantrall & C.S. Carbonell, pp. 165–98.

Johnson, B. (1958). Factors affecting the locomotor and settling responses of alate aphids. *Animal Behaviour*, **6**, 9–26.

Johnson, C.G. (1969). *Migration and Dispersal of Insects by Flight*. London: Methuen.

Johnson, C.G. (1976). Lability of the flight system: a context for functional adaptation. In *Insect Flight*, ed. R.C. Rainey, pp. 217–34. *Symposia of the Royal Entomological Society of London* 7. Oxford: Blackwell Scientific Publications.

Kaster, L.V. & Showers, W.B. (1982). Evidence of spring immigration and autumn reproductive diapause of the adult black cutworm in Iowa. *Environmental Entomology*, **11**, 306–12.

Kehat, M. & Wyndham, M. (1973*a*). Flight activity and displacement in the Rutherglen bug, *Nysius vinitor* (Hemiptera: Lygaeidae). *Australian Journal of Zoology*, **21**, 413–26.

Kehat, M. & Wyndham, M. (1973*b*). The relation between food, age, and flight in the Rutherglen bug, *Nysius vinitor* (Hemiptera: Lygaeidae). *Australian Journal of Zoology*, **21**, 427–34.

Kehat, M. & Wyndham, M. (1974). Differences in flight behaviour of male and female *Nysius vinitor* Bergroth (Hemiptera: Lygaeidae). *Journal of the Australian Entomological Society*, **13**, 27–9.

Kennedy, J.S. (1961). A turning point in the study of insect migration. *Nature*, **189**, 785–91.

Kennedy, J.S. (1985). Migration, behavioural and ecological. In *Migration: Mechanisms and Adaptive Significance*, ed. M.A. Rankin, pp. 5-26. *Contributions in Marine Science*, **27**, Suppl. Port Aransas, Texas: Marine Science Institute, The University of Texas at Austin.

Kennedy, J.S. & Booth, C.O. (1963). Free flight of aphids in the laboratory. *Journal of Experimental Biology*, **40**, 67–85.

Kisimoto, R. (1965). [Studies on polymorphism and its role in the population growth of the brown planthopper, *Nilaparvata lugens* Stål.] *Bulletin of the Shikoku Agricultural Experimental Station*, **13**, 1–106. In Japanese.

Kisimoto, R. (1976). Synoptic weather conditions inducing long distance immigration of planthoppers, *Sogatella furcifera* Horváth and *Nilaparvata lugens* Stål. *Ecological Entomology*, **1**, 95–109.

Kisimoto, R. & Rosenberg, L.J. (1993). Long-distance migration in Delphacid planthoppers. In *Planthoppers: their Ecology and Management*, ed. R.F. Denno & T.J. Perfect, pp. 302–22. New York: Chapman & Hall.

Koerwitz, F.L. & Pruess, K.P. (1964). Migratory potential of the Army Cut-worm. *Journal of the Kansas Entomological Society*, **37**, 234–9.

Lamb, R.J. & MacKay, P.A. (1979). Variability in migratory tendency within and among natural populations of the pea aphid, *Acyrthosiphon pisum*. *Oecologia*, **39**, 289–99.

Li, G.-Z., Li, S.-L. & Liu, S.-F. (1985). [A preliminary analysis of the northern population sources of the rice skipper, *Parnara guttata*.] *Jiangsu Agricultural Sciences*, **2**, 17–18. In Chinese.

Li, K.-P., Wong, H.-H. & Woo, W.-S. (1964). Route of the seasonal migration of the oriental armyworm moth in the eastern part of China as indicated by a three-year result of releasing and recapturing of marked moths. *Acta Phytophylacica Sinica*, **3**, 101–10. In Chinese, English summary.

Loxdale, H.D., Hardie, J., Halbert, S., Foottit, R., Kidd, N.A.C. & Carter, C.A. (1993). The relative importance of long- and short-range movement of flying aphids. *Biological Reviews*, **68**, 291–311.

Malcolm, S.B., Cockrell, B.J. & Brower, L.P. (1993). Spring recolonisation of eastern North America by the Monarch butterfly: successive brood or single sweep migration? In *Biology and Conservation of the Monarch Butterfly*, ed. S.B. Malcolm & M.P. Zalucki, pp. 253–67. Los Angeles: Natural History Museum of Los Angeles County.

McAnelly, M.L. (1985). The adaptive significance and control of migratory behaviour in the grasshopper *Melanoplus sanguinipes*. In *Migration: Mechanisms and Adaptive Significance*, ed. M.A. Rankin, pp. 687–703. *Contributions in Marine Science*, **27**, Suppl. Port Aransas, Texas: Marine Science Institute, The University of Texas at Austin.

McAnelly, M.L. & Rankin, M.A. (1986a). Migration in the grasshopper *Melanoplus sanguinipes* (Fab.). I. The capacity for flight in non-swarming populations. *Biological Bulletin*, **170**, 368–77.

McAnelly, M.L. & Rankin, M.A. (1986b). Migration in the grasshopper *Melanoplus sanguinipes* (Fab.). II. Interactions between flight and reproduction. *Biological Bulletin*, **170**, 378–92.

McDonald, G. & Cole, P.G. (1991). Factors influencing oocyte development in *Mythimna convecta* (Lepidoptera: Noctuidae) and their possible impact on migration in eastern Australia. *Bulletin of Entomological Research*, **81**, 175–84.

McDonald, G. & Farrow, R.A. (1988). Migration and dispersal of the Rutherglen bug, *Nysius vinitor* Bergroth (Hemiptera: Lygaeidae), in eastern Australia. *Bulletin of Entomological Research*, **78**, 493–509.

McNeil, J. N. (1986). Calling behaviour: can it be used to identify migratory species of moths? *Florida Entomologist*, **69**, 78–84.

McNeil, J. N. (1987). The true armyworm, *Pseudaletia unipuncta*: a victim of the pied piper or a seasonal migrant? In *Recent Advances in Research on Tropical Entomology*, ed. M.F.B. Chaudhury, pp. 591–7. *Insect Science and its Application*, vol. 8, Special Issue. Nairobi: ICIPE Science Press.

Mikkola, K. (1971). The migratory habit of *Lymantria dispar* (Lep.: Lymantriidae) adults of continental Eurasia in the light of a flight to Finland. *Acta Entomologia Fennica*, **28**, 107–20.

Mikkola, K. (1986). Direction of insect migrations in relation to the wind. In *Insect Flight: Dispersal and Migration*, ed. W. Danthanarayana, pp. 152–71. Berlin: Springer-Verlag.

Mochida, O. (1970). A red-eyed form of the brown planthopper, *Nilaparvata lugens* (Stål) (Hom., Auchenorrhyncha). *Bulletin of the Kyushu Agricultural Experimental Station*, **15**, 141–273. In Japanese.

Mulder, P.G., Showers, W.B., Kaster, L.V. & Vanschaik, J. (1989). Seasonal activity and response of the black cutworm (Lepidoptera: Noctuidae) to virgin females reared for one or multi-generations in the laboratory. *Environmental Entomology*, **18**, 19–23.

Myers, T.V. & Pedigo, L.P. (1977). Seasonal fluctuations in abundance, reproductive status, sex ratio and mating of the adult green cloverworm. *Environmental Entomology*, **6**, 225–8.

Nakasuji, F. & Isii, M. (1988). [*Butterfly Migration across the Sea.*] Tokyo: Winter Tree Publishing House. In Japanese.

National Cooperative Research Group on Rice Leaf Roller (1981). Advances in studies on the migration of the rice leaf roller in China. *Scientia Agricultura Sinica*, **5**, 1–8. In Chinese, English summary.

Noda, H. (1986). Pre-mating flight of rice planthopper migrants (Homoptera: Delphacidae) collected on the East China Sea. *Applied Entomology and Zoology*, **21**, 175–6.

Nottingham, S.F. & Hardie, J. (1989). Migratory and targeted flight in seasonal forms of the black bean aphid, *Aphis fabae. Physiological Entomology*, **14**, 451–8.

Nottingham, S.F., Hardie, J. & Tatchell, G.M. (1991). Flight behaviour of the bird cherry aphid, *Rhopalosiphum padi. Physiological Entomology*, **16**, 223–9.

Ohkubo, N. (1967). Study on the density effect of the adult of the brown planthopper, *Nilaparvata lugens* Stål. *Japanese Journal of Ecology*, **17**, 230–3. In Japanese, English summary.

Ohkubo, N. & Kisimoto, R. (1971). Diurnal periodicity of the flight behaviour of the brown planthopper, *Nilaparvata lugens* Stål, in the 4th and 5th emergence periods. *Japanese Journal of Applied Entomology and Zoology*, **15**, 8–16.

Oku, T. (1982). Aestivation and migration in noctuid moths. In *Diapause and Life Cycle Strategies in Insects*, ed. V.K. Brown & I. Hodek, pp. 219–31. The Hague: Junk.

Padgham, D.E., Perfect, T. J. & Cook, A.G. (1987). Flight behaviour in the brown planthopper, *Nilaparvata lugens* (Stål) (Homoptera: Delphacidae). *Insect Science and its Application*, **8**, 71–5.

Palmer, J.O. & Dingle, H. (1986). Direct and correlated responses to selection among life-history traits in milkweed bugs (*Oncopeltus fasciatus*). *Evolution*, **40**, 767–77.

Palmer, J.O. & Dingle, H. (1989). Responses to selection on flight behaviour in a migratory population of milkweed bug (*Oncopeltus fasciatus*). *Evolution*, **43**, 1805–8.

Parker, J.R., Newton, R.C. & Shotwell, R.L. (1955). *Observations on Mass Flights and Other Activities of the Migratory Grasshopper*. Technical Bulletin 1109. Washington, DC: United States Department of Agriculture.

Parker, W.E. (1983). An experimental study on the migration of the African armyworm moth, *Spodoptera exempta* (Walker) (Lepidoptera: Noctuidae). PhD thesis, University of Wales.

Parker, W.E. & Gatehouse, A.G. (1985*a*). The effect of larval rearing conditions on flight performance in females of the African armyworm *Spodoptera exempta* (Walker) (Lepidoptera: Noctuidae). *Bulletin of Entomological Research*, **75**, 35–47.

Parker, W.E. & Gatehouse, A.G. (1985*b*). Genetic factors controlling flight performance and migration in the African armyworm moth, *Spodoptera exempta* (Walker) (Lepidoptera: Noctuidae). *Bulletin of Entomological Research*, **75**, 49–63.

Pedgley, D.E. (1982). *Windborne Pests and Diseases. Meteorology of Airborne Organisms*. Chichester: Ellis Horwood.

Perfect, T.J., Cook, A.J., Padgham, D.E. & Crisostomo, J.M. (1985). Interpretation of the flight activity of *Nilaparvata lugens* (Stål) and *Sogatella furcifera* (Horváth) (Hemiptera: Delphacidae) based on comparative trap catches and marking with rubidium. *Bulletin of Entomological Research*, **75**, 93–106.

Rankin, M. A. & Burchsted, J. C. A. (1992). The cost of migration in insects. *Annual Review of Entomology*, **37**, 533–59.

Rankin, M. A., McAnelly, M. L. & Bodenhamer, J. E. (1986). The oogenesis-flight syndrome revisited. In *Insect Flight, Dispersal and Migration*, ed. W. Danthanarayana, pp. 27–48. Berlin: Springer-Verlag.

Rankin, M.A. & Riddiford, L.M. (1977). Hormonal control of migratory flight in *Oncopeltus fasciatus*: effects of the corpus cardiacum, corpus allatum and starvation on migration and reproduction. *General and Comparative Endocrinology*, **33**, 309–21.

Rao, Y.R. (1942). Some results of studies on the Desert Locust (*Schistocerca gregaria* Forsk.) in India. *Bulletin of Entomological Research*, **33**, 241–65.

Rao, Y.R. (1960). *The Desert Locust in India*. New Delhi: Indian Council of Agricultural Research.

Reynolds, D.R. & Riley, J.R. (1988). A migration of grasshoppers, particularly *Diabolocatantops axillaris* (Thunberg) (Orthoptera, Acrididae), in the West African Sahel. *Bulletin of Entomological Research*, **78**, 251–71.

Riley, J.R., Cheng, X.-N., Zhang, X.-X., Reynolds, D.R., Xu, G.-M., Smith, A.D., Cheng, J.-Y., Bao, A.-D. & Zhai, B.-P. (1991). The long-distance migration of *Nilaparvata lugens* (Stål) (Delphacidae) in China: radar observations of mass return flight in the autumn. *Ecological Entomology*, **16**, 471–89.

Riley, J.R. & Reynolds, D.R. (1983). A long-range migration of grasshoppers observed in the Sahelian zone of Mali by two radars. *Journal of Animal Ecology*, **52**, 167–83.

Riley, J.R., Reynolds, D.R. & Farmery, M.J. (1981). Radar observations of *Spodoptera exempta* in Kenya, March–April 1979. *Centre for Overseas Pest Control Miscellaneous Report 54*. London: Overseas Development Administration.

Riley, J.R., Reynolds, D.R. & Farmery, M.J. (1983). Observations of the flight behaviour of the armyworm moth, *Spodoptera exempta*, at an emergence site using radar and infra-red optical techniques. *Ecological Entomology*, **8**, 395–418.

Riley, J.R., Reynolds, D.R. & Farrow, R.A. (1987). The migration of *Nilaparvata lugens* (Stål) (Delphacidae) and other Hemiptera associated with rice, during the dry season in the Philippines: a study using radar, visual observations, aerial netting and ground trapping. *Bulletin of Entomological Research*, **77**, 145–69.

Roff, D.A. (1992). *The Evolution of Life Histories. Theory and Analysis*. New York: Chapman & Hall.

Roffey, J. (1963). *Observations on Night Flight in the Desert Locust* (Schistocerca gregaria *Forskål*). Anti-Locust Bulletin 39. London: Anti-Locust Research Centre.

Roffey, J. & Popov, G.B. (1968). Environmental and behavioural processes in a desert locust outbreak. *Nature*, **219**, 446–50.

Roome, R.E. (1972). *Annual Report 1971–1972. Entomologist, Botswana*

*Dryland Farm Research Scheme*. London: Overseas Development Administration.

Rose, D.J.W. (1972). Dispersal and quality in populations of *Cicadulina* species (Cicadellidae). *Journal of Animal Ecology*, **41**, 589–609.

Rose, D.J.W. (1973). Field studies in Rhodesia on *Cicadulina* spp. (Hem., Cicadellidae), vectors of maize streak disease. *Bulletin of Entomological Research*, **62**, 477–95.

Rose, D.J.W. (1979). The significance of low density populations of the African armyworm *Spodoptera exempta* (Walk.). *Philosophical Transactions of the Royal Society of London B*, **287**, 393–402.

Rose, D.J.W., Dewhurst, C.F., Page, W.W. & Fishpool, L.D.C. (1987). The role of migration in the life system of the African armyworm *Spodoptera exempta*. In *Recent Advances in Research on Tropical Entomology*, ed. M.F.B. Chaudhury, pp. 561–9. *Insect Science and its Application*, vol. 8, Special Issue. Nairobi: ICIPE Science Press.

Sappington, T.W. & Showers, W.B. (1992). Reproductive maturity, mating status, and long-duration flight behaviour of *Agrotis ipsilon* (Lepidoptera: Noctuidae) and the conceptual misuse of the oogenesis-flight syndrome by entomologists. *Environmental Entomology*, **21**, 677–88.

Schaefer, G.W. (1976). Radar observations of insect flight. In *Insect Flight*, ed. R.C. Rainey, pp. 157–97. *Symposia of the Royal Entomological Society of London 7*. Oxford: Blackwell Scientific Publications.

Shaw, M.P.J. (1970*a*). Effects of population density on alienicolae of *Aphis fabae* Scop. II. The effects of crowding on the expression of the migratory urge among alatae in the laboratory. *Annals of Applied Biology*, **65**, 197–203.

Shaw, M.P.J. (1970*b*). Effects of population density on alienicolae of *Aphis fabae* Scop. III. The effect of isolation on the development of form and behaviour of alatae in a laboratory clone. *Annals of Applied Biology*, **65**, 205–12.

Showers, W.B., Keaster, A.J., Raulston, J.R., Hendrix, W.H., Derrick, M.E., McCorcle, M.D., Robinson, J.F., Way, M.O., Wallendorf, M.J. & Goodenough, J.L. (1993). Mechanism of southward migration of a noctuid moth (*Agrotis ipsilon* (Hufnagel)): a complete migrant. *Ecology*, **74**, 2303–14.

Showers, W.B., Smelser, R.E., Keaster, A.J., Whitford, F., Robinson, J.F., Lopez, J.D. & Taylor, S.E. (1989*a*). Re-capture of marked black cutworm (Lepidoptera: Noctuidae) males after long-range transport. *Environmental Entomology*, **18**, 447–58.

Showers, W.B., Whitford, F., Smelser, R.E., Keaster, A.J., Robinson, J.F., Lopez, J.D. & Taylor, S.E. (1989*b*). Direct evidence for meteorologically driven long-range dispersal of an economically important moth. *Ecology*, **70**, 987–92.

Slansky, F. (1980). Food consumption and reproduction as affected by tethered flight in female milkweed bugs (*Oncopeltus fasciatus*). *Entomologia Experimentalis et Applicata*, **28**, 277–86.

Smith, A.M. & McDonald, G. (1986). Interpreting ultra-violet light and fermentation trap catches of *Mythimna convecta* (Walker) (Lepidoptera: Noctuidae) using phenological simulation. *Bulletin of Entomological Research*, **76**, 419–31.

Solbreck, C. (1978). Migration, diapause, and direct development as alternative life histories in a seed bug, *Neacoryphus bicrucis*. In *The*

*Evolution of Insect Migration and Diapause*, ed. H. Dingle, pp. 195–217. New York: Springer-Verlag.

Solbreck, C. (1979). Induction of diapause in a migratory seed bug, *Neacoryphus bicrucis* (Say) (Heteroptera: Lygaeidae). *Oecologia*, **43**, 41–9.

Southwood, T.R.E. (1977). Habitat, the templet for ecological strategies? *Journal of Animal Ecology*, **46**, 337–65.

Southwood, T.R.E. (1981). Ecological aspects of insect migration. In *Animal Migration*, ed. D.J. Aidley, pp. 197–208. Cambridge: Cambridge University Press.

Sparks, A.N., Jackson, R.D. & Allen, L.C. (1975). Corn earworms: capture of adults in light-traps on unmanned oil platforms in the Gulf of Mexico. *Journal of Economic Entomology*, **68**, 431–2.

Spitzer, K. (1972). Seasonal adult activity of *Scotia ipsilon* Hfn. (Lepidoptera, Noctuidae) in Bohemia. *Acta Entomologica Bohemoslovaca*, **69**, 396–400.

Stearns, S.C. (1992). *The Evolution of Life Histories*. Oxford: Oxford University Press.

Swier, S.R., Rings, R.W. & Musick, G.J. (1976). Reproductive behaviour of the black cutworm, *Agrotis ipsilon. Annals of the Entomological Society of America*, **69**, 546–50.

Symmons, P.M. & McCullogh, L. (1980). Persistence and migration of *Chortoicetes terminifera* (Walker) (Orthoptera, Acrididae) in Australia. *Bulletin of Entomological Research*, **70**, 197–201.

Symmons, P.M. & Wright, D.E. (1981). The origins and course of the 1979 plague of the Australian plague locust *Chortoicetes terminifera* (Walker) (Orthoptera: Acrididae) including the effect of chemical control. *Acrida*, **10**, 159–90.

Tauber, M.J., Tauber, C.A. & Masaki, S. (1984). Adaptations to hazardous seasonal conditions: dormancy, migration and polyphenism. In *Ecological Entomology*, ed. C.B. Huffaker & R.L. Rabb, pp. 149–83. New York: John Wiley.

Taylor, L.R. (1958). Aphid dispersal and diurnal periodicity. *Proceedings of the Linnean Society of London*, **169**, 67–73.

Taylor, L.R. (1974). Insect migration, flight periodicity and the boundary layer. *Journal of Animal Ecology*, **43**, 225–38.

Teraguchi, S.E. (1986). Migration patterns of leafhoppers (Homoptera: Cicadellidae) in an Ohio old field. *Environmental Entomology*, **15**, 1199–211.

Têtefort, J.P. & Wintrebert, D. (1967). Ecologie et comportement du criquet nomade dans le Sud-Ouest malagache. *Annales de la Société Entomologique de France*, **3**, 3–30.

Tucker, M.R. (1994). Inter-seasonal and intra-seasonal variation in outbreak distribution of the armyworm, *Spodoptera exempta* (Lepidoptera: Noctuidae), in Eastern Africa. *Bulletin of Entomological Research*, **84**, 275–87.

Tucker, M.R., Mwandoto, S. & Pedgley, D.E. (1982). Further evidence for windborne movement of armyworm moths, *Spodoptera exempta*, in East Africa. *Ecological Entomology*, **7**, 463–73.

Turgeon, J.J. & McNeil, J.N. (1983). Modifications of the calling behaviour of *Pseudaletia unipuncta* (Lepidoptera: Noctuidae), induced by temperature conditions during pupal and adult development. *Canadian Entomologist*, **115**, 1015–22.

Turgeon, J. J., McNeil, J.N. & Roelofs, W. L. (1983). Responsiveness of *Pseudaletia unipuncta* males to the female sex pheromone. *Physiological Entomology*, **8**, 339–44.

Uvarov, B. P. (1977). *Grasshoppers and Locusts*, vol. 2. London: Centre for Overseas Pest Research.

Wada, T. & Kobayashi, M. (1980). Outbreak and ecology of the rice leaf roller. *Shokubutsu-boeki [Plant Protection]*, **34**, 528–32. In Japanese.

Wada, T. & Kobayashi, M. (1982). Mating status depending on the growing stage of the rice plant in the population of *Cnaphalocrocis medinalis* Guénée (Lepidoptera: Pyralidae). *Applied Entomology and Zoology*, **17**, 278–81.

Wada, T. & Kobayashi, M. (1985). Seasonal changes of wing wear of *Cnaphalocrocis medinalis* Guénée in paddy fields. *Japanese Journal of Applied Entomology and Zoology*, **29**, 41–4. In Japanese, English summary.

Wada, T., Ogawa, Y. & Nakasuga, T. (1988). Geographical differences in mated status and the autumn migration in the rice leaf roller moth, *Cnaphalocrocis medinalis*. *Entomologia Experimentalis et Applicata*, **46**, 141–8.

Walker, T.J. (1986). Stochastic polyphenism: coping with uncertainty. *Florida Entomologist*, **69**, 46–62.

Waloff, Z. (1963). *Field Studies on Solitary and transiens Desert Locust in the Red Sea area*. Anti-Locust Bulletin 40. London: Anti-Locust Research Centre.

Wang, Y.-Z. (1991). [Studies on the migratory behaviour of the oriental armyworm *Mythimna separata* (Walker).] MSc thesis, Nanjing Agricultural University. In Chinese, English summary.

Williams, C.B. (1958). *Insect Migration*. London: Collins.

Wilson, K. & Gatehouse, A.G. (1993). Seasonal and geographical variation in the migratory potential of outbreak populations of the African armyworm moth, *Spodoptera exempta*. *Journal of Animal Ecology*, **62**, 169–81.

Wolf, W.W., Sparks, A.N., Pair, S.D., Westbrook, J.K. & Truesdale, F.M. (1986). Radar observations and collections of insects in the Gulf of Mexico. In *Insect Flight: Dispersal and Migration*, ed. W. Danthanarayana, pp. 221–34. Berlin: Springer-Verlag.

Woodrow, K.P., Gatehouse, A.G. & Davies, D.A. (1987). The effect of larval phase on flight performance of African armyworm moths, *Spodoptera exempta* (Walker) (Lepidoptera: Noctuidae). *Bulletin of Entomological Research*, **77**, 113–22.

Zhai, B.-P. & Zhang, X.-X. (1993). Behaviour of airborne insects in their migratory process: a review. *Chinese Journal of Applied Ecology*, **4**, 440–6. In Chinese, English summary.

Zhang, J.-X. & Zhang, X.-X. (1991). [Migration of the white backed planthopper.] In *[Strategy and Techniques for the Control of Insect Pests and Diseases of Rice in China]*, ed. Z.-W. Du, pp. 43–57. Beijing: Agricultural Publishing House. In Chinese.

Zhang, X.-X., Keng, C.-G. & Chen, X.-N. (1979). *[The Principles and Methods for Insect Pest Forecasting.]* Beijing: Agricultural Publishing House. In Chinese, English summary.

Zhang, X.-X., Lo, Z.-C., Keng, C.-G., Li, G.-Z., Chen, X.-L. & Wu, X.-W. (1980). Studies on the migration of the rice leaf roller, *Cnaphalocrocis medinalis* Guénée. *Acta Entomologica Sinica*, **23**, 130–40. In Chinese, English summary.

Zhang X.-X. (1991). [Migration in the rice leaf roller.] In *[Strategy and Techniques for the Control of Insect Pests and Diseases of Rice in China]*, ed. Z.-W. Du, pp. 58–82. Beijing: Agricultural Publishing House. In Chinese.

# 11

## Insect migration in heterogeneous environments

### K. WILSON

### Introduction

Habitats vary in their temporal and spatial stability and, as Southwood (1962, 1977) recognised, this has important consequences for the evolution of migration in insects. The spatiotemporal structure of the habitat, he argued, would act as a 'template' selecting individuals of migratory phenotype appropriate to the prevailing conditions. Based on this assumption, Southwood predicted that insects occupying habitats whose 'favourableness' varied in time and/or space would have relatively high migratory potential, whereas those in more permanent, continuous habitats would have relatively low migratory potential, due to costs generally assumed to be associated with the ability to migrate (Rankin & Burchsted, 1992). Between-species comparisons within a number of insect orders generally support this prediction: migratory species tend to be associated with 'temporary' habitats, non-migratory with 'permanent' ones (Southwood, 1962; Johnson, 1969; Roff, 1990a; Denno et al., 1991).

However, comparative studies that fail to take account of phylogenetic and other constraints on evolution can generate misleading results (Harvey & Pagel, 1991; but see Denno et al., 1991 for an excellent use of the comparative method). An alternative approach to the problem is to make within-species comparisons of populations in environments differing in their spatiotemporal stabilities, the assumption being that the observed levels of migratory potential will reflect the prevailing levels of habitat heterogeneity. However, there is also a problem associated with this method – that the predicted response to habitat heterogeneity may be obscured by the mechanisms regulating migratory potential. An example is the migratory strategy of an insect like the wing-dimorphic weevil *Sitona hispidulus*, which inhabits meadows and other early successional habitats. For this insect, the optimal proportion of long-winged offspring

243

to produce is at its lowest following colonisation of a new habitat and increases over time, as succession proceeds and the habitat becomes less hospitable (Roff, 1990*b*). The observed pattern of offspring production, however, is quite different from that predicted, with the highest proportion of long-winged offspring produced immediately following colonisation (Stein, 1977). The reason for this apparent paradox is that wing length, and hence migratory potential, is regulated by a simple genetic mechanism (a single locus with two alleles – Jackson, 1928) so the long-winged insects that colonised the habitat tend to produce offspring that also have long wings. Therefore, as a number of authors have noted (e.g. Roff, 1975, 1990*b*; Hamilton & May, 1977; Comins, Hamilton & May, 1980), in order to examine the influence of habitat heterogeneity on migratory potential at the *within-species* level, both the ultimate (i.e. adaptive) and proximate (i.e. mechanistic) causes of field variation must be analysed together. This is the approach taken in the present chapter. Specifically, the following questions are addressed:

1 What aspects of habitat heterogeneity are likely to influence the evolution and maintenance of variation in migratory potential?
2 What is the relationship between habitat heterogeneity and the mechanisms regulating migratory potential?
3 Within species, is there any evidence for significant differences in migratory potential within or between field populations?
4 Are these differences correlated with the degree of habitat heterogeneity?

## Terminology and assumptions

Throughout this chapter, the term *migratory potential* is used to describe the level of any traits associated with the potential for migratory flight (Chapter 10, this volume). In monomorphic macropterous species, high migratory potential is often associated with relatively long wings, well-developed flight muscles, a capacity for prolonged tethered flight and an extended pre-reproductive period (see e.g. Palmer & Dingle, 1986, 1989; Rankin & Burchsted, 1992). In migratory wing-dimorphic species, the proportion of macropterous individuals within a population or clone is generally correlated with the average migratory potential of individuals within it, though not all macropters are capable of flight (Fairbairn, 1986; Roff & Fairbairn, 1991). It is assumed that there is at least a small cost associated with the possession of a migratory attribute such as long wings or extended pre-reproductive period and hence that, within stable habitats, insects with low migratory potential are at a selective advantage (Roff & Fairbairn, 1991; Rankin & Burchsted, 1992; Denno, 1993).

*Habitat* is used here to describe the spatial unit delineated by an insect's non-migratory locomotor capabilities ('trivial' movements, *sensu* Johnson, 1969; Hassell & Southwood, 1978; Southwood, 1981), and *environment* to include the area potentially available to it when engaging in migratory flight. *Arena* is used to describe the area encompassed by the environments of all individuals within the population as a whole. Thus, an arena comprises a number of separate habitats, though these may be contiguous and be described as a *habitat aggregation*. Habitats are generally assumed to be either *favourable* for survival and development or *unfavourable*, but no assumptions are made about the causes of variation in favourableness – for example, whether it is due to differences in the level of crowding, food quality, degree of predation or parasitism, etc.

Two types of habitat variation are considered: temporal and spatial. On the temporal scale, habitats may range from *permanent* to *ephemeral*, with the vast area between these two extremes being referred to as *temporary*. On the spatial scale, the distribution of favourable habitats may be either *continuous* or *discontinuous*. When they are discontinuous, their distribution may range from statistically *uniform*, through *random*, to *aggregated*. These habitat classifications can be further divided according to whether the durational stability of habitats, or the migratory effort required to migrate between them, has a predictable component. Spatially 'predictable' habitat distributions are referred to here as *regular*, and 'unpredictable' distributions as *irregular*.

### Factors generating variation in migratory potential

The maintenance of variation in migratory potential can be viewed from two complementary perspectives. The first is the evolutionary perspective, which is concerned with the adaptive significance of variation in migratory potential; the second is the mechanistic perspective, which attempts to determine the proximate causes of that variation. This section briefly examines variation in migratory potential from both these perspectives.

Many theoretical studies have analysed the factors important in the evolution of dispersal and migration strategies (e.g. Cohen, 1967; Gadgil, 1971; Roff, 1974*a,b*, 1975; Hamilton & May, 1977; McPeek & Holt, 1992 and references therein). These studies conclude that temporal variation in habitat favourableness is generally required for migration to evolve. However, under some circumstances, migration may evolve in temporally constant, but spatially varying, environments (see e.g. Hastings, 1983), or even in environments that are both spatially *and* temporally constant (see e.g. Hamilton & May, 1977). Not surprisingly, perhaps, most authors conclude that migration is most likely to evolve in environments that are both spatially and temporally heterogeneous.

Although temporal heterogeneity mostly influences the prevalence and timing of migration, and spatial heterogeneity mainly its duration and extent, it is often the *interaction* between the temporal and spatial components of a species' arena that determines its migratory character. Evolutionarily, within-species variation in migratory potential may be maintained by local adaptation at the population level; by frequency-dependent selection, in which the optimal strategy depends on that adopted by other members of the population; and by the stochastic nature of the selection process.

There are a number of proximate mechanisms capable of generating this variation (see Chapter 10, this volume) and the ones observed are likely to reflect the relative predictability of the insects' environment. When the durational stability or spatial distribution of favourable habitats is predictable, mechanisms will evolve to take advantage of that predictability, such as adaptive responses to environmental cues (e.g. crowding, food quality/quantity, photoperiod, temperature, etc.) and genetic or non-genetic maternal effects. However, when the persistence and distribution of habitats cannot be predicted, genetic differentiation and 'bet-hedging' maternal-age effects can be expected. Whilst all of these mechanisms may generate variation in migatory potential, it is only when there is additive genetic variation that contemporary natural selection, imposed by the current habitat template, can contribute to the observed patterns.

### Empirical evidence for field variation in migratory potential

In this section, the empirical evidence for field variation in migratory potential is reviewed. The review is far from exhaustive and concentrates on species for which variation in migratory potential has been quantified, characteristics of their habitats documented, and details of the mechanisms regulating migratory potential determined. The approach taken is to assign variation in the various components of migratory potential primarily to the spatial or the temporal components of environmental variation, and to the predictability of that variation. However, although this approach provides some useful insights into how the habitat template shapes the evolution of migration, it is necessary, again, to emphasise its limitations. For example, spatial as well as temporal heterogeneity may be involved in determining the prevalence and timing of migration, as can be illustrated by considering the effects of habitat aggregation and isolation.

### Habitat aggregation and isolation

As defined here, habitats are of similar size (determined by the range of the non-migratory movements of the species) but may vary in their degree of clumping. Within a habitat aggregation (a number of contiguous habitats), the average migratory potential of the insects is determined by both its size and degree of isolation. In general, when habitat aggregations are large, natural selection will favour individuals that migrate between habitats *within* aggregations and hence that have relatively low migratory potential; whereas when they are small, selection will tend to favour individuals with the high migratory potential necessary to migrate *between* aggregations. An exception to this rule is when habitat aggregations are both small and spatially isolated. In these circumstances, the prevalence of migratory phenotypes may be reduced, because the rate of exchange of migrants between aggregations may be very low, and the cost of migration to the individual high. Hence, the average migratory potential of insect populations, including the prevalence of migratory phenotypes, may be a function of the spatial distribution, as well as the temporal stability, of their habitats.

Such arguments have been invoked to explain why insects with low migratory potential are often associated with small, isolated habitat aggregations, such as islands and mountaintops (e.g. Darwin, 1859; but see Roff, 1990*a*). Cultivated land often assumes an aggregated distribution, due to the concentration of crops on fertile soil. Lamb & MacKay (1979) studied populations of the aphid *Acyrthosiphon pisum* exploiting fields of alfalfa and found that migratory potential (as determined by the numbers of winged and non-winged offspring produced by parthenogenetic females after a crowding test) was lowest for populations occupying the most isolated fields. These differences appear to be genetic in origin, as they were observed even when maternal-age and environmental parameters were kept constant (Lamb & MacKay, 1979). The authors attributed the lower migratory potential of isolated populations to 'a lower ratio of immigrants to emigrants and a greater loss of the genes for migratory tendency'.

Therefore, the migratory potential of individuals within a population is determined by both the spatial and temporal components of the environment they occupy. Nevertheless, it is still appropriate, and instructive, to assign variation in migratory potential predominantly to one or other of these causes. The remainder of this section is concerned first, with the evolutionary and mechanistic responses to predictable variation in temporal stability; second, with responses to unpredictable variation in temporal stability; third, with responses to spatially regular habitat distributions; and finally, with responses to spatially irregular habitat distributions (as defined above).

### *Temporally predictable habitats*

When habitats are temporary, migration may be the optimal strategy for an insect to adopt. There is evidence from a number of monomorphic macropterous species that migration is induced in response to environmental cues predicting the future deterioration of the current habitat (Dingle, 1985). However, the best evidence concerning within-species variation in the production of migrants and non-migrants comes from studies of wing-dimorphic species (see Harrison, 1980; Roff, 1986*a,b*; Gatehouse, 1989; Roff & Fairbairn, 1991 for reviews) and it is therefore on these species that this part of the review concentrates.

When habitats vary in their durational stability in a *predictable* manner, regulatory mechanisms will evolve to take advantage of that predictability. These mechanisms can be divided into two main types: environmental regulation, in which the production of migrants and non-migrants is determined by direct responses to environmental cues, such as crowding, photoperiod, temperature or food supply (Chapter 10, this volume); and maternal regulation, in which genetic or environmental differences in the maternal generation are expressed as phenotypic differences in the offspring generation (Mousseau & Dingle, 1991).

An example of a monomorphic macropterous species in which the production of migratory and non-migratory phenotypes appears to be determined by facultative responses to environmental cues is the Cotton-Bollworm moth *Helicoverpa armigera*. Colvin (Chapter 12, this volume) demonstrates that most individuals become sexually mature within a day or two of feeding on a carbohydrate source (sugar solution in the laboratory; nectar in the field). Thus, as flight capacity is at its highest in pre-reproductive individuals (Colvin & Gatehouse, 1993), most *H. armigera* migration can be expected when flowering plants are unavailable, as during the dry season. The adaptive significance of using nectar availability as a cue to habitat suitability is obvious for a species whose larvae feed on the reproductive structures of plants (Chapter 12, this volume).

When migratory phenotype is environmentally regulated, an optimum response curve or *norm of reaction* can be expected to evolve for each environmental cue. For some cues a single norm of reaction will be more or less optimal in all environments but, for others, there may be several optima, because of variation between environments in the accuracy with which the cues predict the timing or extent of habitat deterioration (Chapter 10, this volume). For example, a higher threshold of response to crowding will be favoured in relatively permanent habitats than in relatively temporary ones (Gadgil, 1971), because, for a given level of crowding, the probability of subsequent habitat deterioration is higher in temporary habitats. Geographical variation of this kind is observed in the crowding response of several planthopper species, including the Brown Planthopper *Nilaparvata lugens* (Iwanaga , Nakasuji & Tojo, 1987; Tojo, 1991). In South Asia (Philippines, Malaysia and Indonesia), where the insect's host plant (rice) is grown throughout the year, the proportions of macropterous adults produced are small and largely independent of nymphal densities. In China, where habitats remain favourable for shorter periods of time, the proportion of nymphs developing into macropters is much higher and, to some extent, dependent upon nymphal density. In Japan, where planthoppers immigrate each year from mainland China or South Asia (Iwanaga *et al.*, 1987), most individuals develop as macropters and the proportion doing so increases with nymphal density (Tojo, 1991). The response to nymphal crowding is genetically regulated in this species, and controlled by genes at several loci (Tojo, 1991).

One of very few studies to have quantified the relationship between habitat stability and the production of migratory phenotypes is that by Denno & Grissell (1979) on the wing-dimorphic, multivoltine planthopper *Prokelisia marginata*. This species inhabits salt marshes on the Atlantic Coast of North America, and its host plant (an intertidal, perennial grass species) exhibits predictable variation in its durational stability. 'High marsh' habitats persist throughout the year, but are of relatively poor quality, whereas 'streamside' habitats are of much higher quality but disappear completely during the winter. Macroptery is induced in both habitats by cues such as host-plant quality and nymphal density, but there is also significant non-additive genetic variation in the response to these cues (Denno & Grissell, 1979; Denno *et al.*, 1991). Denno & Grissell produced an index of the prevalence of the temporary habitats in an arena, CVSH (coefficient of variation in stand height; see Denno & Grissell (1979) for a full definition) and, for a number of arenas throughout the species' range, correlated this with the proportion of macropters in the population. As predicted, they found that the prevalence of macroptery increased with CVSH (Fig. 11.1): in environments comprising large amounts of temporary, streamside vegetation, the proportion of macropters was extremely high (>90%), whereas in environments comprising large amounts of permanent, high marsh habitat, the proportion of macropterous individuals was much lower (<10%).

Maternal effects are most likely to evolve in species for which ontogenetic constraints require that migratory phenotype is determined

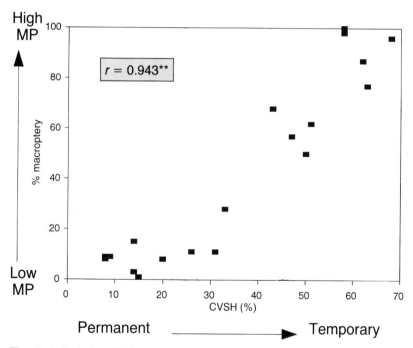

Fig. 11.1. Relationship between percentage macroptery in populations of the planthopper *Prokelisia marginata* and the coefficient of variation in stand height (CVSH) of its host plant *Spartina alterniflora*. (**$p<0.01$.) As the proportion of temporary streamside habitats increases (i.e. as CVSH increases), so too does the average migratory potential (MP) of the population (as indicated by percentage macroptery). Redrawn from Denno & Grissell (1979).

early in development and/or for which cues forecast conditions well in advance. The most common maternal effect is one in which an environmental cue perceived by the maternal generation is manifested in individuals of the offspring generation. For example, in many species of aphids, the environmental conditions experienced by virginoparous females is reflected in the proportion of winged offspring they produce (for reviews see Lees, 1966; Mousseau & Dingle, 1991; Moran, 1992). Maternal effects may not be limited to expression in just the offspring generation: if environmental conditions can be predicted further in advance of this, then they may also extend into subsequent generations. For example, in the aphid *Megoura viciae*, not only do parent alatae produce fewer alate offspring, but so too do their apterous daughters (Lees, 1966, 1967).

Just as with environmental regulation, geographical variation in maternal effects can be expected when the optimal response varies between environments. A number of studies have demonstrated genetic variation between clones with respect to maternal effects (Moran, 1992 and references therein) and geographical variation in clonal frequencies (see e.g. Smith & MacKay, 1989*a*,*b*). Whilst this suggests possible local adaptation to the habitat template (Mousseau & Dingle, 1991), no studies have yet collected the data required to correlate geographical variation in clonal frequencies with variation in habitat temporal stability.

### Temporally unpredictable habitats

When habitats persist for unpredictable lengths of time, a proportion of migratory phenotypes is always favoured. These can be generated by genetic polymorphisms (i.e. by a mixed strategy being employed by the population as a whole) and/or by individuals producing offspring of both phenotypes via maternal influences such as maternal-age effects (i.e. by a mixed strategy being employed by each female within the population – Mousseau & Dingle, 1991). A consequence of both of these mechanisms is that, although the average proportion of migrants in a population will reflect the average persistence of all habitats in the arena, within favourable habitats there may be a change in migratory potential over time. Thus, geographical variation in migratory potential may be observed in field samples, even between arenas comprising habitats with the same average durational stabilities, due to variation in their colonisation times.

As mentioned in the Introduction, when the production of migratory phenotypes is genetically regulated, a temporal decline in migratory potential may result (Roff, 1990*b*). Stein (1977) found that, following colonisation of a recently-seeded meadow, the proportion of macropters in four species of wing-dimorphic weevils fell rapidly over time. For example, a decline from 100% macroptery to 75% in four generations was observed for *Sitona hispidulus* and from 100% to 60% in six generations for *Apion virens*. For these two species at least, wing-morph is under genetic control (a single locus with two alleles, with brachyptery dominant – Roff, 1986*b*) and the temporal decline in migratory potential is likely to be due to both the higher reproductive rate of the brachypterous morphs and the net loss of genes for macroptery via emigration (which is a function of habitat size and isolation – see below). The trajectory of the temporal decline in migratory potential will depend upon the relative contributions of these two factors and the genetic mechanism involved (Roff, 1990*b*).

In some insects, offspring produced early in a female's reproductive life have lower migratory potential than those produced later in life, whereas in others the temporal trend is reversed. For example, females of the bruchid beetle *Callosobruchus maculatus* produce more migratory, 'active' morph offspring as they get older (Sano-Fujii, 1979). In the Pea Aphid *Acyrthosiphon pisum*, however, early-born female offspring of apterous parthenogenetic females produce much larger proportions of alates in response to crowding than do their late-born sisters (MacKay & Wellington, 1977). At the population level, the impact of these maternal-age effects is seen most clearly in a study of *A. pisum* on fields of alfalfa in Canada. MacKay & Lamb (1979) found that the average migratory potential of apterous females (measured as the proportion of winged offspring produced in response to a standard stimulus) was negatively correlated with the age of the source population. They argue that the reason for these differences is that, as the populations age, their age-structures become increasingly biased towards older females. That differences between these populations are due to transient maternal-age effects, rather than genetic differences, is indicated by the fact that they disappear after the females have been maintained under laboratory conditions for two generations.

### Spatially regular habitat distributions

Temporal heterogeneity may explain variation in the proportions of migrants and non-migrants in a population, but in order to account for quantitative variation in traits such as flight capacity and pre-reproductive period, which exhibit variation within macropters, the influence of spatial heterogeneity must also be considered.

When favourable habitats exhibit a *regular* distribution, i.e. are located a fixed distance (or migratory effort) from each other, natural selection will tend to favour individuals with the migratory potential required to reach the nearest favourable habitat. Thus, when the habitat distribution is more or less uniform, or when seasonal migrations are between specific locations (e.g. to and from specific diapause sites), little variation in migratory potential within the migrants of a population can be expected. Geographical variation in migratory potential would result, however, if the distance between favourable habitats varied consistently between environments. In this situation, local adaptation to the habitat template can be expected such that, for example, populations occupying habitats close to suitable diapause sites would comprise individuals adapted for shorter-range migrations than those from populations far from them (Oku, 1982). Although the data required to test this hypothesis are currently lacking, variation in this respect might be found in populations of noctuid moths, such as the Army Cutworm *Chorizagrotis auxiliaris*, which migrate between extensive, low-altitude breeding grounds (such as the Great Plains of North America) and high-altitude aestivation sites (like the Rocky Mountains – see Pruess, 1967 for details of *C. auxiliaris* migrations).

Variation in migratory potential may also be maintained in spatially regular environments if there is a gradient in habitat favourableness. One way in which such a gradient may be produced is by the seasonal appearance of new habitats at high latitudes which, when they first appear, are likely to be of higher quality and/or to have lower levels of competition or natural enemies. In such a situation, insects with high migratory potential might incur greater costs than those with lower migratory potential, but would also recoup greater benefits because of the higher favourableness of the habitats they reach. Thus, at each generation, insects with the highest migratory potential would tend to be those located at the highest latitudes (though stochastic meteorological processes can be expected to limit the integrity of such clines).

Environmental cues forecasting the imminent deterioration of these temporary high-latitude habitats (e.g. photoperiod and temperature) may also be used to indicate the migratory effort required to reach the more permanent habitats at lower latitudes. However, a significant genetic component to migratory potential will ensure that those individuals emerging at high latitudes have the migratory potential required to reach lower-latitude habitats (for a discussion of such *genetic partitioning*, see Han & Gatehouse, 1991).

Several examples of clinal variation in migratory potential have been reported. For example, in the Oriental Armyworm moth *Mythimna separata* in China, and the Silver Y moth *Autographa gamma* in North Africa and western Europe, populations at higher latitudes have significantly longer pre-reproductive periods than those at lower latitudes (Han & Gatehouse, 1991; Hill & Gatehouse, 1993). Thus, individuals in these populations have longer to express their flight capacity than those at lower latitudes.

The best example of this kind of latitudinal cline in migratory potential is a study of tethered-flight performance in the North American milkweed bug *Oncopeltus fasciatus*. Both temperature and photoperiod are used as cues for migration in this species, but Dingle, Blakely & Miller (1980) showed that, in insects reared and tested under similar conditions of temperature and photoperiod (23 °C, 14L : 10D), the proportion of insects making flights in excess of 30 min increased from 0% in Mexico to 6–18% in southern USA (Georgia and Florida) and 24–25% in northern USA (Iowa and Maryland). For all three of these species, genetic partitioning has been shown to contribute to variation in migratory potential and the trait in question is under polygenic control (Han & Gatehouse, 1991; Hill & Gatehouse, 1992; Dingle, 1968, respectively; Table 11.1 below).

### Spatially irregular habitat distributions

When the distribution of favourable habitats is *irregular*, the migratory effort required to locate new habitats is unpredictable and mixed migration strategies can be expected through genetic differentiation and/or bet-hedging maternal-age effects. When variation is maternally mediated, each female has a chance of producing at least some offspring that will reach favourable habitats. However, in the absence of any additive genetic contribution, the proportions of each of the different phenotypes will remain fixed from one generation to the next. This mechanism may evolve in environments in which the spatial distribution of favourable habitats is temporally stable but, when it varies over time, the maintenance of significant additive genetic variation in migratory potential can be expected. When migratory potential is polygenically inherited, not only will a single female produce offspring with a range of migratory potentials, but also, if there is a correlation between the habitat distribu-

Table 11.1. *Heritability of migratory potential in macropterous insects.*

| Species | Heritability ($h^2$) | Reference |
|---|---|---|
| *Relative wing-length* | | |
| Oncopeltus fasciatus | 0.55 | Hegmann & Dingle (1982) |
| (Hemiptera: Heteroptera) | 0.49–0.87 | Palmer & Dingle (1986) |
| Dysdercus fasciatus | 0.51 | Derr (1980) |
| (Hemiptera: Heteroptera) | | |
| *Tethered-flight capacity* | | |
| Lygaeus kalmii (Hemiptera: Heteroptera) | 0.20–0.41 | Caldwell & Hegmann (1969) |
| Helicoverpa armigera | 0.15–0.39 | Colvin & Gatehouse (1993) |
| (Lepidoptera: Noctuidae) | | |
| Spodoptera exempta | 0.40–0.88 | Parker & Gatehouse (1985) |
| (Lepidoptera: Noctuidae) | | |
| Mythimna separata | 0.27 | Han & Gatehouse (1993) |
| (Lepidoptera: Noctuidae) | | |
| *Pre-reproductive period* | | |
| Oncopeltus fasciatus | 0.25 | Hegmann & Dingle (1982) |
| (Hemiptera: Heteroptera) | 0.20 | Palmer & Dingle (1986) |
| Dysdercus fasciatus | 0.18–0.40 | Derr (1980) |
| (Hemiptera: Heteroptera) | | |
| Helicoverpa armigera | 0.12–0.58 | Colvin & Gatehouse (1993) |
| (Lepidoptera: Noctuidae) | | |
| Autographa gamma | 0.24–0.66 | Hill & Gatehouse (1992) |
| (Lepidoptera: Noctuidae) | | |
| Spodoptera exempta | 0.12–0.43 | Wilson & Gatehouse (1992) |
| (Lepidoptera: Noctuidae) | | |

tions faced by successive generations, the majority of their offspring will tend to be of phenotypes appropriate to the conditions that prevail (Roff, 1986*a*).

In tropical and sub-tropical environments, the spatial distribution of the habitats of many, particularly phytophagous, insects depends upon the temporal and spatial pattern of the rainfall on which the growth of their host plants is dependent. Thus the migratory potential required to track them will vary from region to region and time to time. In areas or at times when rainfall is infrequent, patches of host plants become widely dispersed, favouring insects with high migratory potential. When rainfall is more widespread, however, insects with lower migratory potential will generally be favoured, because suitable habitats will be ubiquitous and these individuals will tend to have higher reproductive rates. Thus, an association between the prevalence of rainfall and migratory potential can be expected. However, because rainfall is not easily predicted, it is likely that there will be a minimal environmental component to the regulation of migratory potential and that the majority of the adaptive variation will be genetically derived (Gatehouse, 1986, 1989).

The leafhopper *Cicadulina mbila* feeds on graminaceous crops and grasses in southern Africa. Migratory phenotypes of this species tend to have relatively longer wings and higher tethered-flight capacity than the non-migratory phenotypes. Rose (1972) found that, as the dry season progressed and habitats became more patchily distributed, mean relative winglength (and hence migratory potential) increased. Rose also demonstrated that, as predicted, body length and tethered-flight capacity were both primarily under genetic control in this species, though host-plant quality played a modulating role.

Another species for which the pattern of rainfall appears to select for variation in migratory potential is the polyphagous grasshopper *Melanoplus sanguinipes*. In Arizona, the wet, green areas suitable for reproductive maturation, oviposition and nymphal development are discontinuous, favouring individuals with relatively high migratory potential. In Colorado, on the other hand, green vegetation is more uniformly distributed, so favouring individuals with lower migratory potential but higher reproductive potential (McAnelly, 1985; McAnelly & Rankin, 1986). As predicted, McAnelly & Rankin (1986) found that a higher proportion of individuals from Arizona (28%) than Colorado (5%) made long tethered flights (> 55 min). Moreover, this difference was maintained in subsequent laboratory-reared generations, indicating a significant genetic component to tethered-flight capacity (the only environmental influence was a weak positive effect of nymphal density).

In the two previous examples, a link between rainfall and migratory potential has been implicated but not clearly demonstrated. The only study to have shown a quantitative association between migration and precipitation is that by Wilson & Gatehouse (1993) on the African Armyworm moth *Spodoptera exempta*. The caterpillars of this noctuid moth feed on graminaceous crops and pasture grasses throughout sub-Saharan Africa and adult moths from outbreak populations migrate at every generation to keep track of the changing distribution of their habitats, as determined by the pattern of rainfall (Rose et al., 1987). *S. exempta* females migrate as immature adults (Rose et al., 1987; Page, 1988; Pedgley et al., 1989), and therefore the duration of the pre-reproductive period may be used as an index of their migratory potential

(Wilson & Gatehouse, 1993). Laboratory studies have demonstrated that pre-reproductive period is primarily under genetic control in females, though the evidence for males is equivocal (Wilson & Gatehouse, 1992). Wilson & Gatehouse (1993) therefore predicted a negative correlation between pre-reproductive period and the prevalence of rainfall at the outbreak site during the time that the parent moths initiating the outbreak were migrating (see also Parker & Gatehouse, 1985; Gatehouse, 1986). It is at this time that the process of natural selection occurs, through the action of rainstorms, in themselves inducing the moths to land and reproduce or in generating the spatial pattern of green habitats available to them for reproduction. A corollary to this was the prediction that as the armyworm season progressed, and habitat aggregations increased in size, there would be a progressive reduction in migratory potential.

Wilson & Gatehouse (1993) found that, as predicted, there was a strong negative correlation between female pre-reproductive period and rainfall prevalence (Fig. 11.2a) and that female pre-reproductive period declined through the armyworm season (Fig. 11.2b; the corresponding correlations for males were in the same direction but non-significant). These results provide strong support for the idea that, at each generation, the habitat template selects migratory phenotypes appropriate to the prevailing distribution of habitats. That these differences were not environmentally induced is indicated by the lack of a consistent association between pre-reproductive period and temperature, humidity, body weight or rainfall prevalence before or following the month of migration by the parent moths (Wilson & Gatehouse, 1993), though maternal effects cannot be entirely discounted.

### Contemporary natural selection and the habitat template

For contemporary natural selection to act via the habitat template, it is necessary that there be additive genetic variation in migratory potential. This chapter has emphasised that an additive genetic contribution to variation in migratory potential can be expected whenever the durational stability of habitats, or the migratory effort required to track their spatial distribution, has an unpredictable component. This assertion appears to be borne out in the literature: in nearly all instances in which it has been examined, significant additive genetic variation in migratory potential has been demonstrated. In one-third of wing-dimorphic species, wing-morph is determined by genes at a single locus (Roff, 1986a,b; Roff &

Fairbairn, 1991). In the remaining two-thirds, it is under polygenic control (i.e. determined by genes located at more than one locus), often with high heritability (Roff & Fairbairn, 1991). In monomorphic winged species, relative wing length, tethered-flight capacity and pre-reproductive period are similarly under polygenic control.

Within a population, the amount of additive genetic variation in a quantitative trait is indicated by its narrow sense heritability ($h^2$). Table 11.1 lists published heritability values for these traits in a variety of species. Their relatively high values indicate that contemporary natural selection may be an important factor generating geographical and temporal variation in migratory potential, even in species that exhibit strong responses to environmental cues (via genotype–environment interactions). If variation in migratory potential is predominantly under genetic rather than environmental control, it may be much more difficult to

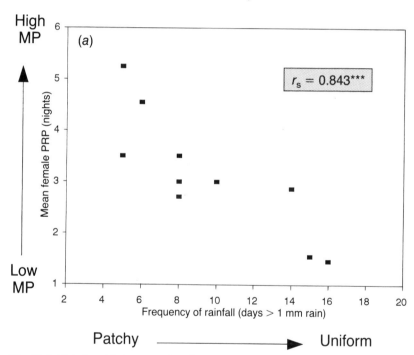

Fig. 11.2. Relationship between female pre-reproductive period (PRP) in populations of the African armyworm moth *Spodoptera exempta* and (*a*) frequency of rainfall at the outbreak sites during the month of migration of the parent moths, (*b*) calendar month. (*p< 0.05;***p<0.001.) As suitable larval habitats become more uniformly distributed (i.e. as the frequency of rainfall increases or the rainy season progresses) so the average migratory potential (MP) of the population increases (as indicated by the duration of the PRP). Redrawn from Wilson &

predict future distributions of migratory insects following environmental perturbations such as changes in land-use practices, pollution episodes or climate changes (see e.g. Southwood, 1971; Stinner *et al.*, 1983). Such considerations are particularly pertinent at the moment, due to the potential impact of global warming.

## Conclusions

This chapter has examined the effects of temporal and spatial habitat heterogeneity on the evolution and maintenance of variation in the migratory potential of insects. It has demonstrated a clear association between the predictability of habitat heterogeneity and the mechanisms regulating migratory potential, and has emphasised the importance of a knowledge of the regulatory mechanisms in interpreting the variation

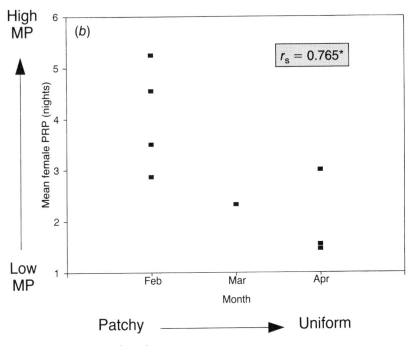

Caption for Fig. 11.2 (*cont.*)
Gatehouse (1993). The rainfall data in (*a*) are taken from archive sources and represent the average frequency of rainfall at the outbreak sites during the calendar month that the moths initiating the outbreaks were migrating; (*b*) includes only those data from outbreaks that occured in Kenya and Tanzania during the long rainy season and, in both figures, data for the Embu sample have been omitted (see Wilson & Gatehouse (1993) for justification).

observed within and between field populations. There is a large body of literature suggesting adaptive variation in migratory potential, but only a small proportion of this has provided a convincing association with habitat heterogeneity. In this respect, studies on the influence of temporal heterogeneity are by far the more numerous and convincing, due mainly to analyses of wing-dimorphic species, though even here the migratory capacity of macropters has often not been directly tested.

There is clearly a need for more studies on the influence of spatial heterogeneity. The largest problem to overcome, in this respect, is the lack of suitable methods for quantifying both habitat heterogeneity and migratory potential. With regard to the former problem, the use of satellite imagery may prove useful (see Robinson, 1992, and in Chapter 16, this volume), but no new methods for measuring migratory potential appear to be forthcoming. Recent studies relating pre-reproductive period to the spatial structure of the environment (McNeil, 1986; Han & Gatehouse, 1991; Hill & Gatehouse, 1993; Wilson & Gatehouse, 1993) provide encouraging results, though this method is not appropriate for all species and has been criticised by some authors (e.g. Sappington & Showers, 1992). Until radio transmitters are developed that are small enough and light enough to be carried by migrating insects, it appears likely that tethered-flight techniques will prevail. Even when these problems are overcome, only the combination of field studies of variation in migratory potential with laboratory studies of the regulatory mechanisms will provide real insight into the effects of habitat heterogeneity on migratory potential. Such studies are likely to require the combined efforts of ecologists, biogeographers, physiologists and geneticists. However, they promise to yield interesting results with applied benefits.

## References

Caldwell, R.L. & Hegmann, J.P. (1969). Heritability of flight duration in the milkweed bug *Lygaeus kalmii*. *Nature*, **223**, 91–2.

Cohen, D. (1967). Optimisation of seasonal migratory behaviour. *American Naturalist*, **101**, 5–7.

Colvin, J. & Gatehouse, A.G. (1993). Migration and genetic regulation of the pre-reproductive period in the cotton-bollworm moth, *Helicoverpa armigera*. *Heredity*, **70**, 407–12.

Comins, H.N., Hamilton, W.D. & May, R.M. (1980). Evolutionarily stable dispersal strategies. *Journal of Theoretical Biology*, **82**, 205–30.

Darwin, C. (1859). *On the Origin of Species*. London: John Murray.

Denno, R.F. (1993). Life history variation in planthoppers. In *Planthoppers: their Ecology and Management*, ed. R.F. Denno & T.J. Perfect, pp. 163–215. New York: Chapman & Hall.

Denno, R.F. & Grissell, E.E. (1979). The adaptiveness of wing-dimorphism in the saltmarsh-inhabiting planthopper *Prokelisia marginata* (Homoptera: Delphacidae). *Ecology*, **60**, 221–36.

Denno, R.F., Roderick, G.K., Olmstead, K.L. & Döbel. H.G. (1991). Density-related migration in planthoppers (Homoptera: Delphacidae): the role of habitat persistence. *American Naturalist*, **138**, 1513–41.

Derr, J.A. (1980). The nature of variation in life history characters of *Dysdercus bimaculatus* (Heteroptera: Pyrrhocoridae), a colonising species. *Evolution*, **34**, 548–57.

Dingle, H. (1968). Life history and population consequences of density, photoperiod, and temperature in a migrant insect, the milkweed bug *Oncopeltus. American Naturalist*, **102**, 149–63.

Dingle, H. (1985). Migration. In *Comprehensive Insect Physiology, Biochemistry and Pharmacology*, vol. 9, ed. G.A. Kerkut & L.I. Gilbert, pp. 375–415. London: Pergamon.

Dingle, H., Blakely, N.R. & Miller, E.R. (1980). Variation in body size and flight performance in milkweed bugs (*Oncopeltus*). *Evolution*, **34**, 371–85.

Fairbairn, D.J. (1986). Does alary dimorphism imply dispersal dimorphism in the waterstrider, *Gerris remigis? Ecological Entomology*, **11**, 355–68.

Gadgil, M. (1971). Dispersal: population consequences and evolution. *Ecology*, **52**, 253–61.

Gatehouse, A.G. (1986). Migration in the African armyworm *Spodoptera exempta*: genetic determination of migratory capacity and a new synthesis. In *Insect Flight: Dispersal and Migration*, ed. W. Danthanarayana, pp. 128–44. Berlin: Springer-Verlag.

Gatehouse, A.G. (1989). Genes, environment and insect flight. In *Insect Flight*, ed. G.J. Goldsworthy & C.H. Wheeler, pp. 115–38. Boca Raton, Florida: CRC Press.

Hamilton, W.D. & May, R.M. (1977). Dispersal in stable habitats. *Nature*, **269**, 578–81.

Han, E.-N. & Gatehouse, A.G. (1991). Genetics of precalling period in the Oriental armyworm, *Mythimna separata* (Walker) (Lepidoptera: Noctuidae), and implications for migration. *Evolution*, **45**, 1502–10.

Han, E.-N. & Gatehouse, A.G. (1993). Flight capacity: genetic determination and physiological constraints in a migratory moth *Mythimna separata. Physiological Entomology*, **18**, 183–8.

Harrison, R.G. (1980). Dispersal polymorphisms in insects. *Annual Review of Ecology and Systematics*, **11**, 95–118.

Harvey, P.H. & Pagel, M.D. (1991). *The Comparative Method in Evolutionary Biology*. Oxford: Oxford University Press.

Hassell, M.P. & Southwood, T.R.E. (1978). Foraging strategies in insects. *Annual Review of Ecology and Systematics*, **9**, 75–98.

Hastings, A. (1983). Can spatial variation alone lead to selection for dispersal? *Theoretical Population Biology*, **24**, 244–51.

Hegmann, J.P. & Dingle, H. (1982). Phenotypic and genetic covariance structure in milkweed bug life history traits. In *Evolution and Genetics of Life Histories*, ed. H. Dingle & J.P. Hegmann, pp. 177–85. New York: Springer-Verlag.

Hill, J.K. & Gatehouse, A.G. (1992). Genetic control of the pre-reproductive period in the silver-Y moth *Autographa gamma* (L.) (Lepidoptera: Noctuidae). *Heredity*, **69**, 458–64.

Hill, J.K. & Gatehouse, A.G. (1993). Phenotypic plasticity and geographical variation in the pre-reproductive period of *Autographa gamma* (L.) (Lepidoptera: Noctuidae) and their implications for migration. *Ecological Entomology*, **18**, 39–46.

Iwanaga, K., Nakasuji, F. & Tojo, S. (1987). Wing polymorphism in Japanese and foreign strains of the brown planthopper, *Nilaparvata lugens*. *Entomologia Experimentalis et Applicata*, **43**: 3–10.

Jackson, D.J. (1928). The inheritance of long and short wings in the weevil, *Sitona hispidulus*, with a discussion of wing reduction among beetles. *Transactions of the Royal Society of Edinburgh*, **55**, 665–735.

Johnson, C.J. (1969). *Migration and Dispersal of Insects by Flight*. London: Methuen.

Lamb, R.J. & MacKay, P.A. (1979). Variability in migratory tendency within and among natural populations of the pea aphid, *Acyrthosiphon pisum*. *Oecologia*, **39**, 289–99.

Lees, A.D. (1966). The control of polymorphism in aphids. *Advances in Insect Physiology*, **3**, 207–77.

Lees, A.D. (1967). The production of the apterous and alate forms in the aphid *Megoura viciae* Buckton, with special reference to the role of crowding. *Journal of Insect Physiology*, **13**, 289–318.

MacKay, P.A. & Lamb, R.J. (1979). Migratory tendency in aging populations of the pea aphid, *Acyrthosiphon pisum*. *Oecologia*, **39**, 301–8.

MacKay, P.A. & Wellington, W.G. (1977). Maternal age as a source of variation in the ability of an aphid to produce dispersing forms. *Researches on Population Ecology*, **18**, 195–209.

McAnelly, M.L. (1985). The adaptive significance and control of migratory behaviour in the grasshopper *Melanoplus sanguinipes*. In *Migration: Mechanisms and Adaptive Significance*, ed. M.A. Rankin, pp. 687–703. *Contributions in Marine Science*, vol. 27, Suppl. Port Aransas, Texas: Marine Science Institute, The University of Texas at Austin.

McAnelly, M.L. & Rankin, M.A. (1986). Migration in the grasshopper *Melanoplus sanguinipes* (Fab.). I The capacity for flight in non- swarming populations. *Biological Bulletin*, **170**, 368–77.

McNeil, J.N. (1986). Calling behaviour: can it be used to identify migratory species of moths? *Florida Entomologist*, **69**, 78–84.

McPeek, M.A. & Holt, R.D. (1992). The evolution of dispersal in spatially and temporally varying environments. *American Naturalist*, **140**, 1010–27.

Moran, N.A. (1992). The evolution of aphid life cycles. *Annual Review of Entomology*, **37**, 321–48.

Mousseau, T.A. & Dingle, H. (1991). Maternal effects in insect life histories. *Annual Review of Entomology*, **36**, 511–34.

Oku, T. (1982). Aestivation and migration in noctuid moths. In *Diapause and Life-Cycle Strategies in Insects*, ed. V.K. Brown & I. Hodek, pp. 219–31. Amsterdam: Junk.

Page, W.W. (1988). Varying durations of arrested oocyte development in relation to migration in the African armyworm moth, *Spodoptera exempta* (Walker) (Lepidoptera: Noctuidae). *Bulletin of Entomological Research*, **79**, 181–98.

Palmer, J.O. & Dingle, H. (1986). Direct and correlated responses to selection among life-history traits in milkweed bugs (*Oncopeltus fasciatus*). *Evolution*, **40**, 767–77.

Palmer, J.O. & Dingle, H. (1989). Responses to selection on flight behaviour in a migratory population of milkweed bug (*Oncopeltus fasciatus*). *Evolution*, **43**, 1805–8.

Parker, W.E. & Gatehouse, A.G. (1985). Genetic factors controlling flight performance and migration in the African armyworm moth, *Spodoptera exempta* (Walker) (Lepidoptera: Noctuidae). *Bulletin of Entomological Research*, **75**, 49–63.

Pedgley, D.E., Page, W.W., Mushi, A., Odiyo, P., Amisi, J., Dewhurst, C.F., Dunstan, W.R., Fishpool, L.D.C., Harvey, A.W., Megenasa, T. & Rose, D.J.W. (1989). Onset and spread of an African armyworm upsurge. *Ecological Entomology*, **14**, 311–33.

Pruess, K.P. (1967). Migration of the army cutworm, *Chorizagrotis auxiliaris* (Lepidoptera: Noctuidae). I. Evidence for a migration. *Annals of the Entomological Society of America*, **60**, 910–20.

Rankin, M.A. & Burchsted, J.C.A. (1992). The cost of migration in insects. *Annual Review of Entomology*, **37**, 533–59.

Robinson, T.P. (1992). Modelling the seasonal distribution of habitat suitability for armyworm population development in East Africa using GIS and remote sensing techniques. PhD thesis, University of Reading.

Roff, D.A. (1974*a*). Spatial heterogeneity and the persistence of populations. *Oecologia*, **15**, 245–58.

Roff, D.A. (1974*b*). The analysis of a population model demonstrating the importance of dispersal in a heterogeneous environment. *Oecologia*, **15**, 259–75.

Roff, D.A. (1975). Population stability and the evolution of dispersal in a heterogeneous environment. *Oecologia*, **19**, 217–37.

Roff, D.A. (1986*a*). Evolution of wing polymorphism and its impact on life cycle adaptation in insects. In *The Evolution of Insect Life Cycles*, ed. F. Taylor & R. Karban, pp. 204–21. New York: Springer-Verlag.

Roff, D.A. (1986*b*). The evolution of wing dimorphism in insects. *Evolution*, **40**, 1009–20.

Roff, D.A. (1990*a*). The evolution of flightlessness in insects. *Ecological Monographs*, **60**, 389–421.

Roff, D.A. (1990*b*). Understanding the evolution of insect life-cycles: the role of genetic analysis. In *Insect Life Cycles: Genetics, Evolution and Coordination*, ed. F. Gilbert, pp. 5-27. London: Springer-Verlag.

Roff, D.A. & Fairbairn, D.J. (1991). Wing dimorphisms and the evolution of migratory polymorphisms among the Insecta. *American Zoologist*, **31**, 243–51.

Rose, D.J.W. (1972). Dispersal and quality in populations of *Cicadulina* species (Cicadellidae). *Journal of Animal Ecology*, **41**, 589–609.

Rose, D.J.W., Dewhurst, C.F., Page, W.W. & Fishpool, L.D.C. (1987). The role of migration in the life system of the African armyworm *Spodoptera exempta*. In *Recent Advances in Research on Tropical Entomology*, ed. M.F.B. Chaudhury, pp. 561–9. *Insect Science and its Application*, vol. 8, Special Issue. Nairobi: ICIPE Science Press.

Sano-Fujii, I. (1979). Effect of parental age and developmental rate on the production of the active form of *Callosobruchus maculatus* (F.) (Coleoptera: Bruchidae). *Mechanisms of Ageing and Development*, **10**, 283–93.

Sappington, T.W. & Showers, W.B. (1992). Reproductive maturity, mating status, and long-duration flight behaviour of *Agrotis ipsilon* (Lepidoptera:

Noctuidae) and the conceptual misuse of the oogenesis-flight syndrome by entomologists. *Environmental Entomology*, **21**, 677–88.

Smith, M.A.H. & MacKay, P.A. (1989*a*). Seasonal variation in the photoperiodic responses of a pea aphid population: evidence for long-distance movements between populations. *Oecologia*, **81**, 160–5.

Smith, M.A.H. & MacKay, P.A. (1989*b*). Genetic variation in male alary dimorphism of the pea aphid *Acyrthosiphon pisum*. *Entomologia Experimentalis et Applicata*, **51**, 125–32.

Southwood, T.R.E. (1962). Migration of terrestrial arthropods in relation to habitat. *Biological Reviews*, **37**, 171–214.

Southwood, T.R.E. (1971). The role and measurement of migration in the population systems of an insect pest. *Tropical Science*, **13**, 275–8.

Southwood, T.R.E. (1977). Habitat, the templet for ecological strategies? *Journal of Animal Ecology*, **46**, 337–65.

Southwood, T.R.E. (1981). Ecological aspects of insect migration. In *Animal Migration*, ed. D.J. Aidley, pp. 197–208. Cambridge: Cambridge University Press.

Stein, W. (1977). Die Beziehung zwischen Biotop-Alter und Auftreten der Kurzflügeligkeit bei populationen dimorpher Rüsselkäfer-Arten (Col., Curculionidae). *Zeitschrift für Angewandte Entomologie*, **83**, 37–9.

Stinner, R.E., Barfield, C.S., Stimac, J.L. & Dohse, L. (1983). Dispersal and movement of insect pests. *Annual Review of Entomology*, **28**, 319–35.

Tojo, S. (1991). Genetic background of insect migration. In *Proceedings of the International Seminar on Migration and Dispersal of Agricultural Insects, Tsukuba, Japan, September 25–28, 1991*, pp. 21–9. Tsukuba: National Institute of Agro-Environmental Sciences.

Wilson, K. & Gatehouse, A.G. (1992). Migration and genetics of pre-reproductive period in the moth, *Spodoptera exempta* (African armyworm). *Heredity*, **69**, 255–62.

Wilson, K. & Gatehouse, A.G. (1993). Seasonal and geographical variation in the migratory potential of outbreak populations of the African armyworm moth, *Spodoptera exempta*. *Journal of Animal Ecology*, **62**, 169–81.

# 12

# The regulation of migration in *Helicoverpa armigera*

## J. COLVIN

## Introduction

The Cotton Bollworm *Helicoverpa armigera* (Hübner) is an agricultural pest occurring throughout Africa, southern Europe, the Middle East, Asia and Australasia (Commonwealth Institute of Entomology, 1968). An important factor contributing to its pest status is its ability to undertake long-distance migratory flights (review by Farrow & Daly, 1987; see also Chapter 8, this volume), that may occasionally exceed 2000 km (Bowden & Johnson, 1976).

Radar studies and field observations have shown that noctuid migration typically involves a characteristic sequence of behaviours (see e.g. Rose *et al.*, 1985). Takeoff occurs shortly after sunset, with moths climbing at a rate of ~0.5–2.0 m s$^{-1}$ until they reach the top of the temperature inversion where windspeeds are usually at a maximum, often 20–50 km h$^{-1}$ (Drake & Farrow, 1988). The insects then maintain their altitude, continuing their wind-assisted progress until they descend at some time during the night or around dawn (Drake, 1985).

Because of the long distances migratory insects travel and their relatively small size, the study of migration at the ecological level presents major logistic and technological challenges. No less problematic, however, is the laboratory-based approach, in which the behavioural, physiological and genetic mechanisms that regulate migratory behaviour are investigated.

This chapter describes the contribution of laboratory studies to our understanding of the migratory behaviour of *H. armigera*.

## Flight activity and the pre-reproductive period

Two components of migratory potential that are amenable to examination in the laboratory are flight activity and the pre-reproductive period

(PRP). Flight is an obvious aspect of migratory behaviour, but the significance of the PRP is perhaps less readily apparent. It lies in the observation that migration often occurs while adult insects are still reproductively immature (Johnson, 1969; Chapter 10, this volume). Information on the reproductive status of migratory *Helicoverpa* spp. is scarce, largely due to the difficulty of differentiating migratory from local populations in light-trap catches. Coombs *et al.* (1993), however, avoided this problem by examining moths trapped in tower-mounted light traps (40–50 m high) during periods when catches of other migratory species suggested that a migration event had occurred. Of the 132 *H. armigera* and 410 *H. punctigera* examined, 88.6% and 91.0%, respectively, were immature. Of the remaining mature females, 78.6% of the *H. armigera* and 97.1% of the *H. punctigera* contained only a single spermatophore, indicating that they had recently reached maturity. Additional evidence of pre-reproductive migration by *H. armigera* is provided by Roome (1972) who found that, on two occasions in Botswana, peak catches of *H. armigera,* thought to have originated in Zimbabwe, consisted predominantly of immature individuals. The PRP of *H. armigera* can therefore be taken as an approximation to the number of nights available for migratory behaviour.

## Techniques for assessing migratory potential

### *Tethered flight*

Various techniques designed to allow prolonged flight have been developed to study insect migration in the laboratory (see e.g. Gatehouse & Hackett, 1980; Hackett & Gatehouse, 1982; McAnelly & Rankin, 1986; David & Hardie, 1988). With strong fliers such as noctuids, the most widely adopted technique has involved tethering individual moths to a flight mill or balance. Flight activity has then usually been recorded electronically, either with a data logger or by using a personal computer with dedicated software (Woodrow, Gatehouse & Davies, 1987; Armes & Cooter, 1991; Colvin & Gatehouse, 1993*a*).

Flight data recorded in this way, especially with a facultative migrant like *H. armigera*, must be interpreted with caution, because, without additional qualitative data on the responsiveness of the flying moths to vegetative stimuli, migratory flight (*sensu* Kennedy, 1985) cannot be distinguished from other types of flight behaviour described by Farrow & Daly (1987).

## Pre-reproductive period

Various criteria have been used to assess the maturity status of female moths. The beginning of oviposition, for instance, has been taken to indicate reproductive maturity, the time between emergence and first oviposition being termed the pre-oviposition period (Isley, 1935; Hackett, 1980). Reproductively mature female *H. armigera* can also be identified because they 'call' (pheromone release accompanied by extrusion of the ovipositor), a behaviour that occurs prior to mating and oviposition (Kou & Chow, 1987). The pre-calling period, defined as the number of nights between eclosion and the initiation of calling, has therefore been used as a more accurate and reliable measure of the PRP (Colvin & Gatehouse, 1993*a*).

Male PRPs have been recorded by observing the response of males to calling females. The age at which a male first attempted to mate was recorded as its PRP (Colvin & Gatehouse, 1993*a*).

## Relationship between age, flight activity and reproductive status

Armes & Cooter (1991) found that tethered flight by virgin moths of a southern Indian strain increased from emergence to night 4 and remained at the same level until at least night 6. In this strain, 77% of unmated moths had started oviposition by night 3, suggesting a rapid rate of reproductive maturation. Mated females showed a 15–28-fold decrease in flight activity compared to virgin females of similar age.

Colvin & Gatehouse (1993*a*) compared the flight activity of female moths from India and Malawi on nights 2, 4 and 6 after eclosion. Flight activity in the Indian strain declined after night 2, whereas that of the Malawian strain remained at a high level and did not differ significantly over the three nights of flight. The female PRP distributions for the two strains were also determined and the Indian strain was found to reach reproductive maturity ~48 h earlier than the Malawian strain (mode PRP values for the Indian and Malawian strains were night 3 and night 5, respectively). As flight activity remained higher for longer in the late-maturing strain (Malawian), these data suggest that it is the rate of reproductive maturation that influences flight activity, rather than age per se.

This hypothesis was tested more rigorously in another experiment, where males and females of different maturity status were flight tested at the same age (four nights old). The results showed that flight activity was greatest while the adults were still immature; ~50% of immature moths of both sexes flew for more than 7.5 h during the 10-h scotophase (Fig. 12.1). As moths reached reproductive maturity, their flight activity decreased significantly. Additional analysis indicated that mating caused a further reduction in flight activity (Colvin & Gatehouse, 1993*a*), confirming the results of Armes & Cooter (1991).

In spite of the difficulties with the tethered-flight technique in identifying specifically migratory flight, the data produced on *H. armigera* are valuable in that they demonstrate the impressive flight capability, particularly of immature individuals. In addition, the significant correlation

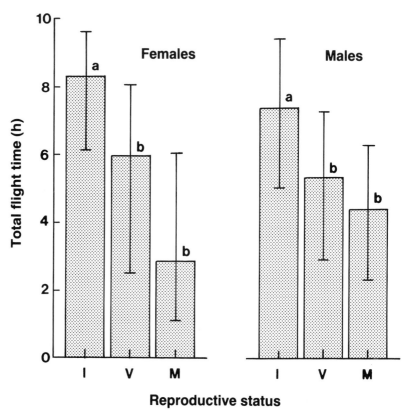

Fig. 12.1. The effect of reproductive status on total flight time (TFT) of 4-night-old Malawian *Helicoverpa armigera*. Median TFT values are indicated by histogram bar heights and interquartile values by thin bars. Distributions followed by a different letter are significantly different at the *p* = 0.05 level. I, immature; V, virgin-mature; M, mated. Sample sizes for females were I, 34; V, 30; M, 22; and for males I, 47; V, 28; M, 25 (from Colvin & Gatehouse (1993*a*), with permission).

between flight activity and reproductive status provides evidence of an oogenesis-flight syndrome in this species (Johnson, 1969; Colvin & Gatehouse, 1993a; Chapter 10, this volume).

### Genetic regulation of migratory potential

A genetic influence on the migratory behaviour of insects has been recognised for some time (review by Gatehouse, 1989). In three noctuid species, loci on both the X-chromosome and the autosomes have been shown to exert an influence on the PRP (Han & Gatehouse, 1991a; Hill & Gatehouse, 1992; Wilson & Gatehouse, 1992). Regulation of flight potential in the African armyworm *Spodoptera exempta* has also been shown to have a genetic component (Parker & Gatehouse, 1985).

#### *Genetic regulation of the PRP and flight activity*

The first indication of a genetic influence on the PRP of *H. armigera* came in a comparison of three strains of different geographical origin. The majority of Malawian moths were found to call on night 5, approximately two nights later than Indian moths and approximately four nights later than Chinese moths. The relatively short PRPs of the Chinese strain were thought to be the result of selection for rapid reproduction, which is often imposed inadvertently in laboratory cultures, as this strain had been in culture for six generations before its PRP distribution was measured (Colvin & Gatehouse, 1993c).

The narrow-sense heritability (Falconer, 1989) of the female PRP has been examined in sib-analysis experiments with Malawian and Chinese *H. armigera*. In both strains, sires were found to contribute a large and significant component of variance to the daughters (*e.g.* Chinese sires $h^2$ = 0.54 ± 0.25 SEM, $p<0.01$) whereas dams did not (Chinese dams $h^2$ = 0.16 ± 0.13, NS). As females are the heterogametic sex in Lepidoptera, this suggests that important genes regulating the female PRP are X-linked (Colvin & Gatehouse, 1993c).

PRP distributions of male and female moths were compared in the progeny of 27 pairs of field-collected Indian moths and were found to be similar (Colvin & Gatehouse, 1993c), suggesting that the same set of genes regulates the PRP in both sexes. This point was further investigated in a selection experiment where early- and late-maturing lines were established using criteria based only on the female PRP. By the second generation of selection, a significant divergence had occurred between the two lines, which was maintained in the subsequent generation. In this third generation, the male PRP distributions in the two lines were also found to have diverged significantly and in the same direction as for females. This result demonstrates that the same group of X-linked loci control the rate of reproductive maturation in both sexes (Colvin & Gatehouse, 1993c).

Heritability of tethered-flight activity has also been estimated (Colvin & Gatehouse, 1993a). With the use of 4–day-old, immature *H. armigera* fed on 10% sugar solution, significant narrow-sense heritability estimates were obtained for both total flying time ($h^2 = 0.39 \pm 0.083$ SEM) and longest flight ($h^2 = 0.15 \pm 0.058$). There is, therefore, a significant genetic influence on the flight activity of *H. armigera*, and inheritance is autosomal.

These results reveal the importance of genetics in the determination of the PRP and flight activity. PRP genes exert their influence in both sexes and appear to be mainly X-linked, while flight-activity genes are inherited autosomally. Due to the pleiotropic effect of PRP genes on flight activity, as revealed by the oogenesis-flight syndrome, flight activity may also be considered as having an X-linked component.

### Theories on X-linkage of the PRP

X-linkage of the PRP has now been demonstrated in four noctuid species, including *H. armigera*, and various reasons have been presented to explain this observation. Han & Gatehouse (1991a) speculated that X-linked inheritance of the PRP might have been a necessary pre-adaptation for the evolution of migration into temperate latitudes by the Oriental Armyworm moth, *Mythimna separata*. They argued that females (XY) migrating northward would not carry genes for early maturation to inappropriate latitudes, increasing the probability that their descendants would have the genetic constitution to return south-ward successfully in the late summer or autumn.

Charlesworth, Coyne & Barton (1987) noted that, in general, traits intimately associated with fitness tend to be X-linked and the PRP of noctuid moths is thought to be another example of this (Colvin & Gatehouse, 1993c). They also showed that X-linked traits, as opposed to autosomally inherited ones, can have a decreased rate of fixation in response to directional selection when certain conditions are fulfilled. X-linkage of the PRP, therefore, may help maintain variation in this trait.

### Genetic variation in migratory potential

Southwood (1977; see also Chpater 11, this volume) argued that the ability to migrate (and/or diapause) would be selected in species occupying temporary habitats and, as these environments are frequently unpredictable, he suggested that maximum reproductive success may often be achieved by partitioning offspring between various strategies. Moreover, theoretical models have shown that even in stable environments, selection can favour the production of some migratory progeny which compete for resources away from their birthplace (Hamilton & May, 1977). In both these situations, the consequent variation in migratory potential can be considered to be the product of selection for the optimal life-history strategy within the species' particular habitat.

If the habitat acts to mould migratory potential (Southwood, 1977; Wilson & Gatehouse, 1993; Chapter 11, this volume), then the differences in the PRP distributions and flight activity between the Indian and Malawian populations suggest that the spatial and/or temporal distribution of resources important to *H. armigera* in Malawi and central India may differ. However, it seems probable that migratory behaviour plays an important part in the life history of *H. armigera* in both regions.

## Environmental factors affecting migratory potential

Environmental factors often provide the stimuli (cues) that trigger migration in insects (see e.g. Johnson, 1969; D.R. Dent & A.B.S. King, unpublished data; Fitt, 1989). This is certainly true for *Helicoverpa* spp., as migration is known to occur in response to the senescence of local host plants (Pair *et al.*, 1987; Farrow & McDonald, 1988; Armes & Cooter, 1991; Chapter 8, this volume).

In temperate regions, photoperiod provides a reliable indicator of future changes in habitat quality (Danks, 1987). In tropical regions, however, other cues may be more important and migration may occur in response to temperature, water stress, food availability and quality, or crowding (Johnson, 1969; Dingle, 1985).

### Effect of temperature and vapour-pressure deficit on PRP

Isley (1935) reported that temperatures both above and below 28 °C resulted in delayed reproduction by *H. zea.* Extended PRPs, due to a reduced metabolic rate, might be expected at low, but not at high, temperatures. When female *H. armigera* were held at a constant vapour-pressure deficit (0 mmHg) at three different temperatures (22 ± 1 °C, 28 ± 1 °C, 34 ± 1 °C), the mean PRP decreased with increasing temperature. In addition, when female moths were held at 22 °C, which was colder than the temperature they had experienced as larvae and pupae (26 ± 1 °C), their rate of reproductive maturation, compared with females kept at 26 °C, was no slower than might be expected due to a reduced metabolic rate with a $Q_{10}$ of 2–3 (Colvin & Gatehouse, 1993b). The disproportionate extensions of the PRP at low temperatures demonstrated in similar experiments with other noctuid species (Turgeon & McNeil, 1983; Han & Gatehouse, 1991b), and the extension of the PRP at high temperatures reported by Isley (1935), evidently do not occur in *H. armigera*.

The effect of the vapour-pressure deficit (VPD) of the environment on the PRP of *H. armigera* was examined by holding female moths from eclosion at 23.7 ± 0.5 mmHg (c. 12% r.h.) and at 6.45 ± 0.15 mmHg (c. 76% r.h.) (Colvin & Gatehouse, 1993b). The PRP distributions of these females, which were supplied with sugar solution *ad lib.*, were not significantly different, showing that, under these conditions, the VPD does not influence PRP. These results indicate that neither suboptimal temperatures nor VPDs are likely to be utilised as cues by *H. armigera* in the regulation of migration.

### *Effect of adult feeding on PRP and on flight activity*

The effect of adult feeding on the PRP was examined by feeding moths either water, or 10% (w/v) sugar solution, or water until they were four nights old, followed by 10% sugar solution for the rest of their lives. In this experiment, all surviving moths fed a sugar solution diet (93%) reached reproductive maturity by night 7 (Fig. 12.2). Survival to maturity in the water/sugar solution treatment was 76%, with the latest moth achieving reproductive maturity on night 11. Most mortality in this treatment occurred while moths were being given only water. Moths in the first treatment, which were given water throughout their lives, showed poor survival to maturity (20%) and all were dead by night 6. Moths in this treatment did not appear to drink and dehydration probably contributed to their high mortality. These results demonstrate that if access to a source of sugar solution is denied to adult *H. armigera*, reproductive development is delayed (Colvin & Gatehouse, 1993b).

Hackett & Gatehouse (1982) investigated the effects of feeding on the

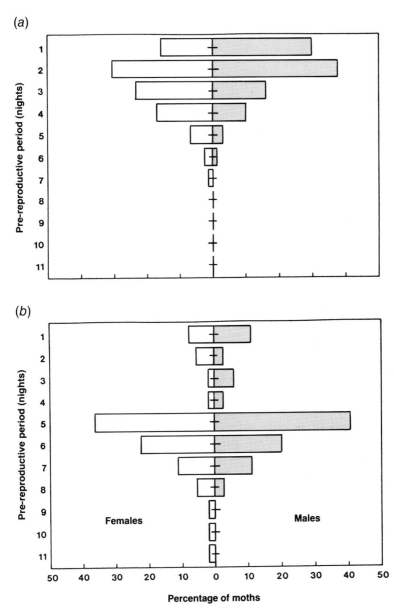

Fig. 12.2. Pre-reproductive period distributions of *Helicoverpa armigera* under two different feeding regimes. Moths were either given 10% (w/v) sugar solution throughout their lives (*a*) or were given water until the end of night 4 and, thereafter, given sugar solution (*b*). Sample sizes for sugar-fed moths were: females, 99; males, 66; and for water/sugar-fed moths: females, 52; males, 34 (from Colvin & Gatehouse (1993*b*), with permission).

tethered-flight activity of female *H. armigera*. They reported that feeding on night 1 did not affect flight performance significantly on night 2, although a small increase in the percentage of fed moths making flights of >120 min duration was observed. They did find, however, that in comparison to starved individuals, moths fed sucrose solution before nightfall on night 2 gave fewer flights of >60 min duration during the first half of the night, which is when migratory flight is thought to begin.

These results show that the availability of sugar solution influences survival (demonstrating that adults have a physiological requirement for carbohydrate), the PRP and flight activity. One possible interpretation is that nectar from flowering host plants may act as an important cue, signalling both the present suitability of the habitat for adult reproduction and its future suitability for larval development (*H. armigera* larvae feed preferentially on the reproductive structures of plants). Field observations suggesting that *H. armigera* populations are sedentary in the presence of flowering host plants are provided by several authors (Parsons, 1940; Roome, 1972; Roome, 1975; Topper, 1987; Riley *et al.*, 1992). To date, however, only anecdotal evidence is available of *H. armigera* emigrating *en masse*, after emergence in an area where host plants were not in flower (Armes & Cooter, 1991). Therefore, field studies on populations emerging into areas where sources of free sugars (nectar, honeydew and plant exudates) are scarce or unavailable, are necessary to establish the validity of this hypothesis.

## Interaction of factors affecting migratory potential

Evidence presented in this chapter shows clearly that *H. armigera* adults have the ability to undertake long-range migratory movements. The regulation of migratory behaviour is complex, however, and involves interactions at several levels. Genes regulating the PRP and flight activity interact both with each other and with the environment, to determine migratory potential. Whether this potential is realised or not probably depends on an ability of newly emerged adults to respond appropriately to environmental conditions, thereby 'fine-tuning' the regulation of migratory behaviour in this species.

## Acknowledgements

I would like to thank Drs Cooter and Gatehouse for constructive criticism of the manuscript.

## References

Armes, N.J. & Cooter, R.J. (1991). Effects of age and mated status on flight potential of *Helicoverpa armigera* (Lepidoptera: Noctuidae). *Physiological Entomology*, **16**, 131–44.

Bowden, J. & Johnson, C.G. (1976). Migrating and other terrestrial insects at sea. In *Marine Insects*, ed. L. Cheng, pp. 97–117. Amsterdam: North Holland Publishing Company.

Charlesworth, B., Coyne, J.A. & Barton, N.H. (1987). The relative rates of evolution of sex chromosomes and autosomes. *American Naturalist*, **130**, 113–46.

Colvin, J. & Gatehouse, A.G. (1993a). The reproduction-flight syndrome and the inheritance of tethered-flight activity in the cotton-bollworm moth, *Heliothis armigera*. *Physiological Entomology*, **18**, 16–22.

Colvin, J. & Gatehouse, A.G. (1993b). Migration and the effect of three environmental factors on the pre-reproductive period of the cotton-bollworm moth, *Helicoverpa armigera*. *Physiological Entomology*, **18**, 109–13.

Colvin, J. & Gatehouse, A.G. (1993c). Migration and the genetic regulation of the pre-reproductive period in the cotton bollworm moth, *Helicoverpa armigera*. *Heredity*, **70**, 407–12.

Commonwealth Institute of Entomology (1968). *Distribution Maps of Pests*, Series A, 15, *H. armigera* (Revised). London: Commonwealth Institute of Entomology.

Coombs, M., Del Socorro, A.P., Fitt, G.P. & Gregg, P.C. (1993). The reproductive maturity and mating status of *Helicoverpa armigera*, *H. punctigera* and *Mythimna convecta* (Lepidoptera: Noctuidae) collected in tower-mounted light traps in northern New South Wales, Australia. *Bulletin of Entomological Research*, **83**, 529–34.

Danks, H.V. (1987). *Insect Dormancy: an Ecological Perspective*. Ottawa: Biological Surveys of Canada Monograph 1.

David, C.T. & Hardie, J. (1988). The visual responses of free-flying summer and autumn forms of the black bean aphid, *Aphis fabae*, in an automated flight chamber. *Physiological Entomology*, **13**, 277–84.

Dingle, H. (1985). Migration. In *Comprehensive Insect Physiology, Biochemistry and Pharmacology*, ed. G.A. Kerkut & L.I. Gilbert, pp. 375–415. New York: Pergamon Press.

Drake, V.A. (1985). Radar observations of moths migrating in a nocturnal low-level jet. *Ecological Entomology*, **10**, 259–65.

Drake, V.A. & Farrow, R.A. (1988). The influence of atmospheric structure and motions on insect migration. *Annual Review of Entomology*, **33**, 183–210.

Falconer, D.S. (1989). *Introduction to Quantitative Genetics*, 3rd edn. London: Longman.

Farrow, R.A. & Daly, J.C. (1987). Long-range movements as an adaptive strategy in the genus *Heliothis* (Lepidoptera: Noctuidae): a review of its occurrence and detection in four pest species. *Australian Journal of Zoology*, **35**, 1–24.

Farrow, R.A. & McDonald, G. (1988). Migration strategies and outbreaks of noctuid pests in Australia. In *Recent Advances in Research on Tropical Entomology*, ed. M.F.B. Chaudhury, pp. 531–42. *Insect Science and its Application*, vol. 8, Special Issue. Nairobi: ICIPE Science Press.

276     *Regulation of migration in* Helicoverpa armigera

Fitt, G.P. (1989). The ecology of *Heliothis* species in relation to agroecosystems. *Annual Review of Entomology*, **34**, 17–52.

Gatehouse, A.G. (1989). Genes, environment and insect flight. In *Insect Flight*, ed. G.J. Goldsworthy & C. Wheeler, pp. 116–38. Boca Raton, Florida: CRC Press.

Gatehouse, A.G. & Hackett, D.S. (1980). A technique for studying flight behaviour of tethered *Spodoptera exempta* moths. *Physiological Entomology*, **5**, 215–22.

Hackett, D.S. (1980). Studies on the biology of *Helicoverpa armigera armigera* (Hb) in the Sudan Gezira. PhD thesis, University of Wales.

Hackett, D.S. & Gatehouse, A.G. (1982). Studies on the biology of *Heliothis* spp. in the Sudan. In *Proceedings of the International Workshop on Heliothis Management*, ed. W. Kumble & V. Kumble, pp. 29–38. Hyderabad: International Crops Research Institute for the Semi-Arid Tropics.

Hamilton, W.D. & May, R.M. (1977). Dispersal in stable habitats. *Nature*, **269**, 578–81.

Han, E.-N. & Gatehouse, A.G. (1991a). Genetics of precalling period in the oriental armyworm, *Mythimna separata* (Walker) (Lepidoptera: Noctuidae), and implications for migration. *Evolution*, **45**, 1502–10.

Han, E.-N. & Gatehouse, A.G. (1991b). Effect of temperature and photoperiod on the calling behaviour of a migratory insect, the oriental armyworm *Mythimna separata*. *Physiological Entomology*, **16**, 419–27.

Hill, J.K. & Gatehouse, A.G. (1992). Genetic control of the pre-reproductive period in the Silver Y moth, *Autographa gamma* (L.) (Lepidoptera: Noctuidae). *Heredity*, **69**, 458–64.

Isley, D. (1935). Relation of hosts to abundance of cotton bollworm. *University of Arkansas Agricultural Experimental Station Bulletin*, **320**, 1–30.

Johnson, C.G. (1969). *Migration and Dispersal of Insects by Flight*. London: Methuen.

Kennedy, J.S. (1986). Migration, behavioural and ecological. In *Migration: Mechanisms and Adaptive Significance*, ed. M.A. Rankin, pp. 1-20. *Contributions in Marine Science*, vol. 27, Suppl. Port Aransas, Texas: Marine Science Institute, The University of Texas at Austin.

Kou, R. & Chow, Y.S. (1987). Calling behaviour of the cotton bollworm *Heliothis armigera* (Lepidoptera: Noctuidae). *Annals of the Entomological Society of America*, **80**, 490–3.

McAnelly, M.L. & Rankin, M.A. (1986). Migration in the grasshopper, *Melanoplus sanguinipes* (Fab.). I. The capacity for flight in non-swarming populations. *Biological Bulletin*, **170**, 368–77.

Pair, S.D., Raulston, J.R., Rummel, D.R., Westbrook, J.K., Wolf, W.W., Sparks, A.N. & Schuster, M.F. (1987). Development and production of corn earworm and fall armyworm (Lepidoptera: Noctuidae) in the Texas High Plains: evidence for reverse fall migration. *Southwestern Entomologist*, **12**, 89–99.

Parker, W.E. & Gatehouse, A.G. (1985). Genetic factors controlling flight performance and migration in the African armyworm moth, *Spodoptera exempta* (Walker) (Lepidoptera: Noctuidae). *Bulletin of Entomological Research*, **75**, 49–63.

Parsons, F.S. (1940). Investigations on the cotton bollworm, *Heliothis armigera* (Hb.). Part III. Relations between oviposition and the flowering curves of food plants. *Bulletin of Entomological Research*, **31**, 147–77.

Riley, J.R., Armes, N.J., Reynolds, D.R. & Smith, A.D. (1992). Nocturnal observations on the emergence and flight behaviour of *Helicoverpa armigera* (Lepidoptera: Noctuidae) in the post-rainy season in central India. *Bulletin of Entomological Research*, **82**, 243–56.

Roome, R.E. (1972). *Annual Report 1971–1972. Entomologist, Botswana Dryland Farm Research Scheme*. London: Overseas Development Administration.

Roome, R.E. (1975). Activity of *Heliothis armigera* (Hb.) with reference to the flowering of sorghum and maize in Botswana. *Bulletin of Entomological Research*, **65**, 523–30.

Rose, D.J.W., Page, W.W., Dewhurst, C.F., Riley, J.R., Reynolds, D.R., Pedgley, D.E. & Tucker, M.R. (1985). Downwind migration of the African armyworm moth, *Spodoptera exempta*, studied by mark-and-capture and by radar. *Ecological Entomology*, **10**, 299–313.

Southwood, T.R.E. (1977). Habitat, the templet for ecological strategies? *Journal of Animal Ecology*, **46**, 337–65.

Topper, C.P. (1987). Nocturnal behaviour of adults of *Heliothis armigera* (Hb.) (Lepidoptera: Noctuidae) in the Sudan Gezira and pest control implications. *Bulletin of Entomological Research*, **77**, 541–54.

Turgeon, J.J. & McNeil, J.N. (1983). Modifications in the calling behaviour of *Pseudaletia unipuncta* (Lepidoptera: Noctuidae) induced by temperature conditions during pupal and adult development. *Canadian Entomologist*, **115**, 1015–22.

Wilson, K. & Gatehouse, A.G. (1992). Migration and genetics of the pre-reproductive period in the moth, *Spodoptera exempta* (African armyworm). *Heredity*, **69**, 255–62.

Wilson, K. & Gatehouse, A.G. (1993) Seasonal and geographical variation in the migratory potential of outbreak populations of the African armyworm moth, *Spodoptera exempta. Journal of Animal Ecology*, **62**, 169–81.

Woodrow, K.P., Gatehouse, A.G. & Davies, D.A. (1987). The effect of larval phase on flight performance of African armyworm moths, *Spodoptera exempta* (Walker) (Lepidoptera: Noctuidae). *Bulletin of Entomological Research*, **77**, 113–22.

# 13

# Physiological integration of migration in Lepidoptera

## J.N. McNEIL, M. CUSSON, J. DELISLE, I. ORCHARD AND S.S. TOBE

### Introduction

Most insect species are subject to either predictable or unpredictable habitat deterioration and have evolved life-history traits that enhance survival when habitat quality is inadequate for normal development and reproduction. In certain species, individuals enter a state of dormancy or quiescence within the habitat and remain in this stage until conditions improve at some later date. In others, individuals emigrate to more favourable sites. These options, broadly defined as the 'here-later' and the 'there-now' strategies (Southwood, 1977; Solbreck, 1978), involve considerable physiological changes, which are initiated by an array of external cues. In cases where habitat deterioration is of a predictable or seasonal nature, parameters such as day-length, temperature or precipitation serve as token stimuli. In contrast, cues such as the absence of suitable food, oviposition sites or mates, as well as sudden changes in abiotic conditions, may serve to induce the physiological modifications necessary to cope with unpredictable changes in habitat quality (see Chapter 10, this volume, for additional information).

The classic scenario relating migration to the reproductive status of the migrant is the 'oogenesis-flight syndrome' (Johnson, 1969), with migration being undertaken by individuals that are sexually immature or between reproductive cycles. Migration and reproduction are perceived as mutually exclusive processes. However, in a number of species known to undertake migratory flight, juvenile hormone (JH) is necessary for both flight and oogenesis (see Rankin, McAnelly & Bodenhamer, 1986, and references therein). Based on their work on the milkweed bug, *Oncopeltus fasciatus*, Rankin & Riddiford (1978) proposed a model to explain the apparent contradiction of having the same hormone controlling reproduction and migration (Fig. 13.1). According to this model,

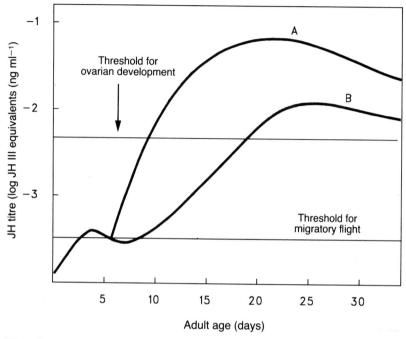

Fig. 13.1. A model describing changes in juvenile hormone (JH) titres in *Oncopeltus fasciatus* under different environmental conditions (Redrawn from Rankin & Riddiford, 1978). Under favourable conditions, JH titres rise and rapidly exceed the threshold for ovarian development, leading to the onset of reproduction (A). In contrast, when temperature and photoperiodic conditions indicate habitat deterioration, migration will occur when JH titres surpass the threshold for migratory flight but remain below that of ovarian development (B).

migratory flight is initiated when the JH titre exceeds a minimum level and continues until the titre surpasses a higher threshold, above which JH promotes sexual maturation.

However, there is now very clear evidence that the relationship between reproduction and migration is considerably more complex than that proposed by the oogenesis-flight syndrome. The strategies deployed vary considerably in different insect orders, not only between but also within species (see Rankin, 1989, 1991; Rankin *et al.*, 1986; Pener, 1985, 1991). As suggested by Rankin *et al.* (1986), there is a need to re-examine the general applicability of the oogenesis-flight syndrome for all situations, an idea recently reiterated by Sappington & Showers (1992). Thus, if we are to advance our understanding, we must not look for an all-encompassing general theory but rather examine the physiological aspects of migration in individual species, within their ecological and phylogenetic contexts. There are a number of excellent reviews dealing with the physiological aspects of migration (see Rankin, 1989, 1991; Rankin *et al.*, 1986; Pener, 1985, 1991) and it is not our intention to duplicate these. Instead, we aim to develop a number of ideas, based on some recent results in our research group and others, to suggest lines of research to further our understanding of the physiology of migratory flight in the Lepidoptera. We are concerned with the physiological adaptations associated with the reproductive state in which individuals initiate migratory flight, and how these may change during the flight period.

## Physiological state at the time migratory flight is initiated

The majority of our own research has been on the True Armyworm *Pseudaletia (Mythimna) unipuncta*, a species that occurs each summer in much of northeastern North America but does not have the ability to survive the prevailing winter conditions in this area (Fields & McNeil, 1984; Ayre, 1985). The populations observed annually in the more northern regions of its distribution are therefore the result of immigration.

Experiments on female calling (the behaviour associated with the emission of the sex pheromone) demonstrated that, under cool, short-day conditions, the age to first calling is greater than under warm, long-day conditions (Turgeon & McNeil, 1983; Delisle & McNeil, 1986, 1987). Similar results were obtained when male responsiveness to the sex pheromone was examined in a wind tunnel (Turgeon, McNeil & Roelofs, 1983; Dumont & McNeil, 1992). In addition, most females caught in light traps during the autumn flight were unmated and very few males were attracted to pheromone lures. The combined laboratory and field data provided the basis for the hypothesis that the observed delay in the onset of reproduction in *P. unipuncta* was associated with a southward autumn migration to suitable overwintering sites (McNeil, 1987). Furthermore, based on data from a five-week spring trapping period in Louisiana, it was proposed that similar temperature and photoperiodic cues serve to initiate the northward spring migration, away from sites where high summer temperatures are unfavourable (McNeil, 1987).

Work on other migrant species has shown that the duration of the pre-calling period may be significantly affected by cues associated with habitat deterioration. When migration is associated with predictable habitat deterioration, cues such as temperature and photoperiod influence the female pre-calling period, as has been shown in two other noctuids, the Oriental Armyworm *Mythimna separata* (Han & Gatehouse, 1991*a*) and *Autographa gamma* (Hill & Gatehouse, 1993). In species where habitat deterioration is unpredictable, the cues to initiate migration may be associated with specific resources. For example, the presence of host-plant pollen, an essential nutrient resource for neonate larvae, significantly reduced the pre-calling period of the Sunflower Moth *Homoeosoma electellum* (McNeil & Delisle, 1989), a species known to undertake long-distance migratory flights when suitable host plants are unavailable (Arthur & Bauer, 1981). Similarly, the presence or absence of migratory behaviour in the Old World bollworm *Helicoverpa armigera* may be associated with the availability of nectar sources for the newly emerged adults (Riley *et al.*, 1992; Colvin & Gatehouse, 1993; Chapter 12, this volume). These examples lend credence to the suggestion that variation in the pre-calling period may identify migrant moth species (McNeil, 1986) and that an extended pre-calling period might provide a longer time-window during which sexually immature individuals could migrate.

We have demonstrated that the onset of calling, pheromone synthesis and oocyte maturation are closely linked in *P. unipuncta*, and that all three processes are JH-dependent (Cusson & McNeil, 1989*a,b*). Therefore, in situations where migration is thought to be undertaken by sexually immature adults, the oogenesis-flight model would predict low but detectable levels of JH production under short-day, low-temperature conditions. The results of experiments carried out to examine JH production under different abiotic conditions provided clear evidence that the delays in ovarian development and the onset of calling in females (Fig. 13.2), as well as in the responsiveness of males, are associated with low levels of production of JH or JH acid (JHA) in *P. unipuncta* (Cusson, McNeil & Tobe, 1990; Cusson *et al.*, 1993). These findings support the hypothesis, derived from field-collected data (McNeil, 1987; Chapter 10, this volume), that adults are sexually immature at the time they initiate migratory flight. We believe that the model proposed for *P. unipuncta* (Fig. 13.3) may hold true for a number of other migrant Lepidoptera, as Gadenne (1993) has shown that JH is essential for ovarian development and the expression of calling behaviour in the Black Cutworm *Agrotis ipsilon*. Furthermore, preliminary data obtained for five noctuid species, considered to be seasonal migrants (Poitout *et al.*, 1974), indicated that most females collected during early October at high altitudes in the Pyrenees were unmated (Table 13.1). The same is true of migrating

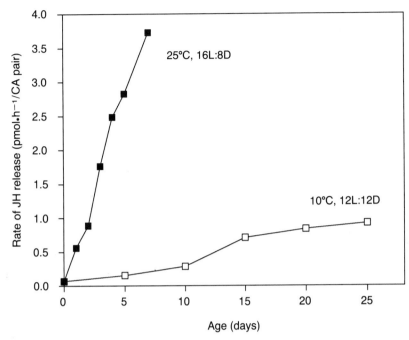

Fig. 13.2. The rate of total juvenile hormone (JH) release *in vitro* by the corpora allata of *Pseudaletia unipuncta* virgin females, as a function of age, when held at either 25 °C, 16L : 8D or 10 °C, 12L : 12D. Data from Cusson *et al.* (1993).

noctuids caught in light traps mounted on towers in eastern Australia (Coombs *et al.*, 1993). However, considerably more data are needed to test this hypothesis.

In insects from most orders, JH III is the only detectable JH homologue, yet moths and butterflies produce at least three homologues (JH I, JH II and JH III). Recent experiments (Cusson *et al.*, 1993) have demonstrated that, in both males and females, the rates of biosynthesis of each homologue produced by the corpora allata (CA) change with age and are affected by prevailing temperature and photoperiodic conditions (Fig. 13.4A,B,D,E). It is, therefore, possible that the reproductive state of Lepidoptera at the time migratory flight is initiated may depend not on the total JH titre but rather on the titre of just one homologue (a modified version of the oogenesis-flight model in Fig. 13.1). Alternatively, migratory flight may require a specific blend, as the ratios of JH I to JH II (or JHA I to JHA II in males), the most abundant homologues in *P. unipuncta*, change as a function of the total amount of JH produced (Fig. 13.4C,F).

Table 13.1. *Reproductive status of southward-moving moths captured in light traps on the Plateau de Lhers[a], France, 8–10 October, 1989*

| Species | Total number captured | Number with some ovarian development | Number with spermatophore |
|---|---|---|---|
| *Agrotis ipsilon* | 30 | 0 | 0 |
| *Autographa gamma* | 2 | 0 | 0 |
| *Peridroma saucia* | 50 | 0 | 0 |
| *Phlogophora meticulosa* | 27 | 6 | 5 |
| *Pseudaletia unipuncta* | 12 | 6 | 1 |

[a]This site is at an altitude of >1000 m in the French Pyrenees.

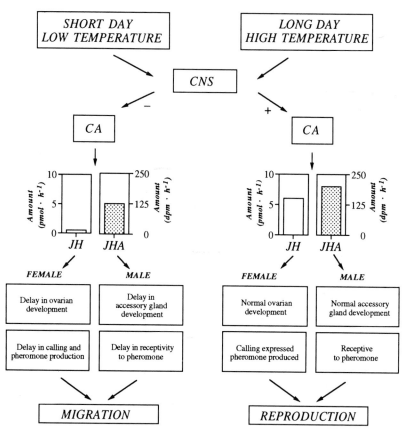

Fig. 13.3. A model for the physiological control of reproduction or the initiation of migratory flight by sexually immature adults of the True Armyworm *Pseudaletia unipuncta*.

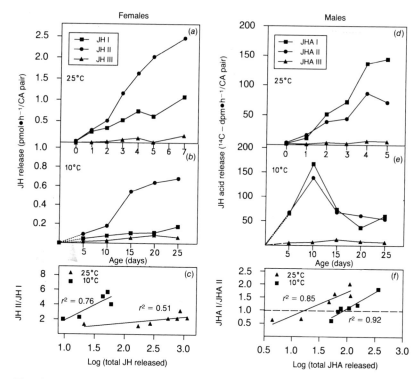

Fig. 13.4. The rate of release *in vitro* of different homologues of juvenile hormone (JH) or JH acid (JHA) by the corpora allata (CA) of virgin *Pseudaletia unipuncta* adults at 25 °C, 16L : 8D (A females; D males) and 10 °C, 12L : 12D (B females; E males). The relationships between JH II/JH I in females or JHA I/JHA II in males to the total JH or JHA released by the CA of virgin *Pseudaletia unipuncta* adults at 25 °C, 16L : 8D and 10 °C, 12L : 12D are presented in C and F, respectively. Data from Cusson *et al.* (1993).

Future work to test the idea that the titres of specific homologues or JH 'blends' are important must examine a diversity of species, as the relative contributions of the different homologues to the total JH titre in *P. unipuncta* differ from those reported for either the non-migrant *Manduca sexta* (Ishizaka, Bhaskaran & Dahm, 1989) or the migrant *Helicoverpa zea* (Satyanarayana *et al.*, 1991). Furthermore, the JH profiles in both sexes should be examined, since, as seen in *P. unipuncta* (Fig. 13.4; Cusson *et al.*, 1993), the ratios of the JHA homologues produced by the CA of males may not be the same as those of the JH homologues produced by the CA of females.

The autumn southward migration of the Monarch butterfly *Danaus plexippus* towards overwintering sites conforms to the oogenesis-flight-syndrome model, as it is associated with low levels of JH and undeveloped reproductive organs (see Herman, 1985; Rankin *et al.*, 1986, and references therein). However, while adults leaving the overwintering site may not be fully mature (Brower, 1985; Herman, 1985), at least part of the northward migration does not appear to conform to the oogenesis-flight-syndrome model, as adults arriving in the more northern parts of the summer range are sexually mature and have high JH titres (Herman, 1985). The implications of these results must now be considered in the context of the recent confirmation that the reinvasion of the summer breeding grounds occurs over several generations (Cockrell & Malcolm, 1993; Malcolm, Cockrell & Brower, 1993). Immigrant *P. unipuncta* females captured in light traps in Quebec are generally mated and males respond to pheromone lures (McNeil, 1987). Similar results have been reported for *A. ipsilon* (Kaster & Showers, 1982; Sappington & Showers, 1992, and references therein). However, immigrant females captured in light traps over the last ten years in Quebec generally contain a full complement of eggs, suggesting that mating has only just occurred. Sexually mature, virgin females of both *P. unipuncta* and *A. ipsilon* call for the first time in the last half of the scotophase (Swier, Rings & Musick, 1977; Turgeon & McNeil, 1982) and thus adults could mate on the same night that they terminate migratory flight. This strategy would ensure rapid exploitation of the new habitat, but could result in an overestimate of the proportion actually migrating as mated individuals (Chapter 10, this volume).

On the other hand, we cannot at present exclude the possibility that a number of moths exhibit a pattern similar to that of *D. plexippus*, with the last part of the northward, spring migration apparently completed by mated individuals. McNeil (1987) pointed out that mean temperature and photoperiodic conditions are similar during the periods when the proposed southward and northward migrations of *P. unipuncta* are initiated and would, therefore, always serve as reliable cues of habitat deterioration. However, several important differences exist between the spring and autumn periods that could affect the reproductive state in which adults terminate their migratory flight.

The first relates to the conditions experienced by migrants during larval and pupal development. Adults emerging at the end of the summer experience declining temperatures and day-length, whereas those emerging during the spring develop when both day-length and temperature are increasing. Under controlled conditions, the pre-calling period of *M. sep-*

*arata* females was significantly longer in individuals reared under decreasing photoperiodic and temperature conditions than under constant conditions (Han & Gatehouse, 1991*a*). Directional changes in these two parameters during pupal development could contribute to the differences observed in the mean age of first calling for *P. unipuncta* females under field conditions at different times of the year (Turgeon & McNeil, 1983).

A second difference is that migrants moving northward in spring will experience a significant and rapid increase in day-length, while, in autumn, migrants moving southward will always remain under short-day conditions (Delisle & McNeil, 1987). Thus, the physiological control of sexual maturation and migration could differ between insects subjected to increasing day-length, both prior to and during migration, and those that are not. Bues *et al.* (1992) reported that the selection of a late-ovipositing line in *A. ipsilon* resulted in delays in both the age of first calling and ovarian maturation, but that these were considerably less than the delay in the onset of oviposition. Is this delay in oviposition, following sexual maturation and mating, associated with the possible migration of mated individuals? It is clear that we do not have sufficient information to answer this question and, as discussed below, more research must be carried out to determine whether or not, under certain conditions, mated female noctuid moths migrate.

Migration has been clearly documented in the Spruce Budworm *Choristoneura fumiferana* (Tortricidae), but its migratory strategy contrasts with that proposed for *P. unipuncta* (McNeil, 1987). The large majority of migrants are females, most of which have mated and deposited a proportion of their egg complement before migrating (Greenbank, Schaefer & Rainey, 1980). Migration is probably initiated by the indirect and direct effects of high densities during the epidemic phase of the population cycle, as there is no evidence of an extended pre-calling period in this species (J. Delisle, unpublished data). *C. fumiferana* is a species in which females are able to detect the sex pheromone of conspecifics (Palaniswamy & Seabrook, 1978; 1985) and, as the flight activity of mated females is significantly higher in pheromone-permeated than in clean air, particularly on the first two days following mating (Sanders, 1987), the high ambient levels of sex pheromone associated with epidemic populations may serve as a cue for migration.

Recent work on the calling behaviour of mated females indicates that ambient pheromone titres may be influenced not only by the density of females but also by the quality of males in the population. *C. fumiferana* females mated with males reared on limited, poor foliage (epidemic conditions) were more likely to resume calling and laid significantly fewer fertile eggs in the 48 h following mating than those paired with males fed unlimited good foliage (endemic conditions) (M. Hardy & J. Delisle, unpublished data). Furthermore, preliminary data on flight behaviour also suggest that females mated with poor-quality males have a greater propensity to fly (J. Delisle, unpublished data). We postulate that male quality, determined by factors such as the availability and quantity of larval resources, may affect female JH titre following mating and influence whether or not *C. fumiferana* females emigrate (Fig. 13.5).

Males transfer resources other than sperm at the time of mating and these substances may affect egg production and oviposition (Friedel & Gillott, 1977; Gillott & Friedel, 1977; Loher *et al.*, 1981), female longevity and/or total fecundity (Gwynne, 1984, 1988; Rutowski, Gilchrist & Terkanian, 1987; Royer & McNeil, 1993), subsequent female receptivity (Svärd, 1985; Young & Downe, 1987; Oberhauser, 1989; Mbata & Ramaswamy, 1990) or the fitness of the progeny (Dussourd *et al.*, 1988). Post-mating changes in female JH titre could result from the direct transfer of JH in the spermatophore (Shirk, Bhaskaran & Roller, 1980). Alternatively, the titre may be modified indirectly through male-derived compounds that alter the activity of the female's corpora allata, or of the

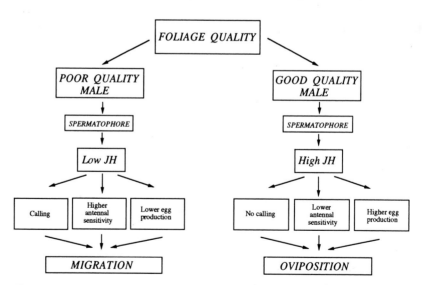

Fig. 13.5. A model for the physiological control of continued oviposition or the onset of migration by mated females of the Spruce Budworm *Choristoneura fumiferana*. JH, juvenile hormone.

enzymes that catabolise JH. If our hypothesis is correct, then females mating with high-quality males will have a higher JH titre than those pairing with poor-quality mates, and will be less likely to emigrate. The idea that post-mating JH titres may influence migration has a certain appeal, given that female sensitivity to *C. fumiferana* sex pheromone decreases following application of JH analogues (Palaniswamy, Seabrook & Sivasubramanian, 1979*a*; Palaniswamy, Sivasubramanian & Seabrook, 1979*b*).

## Changes that could occur once migration has been initiated

Free-flying *P. unipuncta* adults at ambient 20 °C conditions have thoracic temperatures ranging from 35 to 40 °C, with head and abdomen temperatures varying between 27 and 32 °C (T.M. Casey & J.N. McNeil, unpublished data). These increases in body temperature, associated with flight activity, could have a marked effect on various JH-dependent processes that are implicated in both migration and ovarian development.

Females transferred to 25 °C, 12L : 12D or 16L : 8D, after being held from emergence for 5, 10 or 15 days at 10 °C under long- or short-day photoperiodic conditions, became sexually mature within 3–8 days, values that are very similar to controls at 25 °C (Delisle & McNeil, 1987). A similar pattern has also been observed with respect to male responsiveness to the female sex pheromone (Dumont & McNeil, 1992). Furthermore, within 24 h of a switch from 10 to 25 °C there was a marked increase in JH production, to a level higher than that observed in females reared continuously at the higher temperature (Cusson *et al.*, 1990). Therefore, if the extended delay in the onset of reproduction (e.g. a pre-calling period of 21.0 days at 10 °C, 12L : 12D v 5.8 days at 25 °C, 16L : 8D in non-flying *P. unipuncta* females; Delisle & McNeil, 1987) is to provide a greater time-window for migration, then physiological mechanisms must exist to counterbalance the potential effect of high body temperatures on JH biosynthesis and subsequent ovarian development, once individuals initiate migratory flight.

JH biosynthesis may be inhibited by allatostatins (Woodhead *et al.*, 1989; Kramer *et al.*, 1991; Pratt *et al.*, 1991) and we have evidence that compounds present in *P. unipuncta* brains have an allatostatic effect in the cockroach *Diploptera punctata* (M. Cusson, J.N. McNeil & S.S. Tobe, unpublished data). Therefore, if JH biosynthesis is inhibited by allatostatins during migratory flight, one would predict that their production and release would be higher in individuals experiencing cues indicating habitat deterioration (e.g. low temperature, short day-length, low food quality and quantity) than in those reared under conditions favouring normal development. Alternatively, the production and release of allatostatins may remain unchanged, but animals reared under unfavourable conditions may exhibit a greater sensitivity to these compounds. Furthermore, octopamine, a biogenic amine known to play an important role in flight physiology (see below), has been shown to affect JH biosynthesis in *D. punctata* and may modulate the release of other peptidergic allatostatins (Thompson *et al.*, 1990). The possibility that octopamine may modify JH titres during flight merits attention in migratory species.

Lessman & Herman (1983) demonstrated a higher rate of JH degradation in force-flown *D. plexippus* than in inactive individuals, which they associated with the higher temperatures normally encountered during flight. Thus, a higher rate of JH degradation at the relatively high body temperatures attained during flight (35–40 °C in the thorax) might lower the JH titre, slow down the process of ovarian maturation and presumably stimulate continued migratory flight. In addition, the degradation process may not affect all JH homologues equally, as is shown in the house cricket *Acheta domesticus* (Woodring & Sparks, 1987). Such differential degradation could alter the ratio of the different JH homologues which, in turn, may increase or reduce the propensity for migration. If JH esterases are determining either the JH titre or JH ratio during migration, then one would predict differences in the absolute amounts (or possibly activity levels) of JH esterases in individuals reared under different conditions, similar to the differences noted in long- and short-winged morphs of the cricket *Gryllus rubens* (Zera & Tiebel, 1989; Zera *et al.*, 1989).

In the absence of mechanisms to inhibit or counter the rise in JH production, which may operate as a consequence of increased body temperature associated with flight, the time-window for migration would be similar to the pre-calling period observed at high temperatures under short-day conditions (e.g. $9.0 \pm 0.9$ days for *P. unipuncta* at 25 °C; Delisle & McNeil, 1987). There are, however, a number of reasons to believe that migration is associated with an extended pre-calling period.

1  There is evidence that migrants may have longer pre-calling periods than resident species (McNeil, 1986).

2  Under several different temperatures, the pre-calling period in a non-migrant population of *P. unipuncta* from the Azores is significantly shorter than that of the migrant population collected each spring in Quebec (J.N. McNeil, unpublished data).

3 The duration of the pre-calling period is markedly affected by cues indicating either predictable (Turgeon & McNeil, 1983; Delisle & McNeil, 1986; 1987; Han & Gatehouse 1991*a*) or unpredictable (McNeil & Delisle, 1989; Pivnick *et al.*, 1990) habitat deterioration.

4 There is a strong genetic component determining the duration of the pre-calling period (Han & Gatehouse, 1991*b*; Bues *et al.*, 1992; Wilson & Gatehouse, 1992; Hill & Gatehouse, 1993), and individuals selected for a prolonged pre-calling period have a greater propensity for flight than those with a shorter pre-calling period (Han & Gatehouse, 1993).

However, because of the effect of high body temperatures associated with flight, pre-calling periods extended by environmental cues associated with habitat deterioration may not always result in a longer time-window for migration. In such cases, what benefits might be obtained from a delay in the onset of sexual maturation when the insect is exposed to cool temperatures?

One hypothesis is that it is an adaptation associated with the *pre-flight* component of migration. In the absence of adaptations to inhibit sexual maturation in migrants, any delay in the onset of migration must reduce the time available for subsequent, pre-reproductive migratory flight. There is clear evidence, from the composition of pollen grains found on the mouth parts, that several noctuids feed at the site of emergence, prior to migrating (Hendrix *et al.*, 1987; Hendrix & Showers, 1992; Gregg, 1993), probably to accumulate the necessary energy reserves. In such circumstances, strong selection might be expected for mechanisms that prevented sexual maturation during this essential foraging period, especially if the availability of suitable food sources was limited. This, of course, would not apply to species that emigrate in response to the absence of flowering host plants that serve as an essential source of adult food and/or as oviposition sites (Delisle *et al.*, 1989; McNeil & Delisle, 1989; Riley *et al.*, 1992; Chapter 12, this volume).

Alternatively, or in conjunction with the necessity to feed prior to flight, the ability to extend the pre-reproductive period, when prevailing weather conditions are not conducive to migratory flight, would be advantageous for species whose migration includes a specific directional component essential for locating suitable habitats (e.g. north in the spring and south in the autumn). There is little information concerning the degree to which these species actually select appropriate weather conditions. However, work on the autumn migration of the Potato Leafhopper *Empoasca fabae* in North America provides circumstantial evidence that the initiation of flight is not random, as significantly more adults were captured in upper-air masses on days when winds were moving southward than when winds were moving in any other direction (Taylor & Reling, 1986). Again, one would not expect the ability to delay the onset of sexual maturation to be as pronounced in species where a specific directional component is not essential for successful emigration from a deteriorating habitat. The idea that an extended pre-reproductive period, in response to certain environmental cues, may be associated with the pre-flight phase of migration has received little attention to date, and data for a diversity of species that migrate in response to predictable and unpredictable habitat deterioration will be required to test the validity of our hypothesis.

The current status of our understanding of mobilisation and utilisation of fuel during flight has been addressed in two reviews (Candy, 1989; Wheeler, 1989). However, in the context of the supposed costs associated with migratory flight (Rankin & Burchsted, 1992), there are three areas where our understanding of the physiological processes needs to be broadened:

1 The accumulation of lipids prior to the onset of migration;
2 The mobilisation of lipids during flight;
3 Flight muscle efficiency.

Chino *et al.* (1992) examined lipid storage in the locust and observed that the weight of the fat body in gregarious individuals was ≥7 times that of solitarious individuals. Furthermore, the quantity of triacylglycerol per unit mass of fat body tissue in solitarious locusts was <5% of that found in gregarious locusts. Thus a very different pattern of lipid accumulation is seen between the more mobile and sedentary phases, which may account for the higher resting haemolymph lipid levels in gregarious than in solitarious locusts (Ayali & Pener, 1992). There is a need to obtain data on both fat-body weight and lipid content in a number of migratory and resident species, reared under a range of different ecological conditions, to determine whether similar differences in lipid accumulation exist between migratory and non-migratory individuals, within and between species, in the Lepidoptera. Furthermore, it will be important to examine the efficacy of lipid accumulation during larval development, as well as by adult feeding.

In the course of sustained flight, lipids stored in the fat body are released into the haemolymph and transported by lipoproteins to the wing muscles, where they are used as a source of energy. During the initial phase (first 15–20 min) of flight in the locust, release of lipids from the fat body is activated by the monophenolic amine, octopamine (Orchard, 1987; Orchard, Ramirez & Lange, 1993), but this role is soon taken over by the adipokinetic hormones (AKH), synthesised and stored by the corpora cardiaca (CC). More specifically, AKHs activate the conversion of fat-body triacylglycerol (TG) into diacylglycerol (DG), the main lipid used as fuel in insect flight. The lipoproteins that transport DG to the muscles exist in two different forms, high-density (HDLp) and low-density lipophorins (LDLp). The LDLp have a much higher capacity for DG loading and, during flight, HDLp are converted to LDLp. This enhanced capacity appears to result primarily from an increase in the proportion of the smallest of three apoproteins (apo III) forming the protein moiety of lipophorins. In the wing muscles, LDLp-bound DG is unloaded without internalisation, the LDLp being converted back into HDLp to facilitate recycling (for more details see Beenakkers, 1991; Surholt *et al.*, 1992).

In addition to differences in the amounts of lipids that migratory and non-migratory insects may accumulate, it is possible that they also differ in their response to octopamine and/or AKH. For example, gregarious locusts release more lipid than solitarious individuals following injection of the same dose of either AKH or CC extracts (Ayali & Pener, 1992). This could result from the phase-related variation in the amount of lipids available for mobilisation (Chino *et al.*, 1992), but could also be related to differences in the density of AKH receptors in the fat body, an aspect that merits attention in future research. Another avenue that should be pursued relates to the ability to form LDLp, as this appears to vary according to the ability to undertake flights of long duration. The American Cockroach *Periplaneta americana*, in addition to showing only a weak response to AKH, is incapable of forming LDLp, primarily because it lacks the apoprotein III (Chino *et al.*, 1992). Similarly, solitarious locusts are unable to form LDLp, even though apoprotein III is present in their haemolymph. This inability may be associated with the very low amounts of lipids stored in the fat body (Chino *et al.*, 1992). Lastly, although never reported in the literature, it is possible that migratory and non-migratory insects differ either in their sensitivity to AKH or in the amounts of AKH stored in the CC. Certainly, amounts have been shown to change during adult life in locusts (Siegert & Mordue, 1986). Differences in lipid accumulation and in sensitivity to octopamine and/or AKH between the phases in locusts, which presumably reflect differences in their capacities for persistent flight, suggest directions for future research into physiological adaptations for migration. However, some solitaria-phase individuals are known to migrate (see Chapters 10 and 19, this volume) and nothing is as yet known of their physiology.

Changes in the haemolymph lipid during flight may differ with species, as evidenced by the results reported for the Tobacco Hornworm *Manduca sexta* (Ziegler & Schulz, 1986), *D. plexippus* (Dallmann & Herman, 1978) and *P. unipuncta* (Orchard, Cusson & McNeil, 1991) during 60 min of tethered flight (Fig. 13.6). In the non-migratory *Manduca sexta*, levels decreased during flight and only increased once flight had terminated. In contrast, lipid levels in *D. plexippus* and *P. unipuncta*, both migratory species, increased during flight. However, the lipid level rose more slowly in *D. plexippus* and, even after 120 min of tethered flight, had not increased to the same degree as was seen in *P. unipuncta* after 60 min (Fig. 13.6). Orchard *et al.* (1991) suggested that this difference may reflect the different flight strategies used during migratory flight. This should be verified for a number of migratory and non-migratory species,

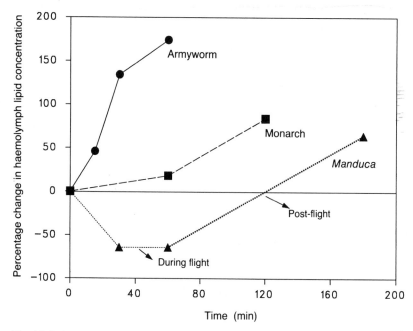

Fig. 13.6. A comparison of the change in haemolymph lipid concentration relative to resting levels after a 60 min bout of tethered flight in a non-migratory species, *Manduca sexta* (data from Dallmann & Herman, 1978) and two migratory species, the Monarch butterfly *Danaus plexippus* (data from Ziegler & Schulz, 1986) and the True Armyworm *Pseudaletia unipuncta* (data from Orchard *et al.*, 1991).

with the understanding that the levels of haemolymph lipids observed at different times during a flying bout will not only depend on lipid mobilisation but also on utilisation by flight muscles.

Recently it has been shown that AKH can inhibit vitellogenesis in locusts (Moshitzky & Applebaum, 1990) and that, at high doses, it may inhibit CA activity (Applebaum *et al.*, 1990). The possibility that the hormone responsible for lipid mobilisation during prolonged flight could, at the same time, have a negative effect on ovarian development merits further consideration.

Adaptations that facilitate the performance of flight muscle may be more prevalent in migrants than in non-migrants, be it at the inter- or intraspecific level. Production of and/or sensitivity to octopamine, a biogenic amine that may potentiate neuromuscular transmission (Claassen & Kammer, 1986) and muscle power output (Malamud, Mizisin & Josephson, 1988), could vary. Data from the only studies that have addressed certain aspects of this question, by comparing the gregarious and solitarious phases of the locust, are somewhat contradictory. Differences between the two phases have been reported (Fuzeau-Braesch & David, 1978; Fuzeau-Braesch, Coulon & David, 1979) but the findings were not confirmed in a subsequent study (Morton & Evans, 1983). A new peptide, belonging to the FMRFamide family, has been isolated from the central nervous system of *Manduca sexta* (Kingan *et al.*, 1990). This peptide increases the force of the neurally evoked contractions of the dorsal longitudinal muscles, and may play an important role in flight (Kingan *et al.*, 1990). Are such compounds common, and do their concentrations vary with respect to local v. long-distance movement?

## Concluding remarks

Migration is not a random act of 'casting one's fate to the winds' but a physiologically coordinated sequence of behaviours, determined by both genetic and environmental factors. Therefore, an interdisciplinary approach is required in future research to elucidate this important life-history strategy. Furthermore, as there is evidently no universal scenario, a number of different species exhibiting a diversity of life histories must be examined under a range of different conditions, in the laboratory and at different times during the season in the field. Finally, in order to resolve the question of whether the long flights undertaken by mature individuals on flight mills actually reflect migratory propensity (see Gatehouse & Hackett, 1980; Gatehouse & Woodrow, 1987; Armes &

Cooter, 1991; Sappington & Showers, 1992; Chapter 10, this volume), the field work must include a geographic component, permitting a comparison of the physiological and reproductive state of individuals at the end of their migration with their status at the time they initiated the migratory flight.

## References

Applebaum, S.W., Hirsch, J., Abd El-Hadi, F. & Moshitzky, P. (1990). Trophic control of juvenile hormone synthesis and release in locusts. In *Progress in Comparative Endocrinology*, ed. A. Epple, C.G. Scanes & M.H. Stetson, pp. 186–92. New York: Wiley-Liss.

Armes, N.J. & Cooter, R.J. (1991). Effects of age and mated status on flight potential of *Helicoverpa armigera* (Lepidoptera: Noctuidae). *Physiological Entomology*, 16, 131–44.

Arthur, A.P. & Bauer, D.J. (1981). Evidence of the northerly dispersal of the sunflower moth by warm winds. *Environmental Entomology*, 10, 523–8.

Ayali, A. & Pener, M.P. (1992). Density-dependent phase polymorphism affects response to adipokinetic hormone in *Locusta*. *Comparative Biochemistry and Physiology A*, 101, 549–52.

Ayre, G.L. (1985). Cold tolerance of *Pseudaletia unipuncta* and *Peridroma saucia* (Lepidoptera: Noctuidae). *Canadian Entomologist*, 117, 1055–60.

Beenakkers, A.M.T. (1991). Adipokinetic hormones and lipoprotein inter-conversions during locust flight. *Insect Science and its Application*, 12, 279–86.

Brower, L.P. (1985). New perspectives on the migration biology of the monarch butterfly, *Danaus plexippus* L. In *Migration: Mechanisms and Adaptive Significance*, ed. M.A. Rankin, pp. 748–85. *Contributions in Marine Science*, vol. 27, Suppl. Port Aransas, Texas: Marine Science Institute, The University of Texas at Austin.

Bues, R., Toubon, J.F., Poitout, S. & Saour, G. (1992). Sélection de souches de *Agrotis ipsilon* Hufn. (Lep., Noctuidae) a durées de préoviposition longue ou courte, comparaison de leur activité reproductrice. *Journal of Applied Entomology*, 113, 41–55. In French, English summary.

Candy, D.J. (1989). Utilization of fuels by the flight muscles. In *Insect Flight*, ed. G.J. Goldsworthy & C.H. Wheeler, pp. 305–19. Boca Raton, Florida: CRC Press.

Chino, H., Lum, P.Y., Nagao, E. & Hiraoka, T. (1992). The molecular and metabolic essentials for long-distance flight in insects. *Journal of Comparative Physiology B*, 162, 101–6.

Claassen, D.E. & Kammer, A.E. (1986). Effects of octopamine, dopamine, and serotonin on production of flight motor output by thoracic ganglia of *Manduca sexta*. *Journal of Neurobiology*, 17, 1–14.

Cockrell, B.J. & Malcolm, S.B. (1993). Time, temperature and latitudinal constraints on the annual recolonisation of eastern North America by the Monarch butterfly. In *Biology and Conservation of the Monarch Butterfly*, ed. S.B. Malcolm & M.P. Zalucki, pp. 233–51. Los Angeles: Natural History Museum of Los Angeles County.

Colvin, J. & Gatehouse, A.G. (1993). Migration and the effect of three environmental factors on the pre-reproductive period of the cotton-bollworm moth, *Helicoverpa armigera*. *Physiological Entomology*, **18**, 109–13.

Coombs, M., Del Socorro, A.P., Fitt, G.P. & Gregg, P.C. (1993). The reproductive maturity and mating status of *Helicoverpa armigera*, *H. punctigera* and *Mythimna convecta* (Lepidoptera: Noctuidae) collected in tower-mounted light traps in northern New South Wales, Australia. *Bulletin of Entomological Research*, **83**, 529–34.

Cusson, M. & McNeil, J.N. (1989*a*). Involvement of juvenile hormone in the regulation of pheromone release in a moth. *Science*, **243**, 210–12.

Cusson, M. & McNeil, J.N. (1989*b*). Ovarian development in female armyworm moths, *Pseudaletia unipuncta*: its relationship with pheromone release activities. *Canadian Journal of Zoology*, **67**, 1380–5.

Cusson, M., McNeil, J.N. & Tobe, S.S. (1990). *In vitro* biosynthesis of juvenile hormone by corpora allata of *Pseudaletia unipuncta* virgin females as a function of age, environmental conditions, calling behaviour and ovarian development. *Journal of Insect Physiology*, **36**, 139–46.

Cusson, M., Yagi, K.J., Tobe, S.S. & McNeil, J.N. (1993). Identification of release products of corpora allata of male and female armyworm moths, *Pseudaletia unipuncta*. *Journal of Insect Physiology*, **39**, 775–83.

Dallmann, S.H. & Herman, W.S. (1978). Hormonal regulation of haemolymph lipid concentration in the monarch butterfly, *Danaus plexippus*. *General and Comparative Endocrinology*, **36**, 142–50.

Delisle, J. & McNeil, J.N. (1986). The effect of photoperiod on the calling behaviour of virgin females of the true armyworm, *Pseudaletia unipuncta* (Haw.) (Lepidoptera: Noctuidae). *Journal of Insect Physiology*, **32**, 199–206.

Delisle, J. & McNeil, J.N. (1987). The combined effect of photoperiod and temperature on the calling behaviour of the true armyworm, *Pseudaletia unipuncta*. *Physiological Entomology*, **12**, 157–64.

Delisle, J., McNeil, J.N., Underhill, E.W. & Barton, D. (1989). *Helianthus annuus* pollen, an oviposition stimulant for the sunflower moth, *Homoeosoma electellum*. *Entomologia Experimentalis et Applicata*, **50**, 53–60.

Dumont, S. & McNeil, J.N. (1992). Responsiveness of *Pseudaletia unipuncta* (Lepidoptera: Noctuidae) males, maintained as adults under different temperature and photoperiodic conditions, to female sex pheromone. *Journal of Chemical Ecology*, **18**, 1797–807.

Dussourd, D.E., Ubik, K., Harvis, C., Resch, J., Meinwald, J. & Eisner, T. (1988). Biparental defensive endowment of eggs with acquired plant alkaloid in the moth *Utetheisa ornatrix*. *Proceedings of the National Academy of Sciences, USA*, **85**, 5992–6.

Fields, P.G. & McNeil, J.N. (1984). The overwintering potential of the true armyworm, *Pseudaletia unipuncta* (Lepidoptera: Noctuidae), populations in Québec. *Canadian Entomologist*, **116**, 1647–52.

Friedel, T. & Gillott, C. (1977). Contribution of mate-produced proteins to vitellogenesis in *Melanoplus sanguinipes*. *Journal of Insect Physiology*, **23**, 145–51.

Fuzeau-Braesch, S., Coulon, J.F. & David, J.C. (1979). Octopamine levels during the moult cycle and adult development in the migratory locust *Locusta migratoria*. *Experientia*, **35**, 1349–50.

Fuzeau-Braesch, S. & David, J.C. (1978). Etude du taux d'octopamine chez *Locusta migratoria* (Insecte: Orthoptère): comparaison entre insectes grégaires, solitaires et traités au gaz carbonique. *Comptes Rendus de l' Academie des Sciences, Paris*, **286D**, 697–9. In French, English summary.

Gadenne, C. (1993). Effects of fenoxycarb, juvenile hormone mimetic, on female sexual behaviour of the black cutworm, *Agrotis ipsilon* (Lepidoptera: Noctuidae). *Journal of Insect Physiology*, **39**, 25–9.

Gatehouse, A.G. & Hackett, D.S. (1980). A technique for studying flight behaviour of tethered *Spodoptera exempta* moths. *Physiological Entomology*, **5**, 215–22.

Gatehouse, A.G. & Woodrow, K.P. (1987). Simultaneous monitoring of flight and oviposition of individual velvetbean caterpillar moths (by Wales, Barfield & Leppla, 1985): a critique. *Physiological Entomology*, **12**, 117–21.

Gillott, C. & Friedel, T. (1977). Fecundity-enhancing and receptivity-inhibiting substances produced by male insects: a review. In *Advances in Invertebrate Reproduction*, vol. 1, ed. K.G. Adiyodi & R.G. Adiyodi, pp. 200–18. Kerala: Peralam-Kenoth.

Greenbank, D.O., Schaefer, G.W. & Rainey, R.C. (1980). *Spruce Budworm (Lepidoptera: Tortricidae) Moth Flight and Dispersal: New Understanding from Canopy Observations, Radar, and Aircraft.* Memoirs of the Entomological Society of Canada 110. Ottawa: Entomological Society of Canada.

Gregg, P.C. (1993). Pollen as a marker for migration of *Helicoverpa armigera* and *H. punctigera* (Lepidoptera: Noctuidae) from western Queensland. *Australian Journal of Ecology*, **18**, 209–19.

Gwynne, D.T. (1984). Courtship feeding increases female reproductive success in bush crickets. *Nature*, **307**, 361–3.

Gwynne, D.T. (1988). Courtship feeding and the fitness of female katydids (Orthoptera: Tettigoniidae). *Evolution*, **42**, 545–55.

Han, E.-N. & Gatehouse, A.G. (1991a). Effect of temperature and photoperiod on the calling behaviour of a migratory insect, the oriental armyworm *Mythimna separata*. *Physiological Entomology*, **16**, 419–27.

Han, E.-N. & Gatehouse, A.G. (1991b). Genetics of precalling period in the oriental armyworm, *Mythimna separata* (Walker) (Lepidoptera: Noctuidae), and implications for migration. *Evolution*, **45**, 1502–10.

Han, E.-N. & Gatehouse, A.G. (1993). Flight capacity: genetic determination and physiological constraints in a migratory moth *Mythimna separata*. *Physiological Entomology*, **18**, 183–8.

Hendrix, W.H. III, Mueller, T.F., Phillips, J.R. & Davis, O.K. (1987). Pollen as an indicator of long-distance movement of *Heliothis zea* (Lepidoptera: Noctuidae). *Environmental Entomology*, **16**, 1148–51.

Hendrix, W.H. III. & Showers, W.B. (1992). Tracing black cutworm and armyworm (Lepidoptera: Noctuidae) northward migration using *Pithecellobium* and *Callandria* pollen. *Environmental Entomology*, **21**, 1092–6.

Herman, W.S. (1985). Hormonally mediated events in adult monarch butterflies. In *Migration: Mechanisms and Adaptive Significance*, ed. M.A. Rankin, pp. 799–815. *Contributions in Marine Science*, vol. 27, Suppl. Port Aransas, Texas: Marine Science Institute, The University of Texas at Austin.

Hill, J.K. & Gatehouse, A.G. (1993). Phenotypic plasticity and geographic variation in the pre-reproductive period of *Autographa gamma* (Lepidoptera: Noctuidae) and its implications for migration in this species. *Ecological Entomology*, **18**, 39–46.

Ishizaka, S., Bhaskaran, G. & Dahm, K.H. (1989). Juvenile hormone production and ovarian maturation in adult *Manduca sexta*. In *Regulation of Insect Reproduction*, vol. 4, ed. T.A. Tonner, pp. 49–57. Berlin: Springer-Verlag.

Johnson, C.G. (1969). *Migration and Dispersal of Insects by Flight*. London: Methuen.

Kaster, L.V. & Showers, W.B. (1982). Evidence of spring immigration and autumn reproductive diapause of the adult black cutworm in Iowa. *Environmental Entomology*, **11**, 306–12.

Kingan, T.G., Teplow, D.B., Phillips, J.M., Riehm, J.P., Rao, K.R., Hilder-brand, J.G., Homberg, U., Kammer, A.E., Jardine, I., Griffin, P.R. & Hunt, D.F. (1990). A new peptide in the FMRFamide family isolated from the CNS of the hawkmoth, *Manduca sexta*. *Peptides*, **1**, 849–56.

Kramer, S.J., Toschi, A., Miller, C.A., Kataoka, H., Quistad, G.B., Li, J.P., Carney, R.L. & Schooley, D.A. (1991). Identification of an allatostatin from the tobacco hornworm *Manduca sexta*. *Proceedings of the National Academy of Sciences, USA*, **88**, 9458–62.

Lessman, C.A. & Herman, W.S. (1983). Seasonal variation in hemolymph juvenile hormone of adult monarchs (*Danaus p. plexippus*: Lepidoptera). *Canadian Journal of Zoology*, **61**, 88–94.

Loher, W., Ganjian, I., Kubo, I., Stanley-Samuelson, D. & Tobe, S.S. (1981). Prostaglandins: their role in egg-laying of the cricket *Teleogryllus commodus*. *Proceedings of the National Academy of Sciences, USA*, **78**, 7835–8.

Malamud, J.G., Mizisin, A.P. & Josephson, R.K. (1988). The effects of octopamine on concentration kinetics and power output of a locust flight muscle. *Journal of Comparative Physiology A*, **162**, 827–35.

Malcolm, S.B., Cockrell, B.J. & Brower, L.P. (1993). Spring recolonisation of eastern North America by the Monarch butterfly: successive brood or single sweep migration? In *Biology and Conservation of the Monarch Butterfly*, ed. S.B. Malcolm & M.P. Zalucki, pp. 253–67. Los Angeles: Natural History Museum of Los Angeles County.

Mbata, G.N. & Ramaswamy, S.B. (1990). Rhythmicity of sex pheromone content in female *Heliothis virescens*: impact of mating. *Physiological Entomology*, **15**, 423–32.

McNeil, J.N. (1986). Calling behaviour: can it be used to identify migratory species of moths? *Florida Entomologist*, **69**, 78–84.

McNeil, J.N. (1987). The true armyworm, *Pseudaletia unipuncta*: a victim of the pied piper or a seasonal migrant? In *Recent Advances in Research on Tropical Entomology*, ed. M.F.B. Chaudhury, pp. 591–7. *Insect Science and its Application*, vol. 8, Special Issue. Nairobi: ICIPE Science Press.

McNeil, J.N. & Delisle, J. (1989). Host plant pollen influences calling behavior and ovarian development of the sunflower moth, *Homoeosoma electellum*. *Oecologia*, **80**, 201–5.

Morton, D.B. & Evans, P.D. (1983). Octopamine distribution in solitarious and gregarious forms of the locust, *Schistocerca americana gregaria*. *Insect Biochemistry*, **13**, 177–83.

300 *Physiological integration in Lepidoptera*

Moshitzky, P. & Applebaum, S.W. (1990). The role of adipokinetic hormone in the control of vitellogenesis in locusts. *Insect Biochemistry*, **20**, 319–23.

Oberhauser, K.S. (1989). Effects of spermatophores on male and female monarch butterfly reproductive success. *Behavioral Ecology and Sociobiology*, **25**, 237–46.

Orchard, I. (1987). Adipokinetic hormones: an update. *Journal of Insect Physiology*, **33**, 451–63.

Orchard, I., Cusson, M. & McNeil, J.N. (1991). Adipokinetic hormone of the true armyworm, *Pseudaletia unipuncta*: immunohistochemistry, amino acid analysis, quantification and bioassay. *Physiological Entomology*, **16**, 439–45.

Orchard, I., Ramirez, J.M. & Lange, A.B. (1993). A multifunctional role for octopamine in locust flight. *Annual Review of Entomology*, **38**, 227–49.

Palaniswamy, P. & Seabrook, W.D. (1978). Behavioral responses of the eastern spruce budworm, *Choristoneura fumiferana*, (Lepidoptera: Tortricidae) to the sex pheromone of her own species. *Journal of Chemical Ecology*, **4**, 649–55.

Palaniswamy , P. & Seabrook, W.D. (1985). The alteration of calling behaviour by female *Choristoneura fumiferana* when exposed to synthetic sex pheromone. *Entomologia Experimentalis et Applicata*, **37**, 13–6.

Palaniswamy, P., Seabrook, W.D. & Sivasubramanian, P. (1979a). Effect of a juvenile hormone analogue on olfactory sensitivity of eastern spruce budworm, *Choristoneura fumiferana* (Lepidoptera: Tortricidae). *Entomologia Experimentalis et Applicata*, **26**, 175–9.

Palaniswamy, P., Sivasubramanian, P. & Seabrook, W.D. (1979b). Modulation of sex pheromone perception in female moths of the eastern spruce budworm, *Choristoneura fumiferana*, by altosid. *Journal of Insect Physiology*, **25**, 571–4.

Pener, M.P. (1985). Hormonal effects on flight and migration. In *Comprehensive Insect Physiology, Biochemistry and Pharmacology*, vol. 8, ed. G.A. Kerkut & L.I. Gilbert, pp. 491–550. Oxford: Pergamon Press.

Pener, M.P. (1991). Locust phase polymorphism and its endocrine relations. *Advances in Insect Physiology*, **23**, 1–79.

Pivnick, K.A., Jarvis, B.J., Gillott, C., Slater, G.P. & Underhill, E.W. (1990). Daily patterns of reproductive activity and the influence of adult density and exposure to host plants on reproduction in the diamondback moth (Lepidoptera: Plutellidae). *Environmental Entomology*, **19**, 587–93.

Poitout, S., Cayrol, R., Causse, R. & Anglade, P. (1974). Déroulement du programme d'études sur la migration des lépidoptères Noctuidae réalisé en montagne et principaux résultats acquis. *Annales de Zoologie-Écologie Animale*, **6**, 585–7. In French, English summary.

Pratt, G.E., Farnsworth, D.E., Fok, K.F., Siegel, N.R., McCormack, A.L., Shabanowitz, J., Hunt, D.F. & Feyereisen, R. (1991). Identity of a second type of allatostatin from cockroach brains: an octadecapeptide amide with a tyrosine-rich address sequence. *Proceedings of the National Academy of Sciences, USA*, **88**, 2412–16.

Rankin, M.A. (1989). Hormonal control of flight. In *Insect Flight*, ed. G.J. Goldsworthy & C.H. Wheeler, pp. 139–63. Boca Raton, Florida: CRC Press.

Rankin, M.A. (1991). Endocrine effects on migration. *American Zoologist*, **31**, 217–30.

Rankin, M.A. & Burchsted, J.C.A. (1992). The cost of migration in insects. *Annual Review of Entomology*, **37**, 533–59.

Rankin, M.A., McAnelly, M.L. & Bodenhamer, J.E. (1986). The oogenesis-flight syndrome revisited. In *Insect Flight, Dispersal and Migration*, ed. W. Danthanarayana, pp. 27–48. Berlin: Springer-Verlag.

Rankin, M.A. & Riddiford, L.M. (1978). Significance of haemolymph juvenile hormone titre changes in timing of migration and reproduction in adult *Oncopeltus fasciatus*. *Journal of Insect Physiology*, **24**, 31–8.

Riley, J.R., Armes, N.J., Reynolds, D.R. & Smith, A.D. (1992). Nocturnal observations on the emergence and flight behaviour of *Helicoverpa armigera* (Lepidoptera: Noctuidae) in the post-rainy season in central India. *Bulletin of Entomological Research*, **82**, 243–56.

Royer, L. & McNeil, J.N. (1993). Male investment in the European corn borer, *Ostrinia nubilalis* (Hübner) (Lepidoptera: Pyralidae): impact on female longevity and reproductive performance. *Functional Ecology*, **7**, 209–15.

Rutowski, R.L., Gilchrist, G.W. & Terkanian, B. (1987). Female butterflies mated with recently mated males show reduced reproductive output. *Behavioral Ecology and Sociobiology*, **20**, 319–22.

Sanders, C.J. (1987). Flight and copulation of female spruce budworm in pheromone-permeated air. *Journal of Chemical Ecology*, **13**, 1749–58.

Sappington, T.W. & Showers, W.B. (1992). Reproductive maturity, mating status, and long-duration flight behavior of *Agrotis ipsilon* (Lepidoptera: Noctuidae) and the conceptual misuse of the oogenesis-flight syndrome by entomologists. *Environmental Entomology*, **21**, 677–88.

Satyanarayana, K., Yu, J.H., Bhaskaran, G., Dahm, K.H. & Meola, R. (1991). Hormonal control of egg maturation in the corn earworm, *Heliothis zea*. *Entomologia Experimentalis et Applicata*, **59**, 135–43.

Shirk, P.D., Bhaskaran, G. & Roller, H. (1980). The transfer of juvenile hormone from male to female during mating in the cecropia silkmoth. *Experientia*, **36**, 682–3.

Siegert, K.J. & Mordue, W. (1986). Quantification of adipokinetic hormones I and II in the corpora cardiaca of *Schistocerca gregaria* and *Locusta migratoria*. *Comparative Biochemistry and Physiology A*, **84**, 279–84.

Solbreck, C. (1978). Migration, diapause, and direct development as alternative life histories in a seed bug, *Neacoryphus bicrucis*. In *Evolution of Insect Migration and Diapause*, ed. H. Dingle, pp. 196–217. Berlin: Springer-Verlag.

Southwood, T.R.E. (1977). Habitat – the templet for ecological strategies? *Journal of Animal Ecology*, **46**, 337–65.

Surholt, B., Van Doorn, J.M., Goldberg, J. & Van Der Horst, D.J. (1992). Compositional analysis of high- and low-density lipophorin of *Acherontia atropos* and *Locusta migratoria*. *Biological Chemistry*, **373**, 13–20.

Svärd, L. (1985). Paternal investment in a monandrous butterfly, *Pararge aegeria*. *Oikos*, **45**, 66–70.

Swier, S.R., Rings, R.W. & Musick, G.J. (1977). Age-related calling behavior of the black cutworm, *Agrotis ipsilon*. *Annals of the Entomological Society of America*, **70**, 919–24.

Taylor, R.A.J. & Reling, D. (1986). Preferred wind direction of long-distance leafhopper (*Empoasca fabae*) migrants and its relevance to the return migration of small insects. *Journal of Animal Ecology*, **55**, 1103–14.

Thompson, C.S., Yagi, K.J., Chen, Z.F. & Tobe, S.S. (1990). The effects of octopamine on juvenile hormone biosynthesis, electrophysiology, and cAMP content of the corpora allata of the cockroach *Diploptera punctata*. *Journal of Comparative Physiology B*, **160**, 241–9.

Turgeon, J.J. & McNeil, J.N. (1982). Calling behaviour of the armyworm *Pseudaletia unipuncta*. *Entomologia Experimentalis et Applicata*, **31**, 402–8.

Turgeon, J.J. & McNeil, J.N. (1983). Modifications of the calling behaviour of *Pseudaletia unipuncta* (Lepidoptera: Noctuidae), induced by temperature conditions during pupal and adult development. *Canadian Entomologist*, **115**, 1015–22.

Turgeon, J.J., McNeil, J.N. & Roelofs, W.L. (1983). Responsiveness of *Pseudaletia unipuncta* males to the female sex pheromone. *Physiological Entomology*, **8**, 339–44.

Wheeler, C.H. (1989). Mobilization and transport of fuels to the flight muscles. In *Insect Flight*, ed. G.J. Goldsworthy & C.H. Wheeler, pp. 273–303. Boca Raton, Florida: CRC Press.

Wilson, K. & Gatehouse, A.G. (1992). Migration and genetics of pre-reproductive period in the moth, *Spodoptera exempta* (African armyworm). *Heredity*, **69**, 255–62.

Woodhead, A.P., Stay, B., Seidel, S.L., Khan, M.A. & Tobe, S.S. (1989). Primary structure of four allatostatins: neuropeptide inhibitors of juvenile hormone synthesis. *Proceedings of the National Academy of Sciences, USA*, **86**, 5997–6001.

Woodring, J.P. & Sparks, T.C. (1987). Juvenile hormone esterase activity in the plasma and body tissue during the larval and adult stages of the house cricket. *Insect Biochemistry*, **17**, 751–8.

Young, A.D.M. & Downe, A.E.R. (1987). Male accessory gland substances and the control of sexual receptivity in female *Culex tarsalis*. *Physiological Entomology*, **12**, 233–9.

Zera, A.J., Strambi, C., Tiebel, K.C., Strambi, A. & Rankin, M.A. (1989). Juvenile hormone and ecdysteroid titers during critical periods of wing morph determination in *Gryllus rubens*. *Journal of Insect Physiology*, **35**, 501–11.

Zera, A.J. & Tiebel, K.C. (1989). Differences in juvenile hormone esterase activity between presumptive macropterous and brachypterous *Gryllus rubens*: implications for the hormonal control of wing polymorphism. *Journal of Insect Physiology*, **35**, 7–18.

Ziegler, R. & Schulz, M. (1986). Regulation of lipid metabolism during flight in *Manduca sexta*. *Journal of Insect Physiology*, **32**, 997–1001.

# 14

# Aerodynamics, energetics and reproductive constraints of migratory flight in insects

## R. DUDLEY

### Introduction

Although flight energetics figure prominently in the ecology of insect migration, direct measurements of metabolism are not available for any migratory insect in free flight. Quantitative analysis of migratory energetics must therefore be approached indirectly. However, there are few data on the aerodynamic and biomechanical processes of migratory flight; even such basic kinematic parameters as airspeeds and wingbeat frequencies are generally unknown. Aerodynamic studies of migratory flight are of particular use in estimating the mechanical and metabolic power required to fly and these biomechanical analyses can elucidate the implications of morphology and flight behaviour for overall migratory performance. For insects that migrate on ambient winds, aerodynamic evaluations of flapping flight are of less significance, but migration within the flight boundary layer (*sensu* Taylor, 1958) is widespread in insects and an understanding of the biomechanics of wing flapping is essential for estimating the energetic costs of powered flight. Ultimately, such studies can be integrated with behavioural and environmental data for interpretation of the physiological ecology of migration.

Pennycuick (1969) presented the first detailed analysis of migration mechanics in birds (see also Pennycuick, 1978) and this treatment has recently been extended by Rayner (1990) to incorporate recent findings on the structure of vortex wakes for flying vertebrates. In insects, aerodynamic analyses of migratory flight are less well developed, primarily because the requisite kinematic and morphological data have been lacking. Fortunately, biomechanical data for flying insects are now becoming more available, allowing quantitative analysis of aerodynamic performance and the power requirements of forward flight. This chapter first reviews current approaches to determining the energetic costs of flight in insects. A newly developed experimental method for measuring airspeeds, wing kinematics and thermal physiology of migrating insects is then presented in the context of current studies of neotropical butterfly migrations. Finally, biomechanical analyses of insect migration are extended into more general issues of reproductive constraints and the 'oogenesis-flight syndrome' (Johnson, 1969), the influence of the flight boundary layer in relation to insect size, and strategies of energy utilisation during migratory flight.

## Power requirements of flight

Overall metabolic expenditure during flight is directly proportional to the mechanical work performed by the flight muscles on the surrounding air. Theoretical analysis of these mechanical costs has involved the generation of power curves that relate the rate of energetic expenditure to forward airspeed. Predicted mechanical power requirements can then be compared with experimental measurements of flight metabolism over a range of airspeeds; the two quantities are related by the flight muscle efficiency, which is not necessarily independent of airspeed. The classic aerodynamic model is that of Pennycuick (1975), predicting a 'U-shaped' power curve with high values of energetic expenditure in hovering flight, decreasing to a minimum at intermediate flight speeds, and then increasing dramatically at higher airspeeds. For studies of migratory energetics, knowledge of the power curve can be used to determine the maximum-range speed (i.e. minimum energy expenditure per unit distance, maximising the distance flown) and the minimum-power speed (i.e. minimum energy expenditure per unit time, maximising the time spent aloft).

For insects flying above their flight boundary layer, minimum-power speed is clearly optimal during migration, whereas, for those flying within it, maximum-range speed is appropriate. Maximum-range speed is determined by taking the tangent from the origin to the power curve (Pennycuick, 1975), while the minimum-power speed is self-explanatory. Selection of optimal flight speeds might well be expected for insect migrants because levels of flight metabolism are so high, typically several orders of magnitude greater than those of standard metabolism (Kammer & Heinrich, 1978).

For migratory insects, comparison of theoretically predicted mechanical power curves with their metabolic counterparts has unfortunately been precluded by the absence of energetic data for free forward flight.

However, Ellington, Machin & Casey (1990) have recently made the first measurements of metabolic expenditure as a function of airspeed for any free-flying insect. In bumblebees, metabolic power requirements did not vary over the range from hovering to 4 m s⁻¹. Reviewing the existing metabolic data for bats and birds, as well as the results for bumblebees, Ellington (1991) argued that, in general, the power curve for both vertebrate and insect fliers is better described as 'J-shaped', with power requirements at hovering and intermediate airspeeds being similar but increasing significantly at higher speeds.

Furthermore, Walsberg (1990) discussed certain discrepancies between existing estimates of mechanical power output in flying pigeons and actual metabolic measurements. Experimental errors arise from artifactual problems with wind-tunnel measurements of flight energetics and there is often high variance between the different aerodynamic models used to estimate mechanical power. Moreover, when these estimates are related to metabolic measurements, unreasonably high flight muscle efficiencies are obtained. Methodological problems may therefore adversely influence experimental measurements of flight metabolism, but certain underlying assumptions in estimates of mechanical power (particularly relating to drag forces on the wings and the associated power requirements) may also be suspect. At least in flying vertebrates, direct estimates of metabolic rates from mechanical power requirements must be viewed cautiously.

The mechanical power requirements of insects in forward flight have recently been analysed with the use of quasi-steady aerodynamic theory and detailed kinematic data (Dudley & Ellington, 1990*a,b*; Dudley, 1992). Total mechanical power required at a given airspeed can be determined by examining four individual components: parasite, induced, profile and inertial power. Parasite power represents the power required to overcome drag forces on the body, while induced power is the power required to accelerate air downwards to support body weight. Profile power is the component necessary to overcome drag forces on the wings and inertial power is required to accelerate, in the first half of a half-stroke, the mass of the wings and the wing virtual mass (i.e. the mass of air accelerated along with the wing). Values of these individual power components for a free-flying bumblebee at different airspeeds are shown in Fig. 14.1. If the kinetic energy of the oscillating wing mass and wing virtual mass can be stored as elastic strain energy, then inertial power requirements as averaged over a half-stroke will equal zero. The total mechanical power requirement will then simply be the aerodynamic

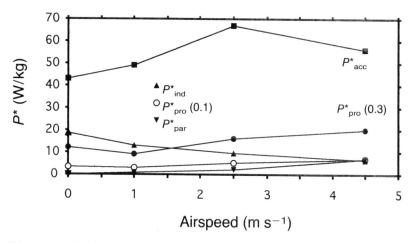

Fig. 14.1. Individual components of the mechanical power curve for a 175-mg bumblebee worker (*Bombus terrestris*) at different airspeeds (modified from Dudley & Ellington, 1990*b*). ■: $P^*_{acc}$, inertial power during the first half of a half-stroke; ●: $P^*_{pro}$ (0.3), profile power assuming a mean profile drag coefficient of 0.3; ○: $P^*_{pro}$ (0.1), profile power given a mean profile drag coefficient of 0.1; ▲: $P^*_{ind}$, induced power; ▼: $P^*_{par}$, parasite power.

power – the sum of the parasite, induced and profile powers. If, on the other hand, there is no elastic energy storage, the additional kinetic energy associated with the reciprocating wings must be supplied for each half-stroke. The energy required to decelerate the wings at the end of a half-stroke is probably negligible (by virtue of energy dissipation in an end stop; Ellington, 1984*b*) and no additional power is necessary in the second half of a half-stroke. Total power requirements for flying insects are therefore typically given as the lower and upper bounds of perfect and zero elastic energy storage, respectively. Unfortunately, the actual extent of elastic energy storage during flight is unknown for any insect.

Values of these components of mechanical power are determined from morphological and kinematic data obtained for individual insects at different airspeeds (Dudley & Ellington, 1990*a,b*) and total power requirements for the two cases of zero and perfect elastic energy storage are then calculated. The resulting power curve (Fig. 14.2) relates the mechanical costs of flight to forward airspeed. For bumblebees, mechanical power requirements show little variation over the range of hovering to 4.5 m s⁻¹, parallelling the relatively constant metabolic power input determined by Ellington *et al.* (1990) over the same speed range. The metabolic rate during flight can be estimated from the mechanical power

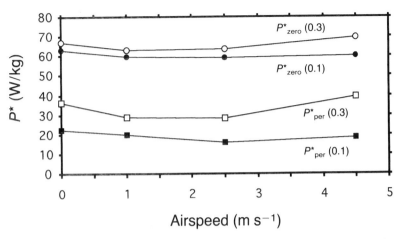

Fig. 14.2. Total specific mechanical power requirements for the bumblebee worker (*Bombus terrestris*) of Fig. 14.1 in free flight at different airspeeds (modified from Dudley & Ellington, 1990*b*). ○: $P^*_{zero}$ (0.3), total power assuming zero elastic energy storage and a mean profile drag coefficient of 0.3; ●: $P^*_{zero}$ (0.1), total power assuming zero elastic energy storage and a mean profile drag coefficient of 0.1; □: $P^*_{per}$ (0.3), total power assuming perfect elastic energy storage and a mean profile drag coefficient of 0.3; ■: $P^*_{per}$ (0.1), total power assuming perfect elastic energy storage and a mean profile drag coefficient of 0.1. Power requirements are relatively constant from hovering to 4–5 m s$^{-1}$.

requirements by dividing total mechanical power by flight muscle efficiency and assuming a standard conversion factor of 20 J ml $O_2^{-1}$. Because of the very high metabolic rates attained during insect flight metabolism (Kammer & Heinrich, 1978), basal metabolic activity, as well as possibly heightened costs of respiration and circulation during flight, are ignored in these energetic estimates. Actual flight muscle efficiencies in free flight are unknown but a value of 10% is in general agreement with experimental results. Mizisin & Josephson (1987) determined a muscle efficiency of 7.7–10.5% for locust muscle in a semi-isolated preparation, while Stevenson & Josephson (1990) estimated a value of 10% for the moth *Manduca sexta* in hovering flight. Neither of these results were dependent upon assumptions concerning elastic energy storage, and similar efficiencies characterise asynchronous flight muscle of insects in both hovering (Ellington, 1984*b*) and forward flight (Casey & Ellington, 1989).

Mechanical power curves have not yet been constructed for a diversity of migratory insects but a general hypothesis about their shape can be proposed. Compared with bumblebees over the same speed range, the mechanical power curve for a migrating diurnal moth (*Urania fulgens*; Dudley & DeVries, 1990) exhibits a much steeper rise with airspeed (Fig. 14.3). Although the curve for *U. fulgens* is composed of data from different insects, not one individual flying at different airspeeds, intra-specific morphological and kinematic variation is small relative to the overall effect of changes in airspeed on the power curve. In bumblebees, the flapping velocity of the wings is high relative to the forward velocity over the airspeed range considered. The ratio of forward velocity to mean flapping velocity is termed the advance ratio (Ellington, 1984*a*); bumblebees thus have advance ratios substantially lower than those of *U. fulgens*. Because bumblebee flapping velocities are higher, changes in forward airspeed have little effect on the relative air velocity experienced by wing sections. The mechanical work performed by the wing is thus less sensitive to variation in forward airspeed and total power requirements

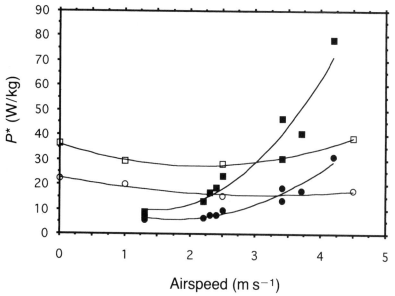

Fig. 14.3. Mechanical power requirements of flight for the bumblebee worker (*Bombus terrestris*) of Figs. 14.1 and 14.2 and for the moth *Urania fulgens*. Data are modified from Dudley & Ellington (1990*b*) and Dudley & DeVries (1990); quadratic equations are fitted to the data to indicate general trends. ■: *U. fulgens*, mean profile drag coefficient of 0.3; ●: *U. fulgens*, mean profile drag coefficient of 0.1; □: *B. terrestris*, mean profile drag coefficient of 0.3; ○: *B. terrestris*, mean profile drag coefficient of 0.1. Perfect elastic energy storage is assumed, but the qualitative trend is unchanged if there is no elastic energy storage. Mechanical power requirements increase much more rapidly with airspeed in *U. fulgens* than in *B. terrestris*.

are little changed. The much higher advance ratios of *U. fulgens* are attributable, principally, to the much lower wingbeat frequencies associated with synchronous flight muscle (Dudley & DeVries, 1990). An increase in the forward airspeed will therefore increase dramatically the relative air velocities experienced by wing sections. In particular, profile power is proportional to the cube of the relative velocity and will increase sharply with forward airspeed, contributing to a steep overall rise in total power requirements.

For migrating insects, this effect of wingbeat frequency on the shape of the power curve should become apparent in comparisons of insect taxa with synchronous and asynchronous flight muscle. For those insects with high wingbeat frequencies and thus a reduced dependency of power requirements upon forward airspeed, choice of airspeed will be less critical for minimisation of energetic expenditure during flight. A greater variance in airspeeds would therefore be expected for asynchronous fliers during migration. This possibility could be evaluated quantitatively by examining intraspecific and individual variance in airspeeds for migrating synchronous and asynchronous fliers. To date, however, reliable airspeed data are not available for any migrating insect with asynchronous flight muscle.

To summarise, biomechanical models of the costs of flapping flight are becoming increasingly sophisticated but the relevance of generalised models to insect migration is unclear. The great diversity in morphology, wing kinematics and airspeeds of flying insects suggests a corresponding diversity in the relative magnitude of the components of the power curve. It is therefore doubtful that a single generalised model for the mechanical power requirements of flight will be appropriate for all insect groups. Wing kinematics may also vary with airspeed (review by Dudley & Ellington, 1990*a*), further confounding simplified analyses of mechanical power requirements. Because the kinematic and morphological data required to implement aerodynamic analyses are unavailable for the vast majority of insect migrants, power curves and associated predictions concerning optimal flight behaviour during migration are of limited utility. To obtain more specific predictions, it is necessary to move from theoretical aerodynamics and laboratory investigation to studies of the performance of migrants in the field.

### Airspeeds and free-flight physiology of migratory insects

Very little is known of the actual airspeeds of migrating insects. Detailed kinematic data are available for migratory locusts in free flight (Baker & Cooter, 1979; Baker, Gewecke & Cooter, 1981) but it was unfortunately not possible to capture the insects filmed in flight in these studies for morphological analysis, and only groundspeeds (as distinct from airspeeds) could be measured. Free-flight data for other insect migrants are similarly limited (see Johnson, 1969), relying on pooled species means to describe migrant performance so that the often wide range of intraspecific variation, particularly due to sexual dimorphism, is necessarily ignored (see e.g. Gibo & Pallett, 1979). Groundspeed measurements must be corrected by the ambient wind vector, although spatial and temporal variation may compromise use of an average wind velocity. Because of the difficulty of following and capturing free-flying insects, flight performance of individual insect migrants under natural conditions is virtually unstudied.

A newly developed experimental method permits direct airspeed measurements on migrating insects flying over a body of water and allows for subsequent capture of individual insects. The investigator follows an insect in a small motorboat so that trajectories of the insect and boat are nominally parallel. A unidirectional anemometer is held laterally from the prow of the boat so that the probe is in the immediate vicinity and at the same height as the flying insect (see Dudley, 1992, for photograph). Airspeed is thus measured directly and no assumptions concerning the ambient wind vector are necessary. Repeated measurements of airspeed are possible if the insect can be followed for an extended period of time. The same insect can then be captured with a net to obtain all relevant morphological data, and thoracic temperature can be measured immediately following capture. Wing kinematics can also be recorded with a cine or video camera concurrently with airspeed measurements.

DeVries & Dudley (1990) and Dudley & DeVries (1990) used this free-flight method to study the annual migrations of the diurnal moth *Urania fulgens* (Uraniidae) from Central into South America. Moths were followed while they crossed Lake Gatún in the Republic of Panama. Airspeeds, wing kinematics, wing and body morphology, thermal data and lipid reserves were assessed for a set of individual moths, and quasi-steady aerodynamics and the mechanical power requirements of forward flight were analysed with the use of the biomechanical and morphological data. The maximum range of migratory flight in *U. fulgens* using stored energy reserves was predicted from estimates of energetic expenditure during forward flight and the mass of stored lipids in individual insects. A comparison of endogenous lipid reserves with estimated metabolic rates during flight suggested that individual moths can fly only 110 km or less using stored energy reserves. Long-distance migration of the order of several thousand kilometres is likely to occur in *U. fulgens*, so extensive nectar feeding *en route* is essential to obtain the energy for these flights (DeVries & Dudley, 1990).

R.B. Srygley and I have recently extended these techniques to investigate the physiology of migratory butterflies flying over Lake Gatún. To date, we have collected data for three pierid species belonging to the genera *Aphrissa* and *Phoebis* and for three nymphaline nymphalids in the genera *Marpesia* and *Historis*. Wing loading (the ratio of body weight to wing area) and flight speed were similar for the genera *Aphrissa*, *Marpesia* and *Phoebis*, with little difference in airspeed between the sexes (R.B. Srygley & R. Dudley, unpublished data). The butterfly *Historis acheronta*, with a much higher wing loading, flew at significantly higher airspeeds. In addition to obtaining airspeed data, we have also made extensive biophysical measurements in order to estimate total heat production via net energy balance. Ultimately such information will be combined with aerodynamic estimates of mechanical power output to derive the flight muscle efficiency.

The principal advantage of this approach is that the physiology of free forward flight can be evaluated under natural conditions and that complete sets of biomechanical data can be obtained for individual insects. Large sample sizes, accurate kinematic measurements and links with flight metabolism are all possible. By obtaining kinematic, thermoregulatory and morphological data for individual insects, both intra- and interspecific variation in flight physiology and aerodynamics can be investigated. For migratory insects in particular, these studies will reveal patterns of physiological variation in relation to current microclimatic and behavioural data, information hitherto unavailable for individual migrating insects.

## Migration and reproductive constraints

In many insect migrants, an oogenesis-flight syndrome has been described whereby migration typically occurs pre-reproductively (Johnson, 1969; Chapter 10, this volume). The flight apparatus of the migrating imago is necessarily well developed while, in females, the ovaries and other elements of the reproductive system remain immature. The ecological basis for the predominance of pre-reproductive migration in female insects has been extensively reviewed (see e.g. Johnson, 1969; Dingle, 1972; Southwood, 1977; Dingle, 1985; Chapters 10 and 11, this volume). Flight itself and oogenesis immediately following migration appear, at least in some species, to be stimulated by the same endocrinological pathway (Rankin & Burchsted, 1992; Chapter 13, this volume).

However, biomechanical and physiological constraints of reproductive development in migrating insects have been less well investigated. For example, development and maintenance of the morphological features required for flight will impose energetic demands, as yet unquantified (Roff & Fairburn, 1991; Rankin & Burchsted, 1992). Flight metabolism and reproductive development, in particular egg production, may compete directly for biochemical substrate as lipid is the principal metabolic fuel of insects in long-distance flights. However, Rankin & Burchsted (1992) noted that egg constituents are not exclusively lipid and that the proportion of carbohydrate to lipid in flight metabolism may be higher in female insects, relative to males. This observation raises the possibility of facultative modulation in the relative lipid requirements of flight and reproduction according to substrate availability.

An important biomechanical consequence of lipid loading and the consequent increase in mass prior to migratory flight is that the mechanical power required to fly will increase disproportionately. Because wing dimensions do not change with either increased lipid reserves or egg loads, the mechanical power (and by implication metabolic power) required to fly will increase approximately as $mass^{1.5}$ (Pennycuick, 1975) or as $mass^{1.56}$ (Norberg & Rayner, 1987). Dudley & Vermeij (1992) recently proposed that such non-linear increases in power requirements have been a general constraint on the evolution of folivory and related dietary mass loading in flying animals. Moreover, potential increases in abdominal size through lipid loading or oogenesis will increase the cross-sectional area of the insect, thereby increasing the body drag forces and, correspondingly, the parasite and total power (Pennycuick, 1975). An increase in body mass through reproductive development or addition of lipid reserves will thus have a disproportionately negative effect on flight performance. Because the energetic costs of flight are so high, the development and maintenance in migratory phenotypes of structures not essential for flight should be minimised and oogenesis should be postponed until flight activity has terminated.

Flight with extensive lipid reserves or egg loads may also impair locomotor performance and manoeuvrability of insect migrants. Although relevant data are not available for flying insects, examples from the vertebrate literature demonstrate this effect. Videler *et al.* (1988) showed that flight speeds of kestrels carrying added weight were lower, while Pennycuick, Fuller & McAllister (1989) found that climbing rates of Harris' Hawks tended to decrease with added loads. Flight speeds of Long-eared Bats also decreased following addition of artificial loads (Hughes & Rayner, 1991). Mass loading will also reduce the capacity for acceleration and manoeuvrability, possibly increasing vulnerability of insect migrants to aerial predation. Many flying insects respond to attempted capture and bat echolocation with evasive manoeuvres, although the impact of predation on migrating insects is unknown.

Pre-migratory lipid loading will obviously increase the maximum range attainable during migration but the magnitude of this effect will decrease exponentially with increasing mass (Rayner, 1990). As lipid reserves are gradually utilised through the course of a migration, the decrease in mass-specific power required to fly should result in a reduction of the maximum-range speed, in proportion to $mass^{0.5}$ (Pennycuick, 1978). A similar reduction in flight speed might be expected to characterise migrants maximising time spent aloft (minimising energetic expenditure per unit time rather than per unit distance). Detailed surveys of variation in airspeeds through migration would be required to test these predictions and, at present, no such data are available.

The relevance of theoretical models for airspeed selection is, however, seriously compromised if energy reserves can be replenished during migration (Walsberg, 1990). Such activity involves costs in time, expenditure of energy in searching behaviour and possibly heightened exposure to predators. However, replenishment *en route* decouples migratory behaviour from the exigencies of a fixed energy reserve. Sources of nutrients are often available during migration, particularly for migrants using continuous flapping flight within the flight boundary layer (see below) and, in some cases, the magnitude of energy uptake during migration can be greater than pre-migratory lipid loading (Brower, 1985). Knowledge of the energy intake during migration is therefore essential for evaluating overall patterns of migratory energetics. Ecological costs (e.g. exposure to predators) of feeding during migration may also differ from those of pre-migratory lipid loading.

Post-migratory feeding is another important feature of the suite of co-evolved adaptations for migration but its extent, relative to rates of energy utilisation and uptake while migration is in progress, is unknown.

A complete description of migratory energetics must therefore include not only energetic expenditure during flight but also energy intake prior to, during and after migration. Few data relating to this issue are available for insects, but measurement of feeding rates of migrating diurnal Lepidoptera might be possible under field conditions.

### Migration and the flight boundary layer

One of the central concepts in insect migration is that of the flight boundary layer (Taylor, 1958), the height in the atmosphere below which insect airspeed exceeds that of the ambient airflow. Within the flight boundary layer, continuous flapping flight can maintain directionality, whereas above it, displacement is predominantly determined by the prevailing wind. Drake & Farrow (1988) noted that flight within the flight boundary layer is most advantageous for movement in fixed directions and that some insects may be restricted to this zone when ambient winds are in an unfavorable direction. Flight above the flight boundary layer, in contrast, exposes insects to the vagaries of the prevailing wind, although directed flight may to some extent be possible through controlled gliding and soaring (see e.g. Gibo & Pallett, 1979; Gibo, 1986). Downwind flight does not unequivocally confirm that an insect is flying above the flight boundary layer, but only that ambient airflow is contributing to some extent to its displacement. The magnitude of this contribution can only be assessed when accurate flight speed data are available. Because wind-speeds vary continuously in space and time and average and maximum flight speeds during migration are in general unknown, the dimensions of flight boundary layers are dynamic quantities that cannot be specified precisely.

The concept of the flight boundary layer makes clear the overwhelming influence of insect size on strategies of migratory flight. Average insect body lengths are of the order of 4–5 mm (May, 1978) and insects less than 10 mm in length typically fly at airspeeds below 1 m s$^{-1}$ (Lewis & Taylor, 1967; Johnson, 1969). Such speeds are well below average values of daytime ambient winds, even in, for example, tropical forests (see Haddow & Corbet, 1961; Allen, Lemon & Muller, 1972; Thompson & Pinker, 1975; Aoki, Yabuki & Koyama, 1978). For the large majority of insects, therefore, ambient winds will dominate flight trajectories and the best possibility for effective dispersal is movement downwind above the flight boundary layer. Diurnal migration in convective conditions is widespread amongst smaller insects, many of which utilise updrafts to gain height and even some larger insects (e.g. locusts, dragonflies) gain altitude in this way (Johnson, 1969). In contrast, butterfly migrants generally engage in continuous flapping flight within the flight boundary layer (Williams, 1930; Baker, 1978; Walker, 1985). Migrating butterflies can often be aided by the wind (see e.g. Mikkola, 1986) although Walker & Riordan (1981) concluded that neither local nor synoptic winds determined flight directionality in four migrating butterfly species. With the exception of the monarch butterfly (Gibo & Pallett, 1979), extensive soaring and gliding behaviour have rarely been observed in migrating butterflies.

Diurnal insect migrants appear, therefore, to follow two general strategies correlated with body size: powered flight by large insects within the flight boundary layer, or convection-aided ascent and dispersal of small insects on the wind above the flight boundary layer. There are clearly exceptions to these trends and a systematic quantitative survey of insect migratory behaviour in relation to body size and flight speed has not yet been undertaken. Flight during the day exposes the insect migrant to avian insectivores, as well as the possibility of significant thermal stress, particularly in the tropics (DeVries & Dudley, 1990). Flying at night reduces the problem of heat overload, an effect also obtained by flying at higher altitudes, but may substitute chiropteran predators for their diurnal avian counterparts. Because of a more stable planetary boundary layer and the occurrence of temperature inversions (Drake & Farrow, 1988), the depth of the flight boundary layer at night may be greater than during the day. However, night-time winds can still be strong relative to the flight speeds of small insects. Taylor (1974) found that fewer small nocturnal insect taxa, compared with diurnally active insects, escape the flight boundary layer into winds at higher altitudes. Similarly, Drake & Farrow (1988) suggested that nocturnal migration is most common among larger insects (see also Drake, 1984).

Self-powered directional flight, entirely or partially independent of wind direction, is most likely to occur, day or night, in larger insects capable of flying at greater airspeeds and there can be little chance of migration independent of ambient wind direction for the vast majority of insect species. In the evolution of migratory behaviour, the magnitude of insect flight speeds relative to windspeeds must, therefore, have played a major role.

## Conclusions

Given that insect migration is a phenomenon principally effected by flight, the absence of data on flight characteristics of insect migrants in free flight is striking. Even for migratory Orthoptera of economic significance, data on kinematics and flight performance of individual insects are sparse. Our understanding of migratory phenomena in the tropics is particularly limited, even though tropical insect migrants exhibit an extraordinary taxonomic and morphological diversity. For example, spectacular migrations of several hundred species of butterflies and diurnal moths have been described at the montane Venezuelan station Rancho Grande (Beebe, 1949), and numerous Coleoptera, Diptera and Hymenoptera were found migrating through the same site (Beebe, 1951).

Butterfly migrations are particularly amenable to studies of flight performance and energetics and of the physiological ecology of migration. They occur diurnally at heights accessible to observers so that methods can be developed for studying free-flight physiology (see above). Moreover, butterflies are generally large and conspicuous, so that species can be readily identified and censused in the field. While the implications of aerodynamics and flight energetics for insect migration are understood in general terms, specific case studies are scarce and data on spatial and temporal variation in free-flight performance through a migration are nonexistent. For migrating butterflies, detailed information on body morphology, lipid reserves and flight speeds can be obtained through the course of a migratory period, allowing the following (and other) questions to be addressed. Are larger and faster butterfly species more active and persistent in the face of adverse winds? Do migrants fly at the optimal airspeeds predicted by biomechanical models? Does airspeed decrease proportionally with relative energy reserves, possibly indicating that the migrants that have flown greater distances have adjusted flight speeds correspondingly?

Williams (1930) reviewed historical records of migration in more than 150 butterfly species, principally in tropical regions. It is unfortunate that, 60 years after his compilation, the only information relating to these migrations is the fact of their occurrence. Given current rates of habitat destruction in the tropics, it seems probable that few of them will survive another 60 years into the future.

# References

Allen, L.H., Lemon, E. & Muller, L. (1972). Environment of a Costa Rican forest. *Ecology*, **53**, 102–11.

Aoki, M., Yabuki, K. & Koyama, H. (1978). Micrometeorology of Pasoh Forest. *Malayasia Nature Journal*, **30**, 149–59.

Baker, P.S. & Cooter, R.J. (1979). The natural flight of the migratory locust, *Locusta migratoria* L. I. Wing movements. *Journal of Comparative Physiology*, **131**, 79–87.

Baker, P.S., Gewecke, M. & Cooter, R.J. (1981). The natural flight of the migratory locust, *Locusta migratoria* L. III. Wing-beat frequency, flight speed and attitude. *Journal of Comparative Physiology A*, **141**, 233–7.

Baker, R.R. (1978). *The Evolutionary Ecology of Animal Migration*. New York: Holmes & Meier.

Beebe, W. (1949). Insect migration at Rancho Grande in north-central Venezuela. General account. *Zoologica*, **34**, 107–10.

Beebe, W. (1951). Migration of insects (other than Lepidoptera) through Portachuelo Pass, Rancho Grande, north-central Venezuela. *Zoologica*, **36**, 255–66.

Brower, L.P. (1985). New perspectives on the migration biology of the monarch butterfly, *Danaus plexippus* L. In *Migration: Mechanisms and Adaptive Significance*, ed. M.A. Rankin, pp. 748–85. *Contributions in Marine Science*, vol. 27, Suppl. Port Aransas, Texas: Marine Science Institute, The University of Texas at Austin.

Casey, T.M. & Ellington, C.P. (1989). Energetics of insect flight. In *Energy Transformation in Cells and Organisms*, ed. W. Wieser & E. Gnaiger, pp. 200–10. Stuttgart: Georg Thieme Verlag.

DeVries, P.J. & Dudley, R. (1990). Morphometrics, airspeed, thermo-regulation and lipid reserves of migrating *Urania fulgens* (Uraniidae) moths in natural free flight. *Physiological Zoology*, **63**, 235–51.

Dingle, H. (1972). Migration strategies of insects. *Science*, **175**, 1327–35.

Dingle, H. (1985). Migration. In *Comprehensive Insect Physiology, Biochemistry, and Pharmacology*, vol. 9, *Behaviour*, ed. G.A. Kerkut & L.I. Gilbert, pp. 375–415. Oxford: Pergamon Press.

Drake, V.A. (1984). The vertical distribution of macro-insects migrating in the nocturnal boundary layer. *Boundary-Layer Meteorology*, **28**, 353–74.

Drake, V.A. & Farrow, R.A. (1988). The influence of atmospheric structure and motions on insect migration. *Annual Review of Entomology*, **33**, 183–210.

Dudley, R. (1992). Aerodynamics of flight. In *Biomechanics (Structures & Systems): A Practical Approach*, ed. A.A. Biewener, pp. 97–121. Oxford: Oxford University Press.

Dudley, R. & DeVries, P.J. (1990). Flight physiology of migrating *Urania fulgens* (Uraniidae) moths: kinematics and aerodynamics of natural free flight. *Journal of Comparative Physiology A*, **167**, 145–54.

Dudley, R. & Ellington, C.P. (1990a). Mechanics of forward flight in bumblebees. I. Kinematics and morphology. *Journal of Experimental Biology*, **148**, 19–52.

Dudley, R. & Ellington, C.P. (1990b). Mechanics of forward flight in bumblebees. II. Quasi-steady lift and power requirements. *Journal of Experimental Biology*, **148**, 53–88.

Dudley, R. & Vermeij, G.J. (1992). Do the power requirements of flapping flight constrain folivory in flying animals? *Functional Ecology*, **6**, 101–4.

Ellington, C.P. (1984a). The aerodynamics of hovering insect flight. III. Kinematics. *Philosophical Transactions of the Royal Society of London B*, **305**, 41–78.

Ellington, C.P. (1984b). The aerodynamics of hovering insect flight. VI. Lift and power requirements. *Philosophical Transactions of the Royal Society of London B*, **305**, 145–81.

Ellington, C.P. (1991). Limitations on animal flight performance. *Journal of Experimental Biology*, **160**, 71–91.

Ellington, C.P., Machin, K.E. & Casey, T.M. (1990). Oxygen consumption of bumblebees in forward flight. *Nature*, **347**, 472–3.

Gibo, D.L. (1986). Flight strategies of migrating Monarch butterflies (*Danaus plexippus* L.) in southern Ontario. In *Insect Flight: Dispersal and Migration*, ed. W. Danthanarayana, pp. 172–84. Berlin: Springer-Verlag.

Gibo, D.L. & Pallett, M.J. (1979). Soaring flight of monarch butterflies, *Danaus plexippus* (Lepidoptera: Danaidae), during the late summer migration in southern Ontario. *Canadian Journal of Zoology*, **57**, 1393–401.

Haddow, A.J. & Corbet, P.S. (1961). Entomological studies from a high tower in Mpanga Forest, Uganda. II. Observations on certain environmental factors at different levels. *Transactions of the Royal Entomological Society of London*, **113**, 257–69.

Hughes, P.M. & Rayner, J.M.V. (1991). Addition of artificial loads to long-eared bats *Plecotus auritus*: handicapping flight performance. *Journal of Experimental Biology*, **161**, 285–98.

Johnson, C.G. (1969). *Migration and Dispersal of Insects by Flight*. London: Methuen.

Kammer, A.E. & Heinrich, B. (1978). Insect flight metabolism. *Advances in Insect Physiology*, **13**, 133–228.

Lewis, T. & Taylor, L.R. (1967). *Introduction to Experimental Ecology*. London: Academic Press.

May, R.M. (1978). The dynamics and diversity of insect faunas. In *Diversity of Insect Faunas*, ed. L.A. Mound & N. Waloff, pp. 188–204. Oxford: Blackwell Scientific Publications.

Mikkola, K. (1986). Direction of insect migrations in relation to the wind. In *Insect Flight: Dispersal and Migration*, ed. W. Danthanarayana, pp. 152–71. Berlin: Springer-Verlag.

Mizisin, A.P. & Josephson, R.K. (1987). Mechanical power output of locust flight muscle. *Journal of Comparative Physiology A*, **160**, 413–19.

Norberg, U.M. & Rayner, J.M.V. (1987). Ecological morphology and flight in bats (Mammalia: Chiroptera): wing adaptations, flight performance, foraging strategy and echolocation. *Philosophical Transactions of the Royal Society of London B*, **316**, 335–427.

Pennycuick, C.J. (1969). The mechanics of bird migration. *Ibis*, **111**, 525–56.

Pennycuick, C.J. (1975). Mechanics of flight. In *Avian Biology*, ed. D.S. Farner & J.R. King, pp. 1–75. London: Academic Press.

Pennycuick, C.J. (1978). Fifteen testable predictions about bird flight. *Oikos*, **30**, 165–76.

Pennycuick, C.J., Fuller, M.R. & McAllister, M. (1989). Climbing performance of Harris' hawks (*Parabuteo unicinctus*) with added load:

implications for muscle mechanics and for radiotracking. *Journal of Experimental Biology*, **142**, 17–29.

Rankin, M.A. & Burchsted, J.C.A. (1992). The cost of migration in insects. *Annual Review of Entomology*, **37**, 533–59.

Rayner, J.M.V. (1990). The mechanics of flight and bird migration performance. In *Bird Migration: Physiology and Ecophysiology*, ed. E. Gwinner, pp. 283–99. Berlin: Springer-Verlag.

Roff, D.A. & Fairburn, D.J. (1991). Wing dimorphisms and the evolution of migratory polymorphisms among the Insecta. *American Zoologist*, **31**, 243–51.

Southwood, T.R.E. (1977). Habitat, the templet for ecological strategies? *Journal of Animal Ecology*, **46**, 337–65.

Stevenson, R.D. & Josephson, R.K. (1990). Effects of operating frequency and temperature on mechanical power output from moth flight muscle. *Journal of Experimental Biology*, **149**, 61–78.

Taylor, L.R. (1958). Aphid dispersal and diurnal periodicity. *Proceedings of the Linnean Society of London*, **169**, 67–73.

Taylor, L.R. (1974). Insect migration, flight periodicity, and the boundary layer. *Journal of Animal Ecology*, **43**, 225–38.

Thompson, O.E. & Pinker, R.T. (1975). Wind and temperature profile characteristics in a tropical evergreen forest in Thailand. *Tellus*, **27**, 562–73.

Videler, J.J., Vossebelt, G., Gnodde, M. & Groenewegen, A. (1988). Indoor flight experiments with trained kestrels. I. Flight strategies in still air with and without added weight. *Journal of Experimental Biology*, **134**, 173–83.

Walker, T.J. (1985). Butterfly migration in the boundary layer. In *Migration: Mechanisms and Adaptive Significance*, ed. M.A. Rankin, pp. 704–23. *Contributions in Marine Science*, vol. 27, Suppl. Port Aransas, Texas: Marine Science Institute, The University of Texas at Austin.

Walker, T.J. & Riordan, A.J. (1981). Butterfly migration: are synoptic-scale wind systems important? *Ecological Entomology*, **6**, 433–40.

Walsberg, G.E. (1990). Problems inhibiting energetic analyses of migration. In *Bird Migration: Physiology and Ecophysiology*, ed. E. Gwinner, pp. 413–21. Berlin: Springer-Verlag.

Williams, C.B. (1930). *The Migration of Butterflies*. London: Oliver & Tweed.

# Part three
Forecasting migrant pests

# 15

# Operational aspects of forecasting migrant insect pests

R.K. DAY AND J.D. KNIGHT

## Introduction

Forecasting future events is a widespread activity that has been practised for millennia. Many methods have been used, including the reading of naturally occurring 'signs' (e.g. weather patterns), performing special operations (e.g. the inspection of chicken entrails) and relaying messages received from God. The variety of sometimes startling ways that have been used to make forecasts is perhaps indicative of the importance, in many situations, of being able to predict the future.

While it may be a part of the human psyche to wish to see into the future, there are also more specific reasons why it is advantageous to attempt to predict what the future might hold. In pest management, the object of forecasting must be to improve the decisions associated with controlling pests. In the first part of this chapter, we discuss the decision processes by which migrant pests are controlled, in order to define the types of forecast information that are required. This information, together with knowledge of the biology of the specific pest, determines what events must be forecast. We then examine the practical constraints on the production of these forecasts and methods for their evaluation, before suggesting how forecast services can be improved. Our central theme is that more attention must be given to the details of how forecasts are intended to improve control, if the value of migrant pest forecasts is to be improved.

Much of what we have to say about forecasting of migrant pests is also applicable to forecasting of non-migrant pests. This is, in part, a consequence of our approach, in which we make the forecast recipient or decision maker the starting point of our discussion, rather than the biology of the pest, which is more usual. However, in the following sections we emphasise aspects peculiar to the forecasting of migrant rather than non-migrant pests.

These aspects are discussed in relation to African Armyworm *Spodoptera exempta* forecasting in East Africa and aphid forecasting in the UK. Although the views expressed are our own, we acknowledge the influence of the many people involved with these forecasting services, with whom we have worked.

### Decision makers in migrant pest control

Decisions affecting how and with what success migrant pests are controlled are made by many people. They may include farmers, agricultural officers, employees of agrochemical companies, scientists and policy makers, and are sometimes in countries remote from the scene of pest attack. One way of examining these decisions is to look at the options that are available to the decision makers and what their objectives are (Norton & Way, 1983). Table 15.1 gives an example from African Armyworm control in East Africa. The decisions in this example illustrate several points.

1 The objectives of decision makers are sometimes conflicting. For example, the district agricultural officer may have yield targets for the district that have to be met, so their aim might be to maximise yield over the district. However, an individual subsistence farmer is more likely to adopt a 'satisficing' approach, attempting to minimise the risk of a very low yield rather than to seek to maximise yield (Norton, 1976).

2 Decisions affecting the outcome of the control programme can be spread over a long period of time, so that some of them may be made far in advance of the biological events that are the subject of the forecast.

3 Different decisions are made with differing frequencies. Decisions by central government on the importation of insecticides may be made once or a few times in a season, but a district agricultural officer, who has been allocated an annual budget, must decide almost daily how much of the budget to use and how much to retain for use later in the season.

4 Different decisions relate to different spatial scales, from the national level, where decisions affect the whole country, to the farmer level, where individual fields are considered.

Because migrations can span several countries, regional or international organisations are often involved in their forecasting and control, and in

Table 15.1. *An example of three levels of decision makers in armyworm control in East Africa.*

| Decision maker | Example decision/ options | Possible objectives |
| --- | --- | --- |
| Central government | Import of insecticides | Minimise foreign exchange use |
| District agricultural officers | Utilise farm inputs budget | Achieve yield target. Retain funds for use later in the season |
| Farmer | Monitor fields. Choose variety | Minimise risk of poor crop |

some cases organisations have been set up specifically with this objective. The Desert Locust Control Organisation for East Africa is one example, which also provides a regional African Armyworm forecasting service.

Such organisations introduce another layer of decision makers, with another set of objectives and perceptions, increasing the complexity of the system. The potential advantages of a regional approach are clear, but this can also lead to an overemphasis on centralised forecasting, creating constraints on the regional forecaster that would not exist locally.

## Defining the forecast required

Analysis of the decision processes involved in the control of a migrant pest should make it possible to specify more precisely the types of forecast that would be appropriate. A forecast of an infestation by a migrant pest is characterised by its timing, resolution and accuracy.

### *Timing*

The time at which a forecast is required is largely determined by when the decision it is intended to support must be made, allowing time for an appropriate response. Allowance must also be made for the time required to disseminate the forecast and for the decision process itself. Decisions made by individuals require little time, but decisions made by government or regional organisations can take longer, particularly if, for example, funding for a major control operation has to be sought. In general, the best time for a forecast to reach a recipient is just before the decision has to be made. This gives the forecaster the opportunity to use the most up-to-date information available. The forecast is also more likely to be acted on if it is received at the time the information is required.

A further feature of the timing of forecasts is the frequency with which they are required; frequent decisions are likely to require frequent forecasts. In the example given in Table 15.1, insecticides may be purchased, at a national level, only once or a few times a year, well in advance of the season, requiring perhaps one long-term forecast per season. The farmer, on the other hand, may make daily decisions on whether to monitor his crops, for which frequent short-term forecasts are appropriate.

### *Resolution*

The resolution of a forecast defines how it aims to predict future events in terms of location, time-period and magnitude. A high-resolution forecast makes predictions for a small area over a short time-period, forecasting the magnitude or intensity of the event (e.g. the size of the pest population) with precision.

As with timing and frequency, the resolution of a forecast should correspond to the target decision. National decisions require long-term forecasts that can be of relatively low spatial and temporal resolution but of higher resolution on the magnitude of the pest attack. With low spatial and temporal resolution required, it might not even be necessary to forecast specific migrations. In deciding what quantity of pesticide to purchase at the start of an African Armyworm season, the important question is how much will be needed in the country as a whole, rather than the details of where and when the chemical will be required. Such decisions can be taken later, when higher spatial and temporal resolution will be necessary.

In contrast, short-term forecasts for farmers must be of high spatial and temporal resolution. Infestations of migrant pests are often at high densities, so the expected severity of infestation may be of lesser importance to the farmer. Armyworm outbreaks are almost always at densities high enough to cause damage, so a risk-averse farmer is likely to be interested in only whether or not there will be an attack.

## Accuracy

The accuracy of a forecast can be defined as the extent to which it matches subsequent events. It is obviously desirable for a forecast to be as accurate as possible but, in practice, there are several constraints that limit accuracy. The question of the minimum accuracy necessary is examined further below.

## Constraints and trade-offs

There are two main types of constraint under which forecasting services must operate. Firstly, there are constraints that are a result of the nature of the decision processes and methods by which the migrant pest is controlled. These determine the type of forecast likely to be of use to a particular recipient (see above); they can be termed user or recipient constraints.

Secondly, the forecaster is constrained by his own capabilities and by what is technically feasible. Forecasts are based on current data, historical data, knowledge about the behaviour of the systems to which they apply and on forecasts of other variables, such as weather. For various reasons, these four areas of information are imperfect, limiting the achievable accuracy of the forecast so that trade-offs are required. For a given level of knowledge of the system and of the available data, the level of accuracy that can be achieved will be affected by how far ahead the forecast is for, and by the resolution attempted. Fig. 15.1 shows the general relationship between resolution, the interval between issue of the forecast and the events predicted, and forecast accuracy. As we move from short- to long-term forecasts, and from low- to high-resolution forecasts, accuracy is likely to fall.

Lines A and B in Fig. 15.1 denote possible recipient constraints. If the decision the forecast is intended to support requires at least a medium-term forecast, then no forecast below line A will be of use. Similarly, if the level of resolution required is at least 'medium', no forecast to the left of line B will be adequate. In this particular case, the recipient and forecaster constraints mean that the best forecast that could be achieved would be of 'fair' accuracy. If this proved to be inadequate, issuing the forecast would be pointless. The question of the value of inaccurate forecasts is discussed further below.

A particular case where such trade-offs need to be examined is in determining the appropriate degree of centralisation of migrant pest

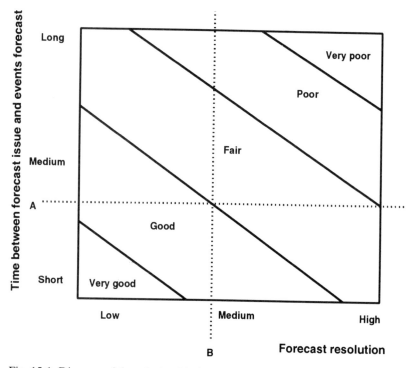

Fig. 15.1. Diagram of the relationship between forecast resolution, accuracy and the interval between the issue of the forecast and the period to which it applies. The longer the term and the higher the resolution of the forecast, the less accurate it becomes (see text for explanation of lines A and B).

forecasting. Centralised forecasting allows the use of data from a wider area and from sources such as remote sensing that would not be available locally. This should allow forecasts of higher resolution to be made. But to allow time for information flow to the forecaster, and for dissemination of the forecast, centrally prepared forecasts may need to be prepared further ahead of the forecast event than locally prepared forecasts, with a possible reduction in the accuracy that can be achieved.

## Evaluating forecasts

Forecasts can be evaluated at several levels. For those funding the forecasting service, it is appropriate to consider whether the cost of providing the service is more than offset by the reduction in crop losses attributable to the forecast. If it is not, it could be argued that the money might be better used as a pest compensation fund, although this is rarely feasible, for political reasons.

An individual farmer receiving the forecast might ask if it is worth his while to act on it. In the simplest case, a forecast predicts that pest outbreaks either will or will not occur. The recipient can either prepare for outbreaks (e.g. by monitoring crops or buying chemicals) or not, and they either do or do not occur. There are thus four potential cost outcomes (A, B, C and 0), as shown in Table 15.2. If a fixed strategy is followed, i.e. always prepare or never prepare, the expected (long-term average) cost is a function of the outbreak frequency, as shown by the lines AB and from the origin to C respectively, in Fig. 15.2. If a forecast that is 100% accurate is issued (and always acted on), then the expected cost is indicated by the lower line from the origin to A. The distance between this and the two lines mentioned above gives an indication of the maximum potential monetary value of the forecast.

However, forecasts are rarely, if ever, 100% accurate. Two parameters are needed to describe accuracy, defined as the proportion of outbreaks correctly predicted ($a1$) and the proportion of outbreak-free periods correctly predicted ($a2$). Fig. 15.2 also shows the expected-cost line for an inaccurate forecast defined by these parameters.

From Fig. 15.2, it is apparent that if the historical outbreak frequency is below X or above Y, with a forecast of the accuracy shown, it is better to follow a fixed strategy than to follow the forecast. Alternatively,

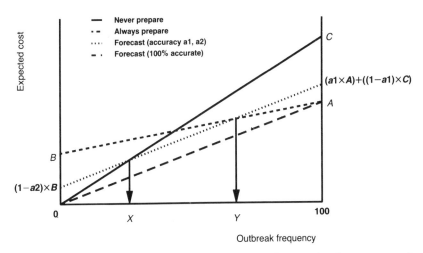

Fig. 15.2. The expected costs of four pest-control strategies for a range of outbreak frequencies (see text for explanation).

Table 15.2. *The four cost outcomes in a simple pest system.*

|  | Outbreaks | No outbreaks |
|---|---|---|
| Preparation | A | B |
| No preparation | C | 0 |

Where cost C > cost A > cost B.

assuming we can quantify the costs, we can specify how accurate the forecast must be to be of value.

This approach was adopted by Cammell & Way (1977) to examine the value of Black Bean Aphid (*Aphis fabae*) forecasts for field beans in the UK. For the period from 1970 to 1975, they calculated that the net loss (from *A. fabae* attack and/or cost of treatment) would have amounted to £16.1 ha$^{-1}$ per annum for no treatment, £8.5 ha$^{-1}$ for routine preventive treatments and £4.7 ha$^{-1}$ if the forecasting scheme had been adopted. The benefits were greater in regions where a large area of beans was grown and attacks were infrequent, than in those where attacks occurred frequently and only a small area of beans was grown. An analysis of the economics of forecasting for the control of the aphid *Sitobion avenae* in the UK (Griffiths & Holt, 1986) showed that, for average or above average yields, control decisions based on a forecast of between 70 and 80% accuracy were more profitable than either prophylactic treatment or no control. However, on farms with low yield expectations, prophylaxis was the only strategy that did not incur a risk of financial loss.

The above analyses assume a cost-minimisation strategy. However, it was noted earlier in this chapter that subsistence farmers might be risk-averse and use a 'satisficing' approach to decision-making (Norton, 1976). In general, this will reduce the space in which a less-than-perfect forecast is worth acting upon. Griffiths & Holt (1986) showed that, if a farmer is willing to tolerate a loss in only one year in 50, then the forecast must be 94% accurate to be of use.

An economic analysis of this nature may not always be possible. In that case, other indicators of value can be assessed. For example, in East Africa, the age at which African Armyworm outbreaks are discovered might be expected to be lower when the outbreak was forecast than when it was not; the earlier an outbreak is detected, the lower the damage and the easier it is to control.

More generally, a forecast is of value only if, at least on some occasions, it influences the action taken by the recipient; an ignored forecast can have no economic value.

In evaluating forecasting services, it is appropriate to conduct practical assessments, comparable to those used for other components of a pest management programme, e.g. chemical trials. Therefore, a forecast should be evaluated by assessing costs and benefits in comparable areas in which it had or had not been acted upon. Alternatively, field situations in which forecasts have or have not been followed could be analysed retrospectively.

## Improving forecasting of migrant pests

The foregoing discussion suggests several ways in which the value of a forecasting service might be improved.

### *More precise targeting*

Forecasts should be more precisely targeted at the appropriate decision makers by consideration of their objectives. This could be achieved by presenting only the relevant information, presenting the forecast in a readily understandable form and, if necessary, specifying the recommended response.

Where one forecast is distributed to all recipients, it may contain only a small amount of information relevant to each, perhaps reducing the impact of the information and the chance of it being acted on. The use of computerised systems affords the opportunity for highly customised forecasts containing only the information the recipient requires.

In designing forecasts more accurately for the recipient's needs, the way the forecast is presented is also of importance. African Armyworm forecasts use a verbal expression of the probability of outbreaks, though an analysis of all the forecasts issued showed that, over the years, a wide range of terms had been used, possibly confusing the recipients. A particular word may also have different meanings to different people (Beyth-Marom, 1982), so if verbal expressions are used, it is desirable that they should be limited in number and have well-defined meanings.

In some cases, it may be appropriate for a forecast to be prescriptive as well as predictive. If the correct response to a forecast is known to the forecaster, this can be added to the forecast and may take precedence over the prediction itself. However, more sophisticated users are likely to require only the forecast, which they will then use, along with other information, in their own decision making.

### *Relaxing forecaster constraints*

One way to reduce the constraints on the forecaster is to minimise the amount of time between the issue of the forecast and the event being predicted. Centralised forecasting requires more information flow than local forecasting, so decentralised forecasting would reduce this time. However, there are other advantages of centralised forecasting which might outweigh this gain. An alternative approach is to improve communication, a greater problem in developing, than in developed, countries. In East Africa, radio links have been established for communicating information to and from the Regional Armyworm Forecaster, though there remains the problem of information flow from the field to national coordinating units. Even within a country, it is not uncommon for less than 50% of moth-trap operators to report their weekly catches in time for use by the forecaster. Similarly, there are also occasions when a forecast sent out by post arrives after the period the forecast covers.

It was noted above that forecasts are based on four types of information, all of which may constrain the forecaster.

Historical data are used in forecasting by analogy, and as a way of calculating the simplest type of forecast, the historical probability of an event. These data can be made more accessible by the development of database systems, such as FLYPAST for aphids in the UK (Knight *et al.*, 1992), and WormBase for African Armyworm in East Africa (Odiyo, 1990; R.K. Day, *WormBase: Armyworm Database and Forecasting System*, unpublished user manual, 1991).

These systems also improve the management and analysis of current data, and incorporate expert-system components that improve the analysis of available information. Other techniques make more and different data available. Remote-sensing methods (Chapter 16, this volume) offer considerable possibilities in this respect, and may lead to a complete reappraisal of the way in which migrant forecasting services operate.

Forecasts are also based on an understanding of the behaviour of the pest system, so increased knowledge of this can lead to improved forecasting (Pedgley *et al.*, 1989). However, incomplete knowledge of the system may not be the major constraint to improving the value of a forecast service, a possibility which is sometimes overlooked.

An important component of migrant pest forecasting is other forecast data, especially weather forecasts, as it is the weather that plays a key role in determining whether a migration occurs, where and when it ends, and how the immigrant population subsequently develops. Substantial resources are already devoted to improving weather forecasting, but it is possible that the use of weather forecasts by pest forecasters could be improved by adopting the approach to forecasting described in this chapter. Migrant pest forecasters are users of the weather forecasting system, and by more precise definition of their requirements, better use could perhaps be made of existing weather forecasting capabilities.

### *Forecast evaluation*

There is a need for the development and application of methods for evaluating forecasts, in order to detect where improvements can be made, and to monitor efforts to implement them. Although it is not always practical, one way of assessing the value of forecasts is to adopt the 'user pays' approach. If recipients have to pay for information, in the same way that they pay for pesticides, they will soon stop purchasing forecasts if the information is not useful.

### *Improving the ability of recipients to respond*

Forecasts may be of great potential value but, for reasons such as inadequate training or insufficient resources, the recipients may be unwilling or unable to respond to them. It is therefore essential that the introduction of any forecasting system is accompanied by appropriate training and provision of the resources necessary to ensure that decisions based upon it can be implemented.

### *Continual appraisal*

Finally it must be remembered that situations change, so a continual appraisal of pest forecasting services is required. User requirements can change, and if organisations are set up, change or close, then new users may be created or former ones lost. In the last few years, computer technology has markedly improved the potential of forecasters to use both historical and new information, a trend that is likely to continue. Care will be required to ensure that attempts to improve migrant pest forecasting services are not simply technology-driven, but are based on the needs of the users.

## References

Beyth-Marom, R. (1982). How probable is probable? A numerical translation of verbal probability expressions. *Journal of Forecasting*, **1**, 257–69.

Cammell, M.E. & Way, M.J. (1977). Economics of forecasting for chemical control of the black bean aphid, *Aphis fabae*, on the field bean, *Vicia faba*. *Annals of Applied Biology*, **85**, 333–43.

Griffiths, E. & Holt, J. (1986). Economics of *Sitobion avenae* (Aphididae) forecasting and control in the U.K. *Crop Protection*, **5**, 238–44.

Knight, J.D., Tatchell, G.M., Norton, G.A., & Harrington, R. (1992). FLYPAST: an information management system for the Rothamsted Aphid Database to aid pest control research and advice. *Crop Protection*, **11**, 419–26.

Norton, G.A. (1976). Analysis of decision making in crop protection. *Agroecosystems*, **3**, 27–44.

Norton, G.A. & Way, M.J. (1983). Forecasting and crop protection decision making – realities and future needs. *10th International Congress of Plant Protection, Brighton*, vol. 1, 131–8.

Odiyo, P.O. (1990). Progress and developments in forecasting outbreaks of the African armyworm, a migrant moth. *Philosophical Transactions of the Royal Society of London B*, **328**, 555–69.

Pedgley, D.E., Page, W.W., Mushi, A., Odiyo, P.O., Amisi, J., Dewhurst, C.F., Dunstan, W.R., Fishpool, L.D.C., Harvey, A.W., Megenasa, T. & Rose, D.J.W. (1989). Onset and spread of an African armyworm upsurge. *Ecological Entomology*, **14**, 311–33.

# 16

# Geographic information systems and remotely sensed data for determining the seasonal distribution of habitats of migrant insect pests

## T.P. ROBINSON

### Introduction

To recognise and understand the environmental processes that lead to outbreaks of migrant pests, many environmental variables may need to be combined in a spatial context. Computer-based geographic information systems (GIS) make it possible to manipulate large spatially referenced data sets, including remotely sensed data on, for example, vegetation, rainfall and surface temperature. In this chapter, the extent to which these techniques have been applied to the ecology of migrant insect pests is discussed and ways in which they may be developed further are considered. The intention is not to provide technical information on the use of GIS and remote sensing, for which the reader is referred to standard texts (e.g. Curran, 1985; Burrough, 1986).

### The use of GIS and remote sensing in entomology

GIS is the name given to the general techniques, and to specific computer packages, that can be used to register, store, manipulate and retrieve spatial data. The function of GIS is to integrate different types of spatial data. Spatial data are commonly held in computers in either vector or raster format and the relative advantages and disadvantages of each format are discussed by Burrough (1986). Vector data are a series of spatial coordinates defining points, lines and polygons that may be generated when maps are digitised, as well as specific point measurements, e.g. from meteorological stations, which are entered as tabular data. Raster data comprise a regular grid of cells, each having a particular value for a given variable. Satellite data are recorded in this format. As remotely sensed data become more widely available, raster and vector data must be managed within the same system.

335

Remote sensing includes all methods of observation of a target by a device some distance from it. This broad definition encompasses a wide range of techniques, including ground-based and aerial photography and videography; satellite-borne photography, multispectral scanning and thermal imaging; ground-based and airborne microwave sensing; acoustic sounding; and low-light optical methods (Riley, 1989). The techniques are becoming more widely applied in entomology and they have been reviewed by Barrett (1980), Nageswara Rao (1988), Hugh-Jones (1989), Riley (1989) and Robinson (1991).

Remotely sensed data have been used to locate insect populations by observing:

1 The insects themselves;
2 The damage they cause;
3 The crops in which they could be pests;
4 The environments favourable for population development.

These observations require remote-sensing systems that gather information at an appropriate resolution. Direct observation of insects requires a resolution of the order of centimetres; identification of a crop or damage within it requires a resolution of the order of metres; the monitoring of larger-scale environmental conditions requires a resolution of the order of tens of metres or kilometres. Entomological radar and optical techniques have a resolving power of the order of centimetres; air photography of the order of metres; the dedicated Earth Observation Satellites (e.g. Landsat and SPOT; a full list of acronyms used in this chapter is given on page 348) of the order of tens of metres; and meteorological satellites (e.g. NOAA and Meteosat) of the order of kilometres.

Observations of insect damage and of crops are of limited use in monitoring migrant pests, because of the insects' mobility and the spatial scale of their populations. However, direct observation of insects using ground-based radar and optical systems has been used to investigate migration in the African Armyworm *Spodoptera exempta* and other migrant insects, particularly grasshoppers and locusts (Riley, 1989).

This chapter is concerned with the monitoring of environmental conditions and with using remotely sensed data to estimate the values of those variables that influence insect population development and movement. Examples of environmental variables that may be measured with the use of remotely sensed data and that influence insect population dynamics are vegetation, rainfall, temperature, saturation deficit, sunshine and soil moisture.

GIS and remote sensing have proved useful in the identification of suitable habitats for a number of non-migratory species, including mosquitoes (Barnes & Cibula, 1979; Linthicum *et al.*, 1987; Wood *et al.*, 1991), ticks (Hugh-Jones *et al.*, 1988; Perry *et al.*, 1990), tsetse flies (Rogers & Randolph, 1991) and the Screw-worm Fly (Arp *et al.*, 1976; Giddings, 1976). Although much can be learned from these studies, migrants pose additional problems, because of their mobility.

Forecasting the distribution and abundance of migrant insect pests depends on estimating, firstly, the probability of a pest occurring in a given area, by using models for pest movement, and, secondly, the probability of its thriving in that area, by using models for habitat suitability. Simple trajectories for insect movement have been used to estimate sources and destinations of a number of pest species (see e.g. Tucker, 1984; Rosenberg & Magor, 1986). These rely on interpolating windspeed and direction at specific times, locations and altitudes, with the use of meteorological data. Models for insect movement are reviewed and compared by Scott & Achtemier (1987) and comprehensive GIS to forecast migrant pests should include such a model.

## Review of three migrant pests

Remotely sensed data are used operationally to forecast outbreaks of three pest species. For the Desert Locust *Schistocerca gregaria* and the Australian Plague Locust *Chortoicetes terminifera*, the techniques are used to identify conditions that lead to a buildup of low-density populations, whereas for the African Armyworm *Spodoptera exempta*, they are used to identify mechanisms that lead to concentration of moths flying from low-density source areas.

### *Desert Locust* **Schistocerca gregaria**

During the time between plagues, *S. gregaria* persists in low-density populations within the generally arid/semi-arid belt extending from Mauritania to northwest India, known as the recession area. Periodically, rainfall or runoff produce ideal conditions for breeding, by moistening the soil sufficiently for oviposition to be initiated and by giving rise to vegetation for food and shelter. If, as a result of several generations of successful breeding, numbers and densities rise, gregarisation may occur and large numbers of highly mobile swarms containing billions of locusts may develop. It is becoming increasingly evident (Skaf, Popov & Roffey,

1990) that outbreaks of locusts arise from low-density populations rather than from undetected swarming populations.

Pedgley (1973) tested the feasibility of using satellite remote sensing for monitoring conditions conducive to *S. gregaria* population development. False-colour composites were made from ERTS-1 imagery for an area in the Red Sea coastal plain of Saudi Arabia. The arrival and subsequent breeding by locusts in a region where the suitability of the vegetation was indicated by satellite imagery demonstrated the potential of these techniques. Roffey (1975) concluded that satellite remote sensing was likely to be the best tool with which to monitor this vast recession area (c. 16 million square kilometres) comprehensively and economically.

Vegetation giving rise to population development occurs either in narrow drainage channels (wadi habitats), of the order of 1–3 km², or in broader areas of low relief, of the order of 20–40 km² (Tucker, Hielkema & Roffey, 1985*a*). Hielkema (1980) developed an automated procedure for detecting the wadi-type habitats using Landsat TM imagery. Comparing various ratios of reflectance values in the red and near-infrared (NIR) wavelengths, he concluded that the normalised difference vegetation index (NDVI) was optimal for monitoring vegetation. A 'potential breeding activity factor' (PBAF) was then developed to weight, progressively, successive NDVI classes in an exponential manner. The NDVI is the most commonly used estimate of vegetation activity and is computed as:

$$NDVI = (NIR-R)/(NIR+R)$$

where R is the reflectance in the red, and NIR is the reflectance in the near-infrared portion of the spectrum. Photosynthetically active vegetation has a high level of chlorophyll, which absorbs strongly in the red wavelengths, compared to the soil background, and reflects strongly in the near-infrared, due to intra-leaf scattering and an absence of absorption. This selective absorption is the basis of satellite monitoring of vegetation and the processes involved have been extensively reviewed by, for example, Knipling (1970) and Woolley (1971).

Tucker *et al.* (1985*a*) concluded that the wadi-type habitats were best identified with the use of Landsat TM, with a resolution of 30 m, but could also be detected by using local area coverage (LAC) 1.1 km resolution AVHRR data from the NOAA series of satellites. The larger areas of vegetation development were best monitored with LAC data but resampled global area coverage (GAC) data, derived by subsampling LAC data at a resolution of 4 km, were adequate.

Some of these techniques were tested in a study of the 1980–81 *S. gregaria* plague upsurge in West Africa (Hielkema, Roffey & Tucker, 1986*b*) by comparing NDVI GAC data with reports of the occurrence of the locusts. Qualitative analysis, comparing the coincidence of locusts and suitable vegetation on a one-degree-square basis, looked promising. A quantitative analysis which compared, for three study areas on various dates, the number of locust reports with the PBAF of that area indicated a significant positive relationship. These were, however, very large areas (up to 500 km × 400 km) and although averages over such large areas give significant correlations, populations must be located much more precisely for strategic control. These data may, however, indicate areas where ground teams should concentrate searches and for which higher-resolution satellite data should be used.

Concurrently with Hielkema's (1980) studies on monitoring vegetation, Barrett (1980) investigated the potential for monitoring rainfall using the 'Bristol method of satellite-improved rainfall mapping' in northwest Africa. Rainfall-hazard-potential maps of the recession area were produced which identified five hazard classes, ranging from no hazard (no rain) to very high hazard (>30 mm rain). These were used, in conjunction with a knowledge of surface characteristics, to recommend Landsat scenes for detailed study of the vegetation. More recently, Milford & Dugdale (1987) have developed an operational technique for rainfall estimation based exclusively on the use of Meteosat thermal-infrared data. This involves the selection of pixels representing rain-bearing clouds by applying temperature thresholds that have been derived through calibration studies in Africa over an extensive period. Linear regression coefficients have been determined to convert cold-cloud-duration (CCD) values into rainfall estimates, usually over 10-d intervals. Both the optimal temperature thresholds and the regression coefficients vary geographically and seasonally (Milford & Dugdale, 1990).

Following early pilot studies, the development of the ARTEMIS system was initiated (Hielkema *et al.*, 1986*a*) as a collaborative project between FAO, NASA, NOAA, ESA and NLR. This system uses hourly Meteosat and daily AVHRR (GAC) data to derive a number of products on a ~7.6-km pixel grid. They include 10-d images of rainfall estimation, with the use of Milford & Dugdale's (1987) techniques, and vegetation activity, with the use of the NDVI (NDVI images from September 1981 are available). The current ARTEMIS system characteristics are described by Hielkema (1990), although research continues to improve the rainfall estimations (Milford & Dugdale, 1990).

Operationally, it has not been possible to identify the very low (5%) vegetation cover typical of suitable *S. gregaria* breeding habitats by using the 7.6-km NDVI data, although the rainfall estimation maps have been useful in forecasting. Since August 1989, 1.1 km AVHRR (LAC) NDVI images have been available for much of the *S. gregaria* recession area, and these are proving more useful in detecting this sparse but critical vegetation (Cherlet & Di Gregorio, 1991).

An integrated approach is proposed by Cherlet & Di Gregorio (1991) as a result of a detailed calibration study in the Tamesna region of Niger. Following comparison of a variety of vegetation indices, the NIR/R ratio was recommended as preferable to the NDVI, because of its greater sensitivity in detecting very low levels of the major plant associations (*Schouwia* ass. and *Tribulus* ass.) important for *S. gregaria* breeding in the study area. They suggested the integration of detailed land classification maps, vegetation index data and rainfall information within a GIS, to improve estimates of potential *S. gregaria* breeding areas. Work is also under way to produce evapotranspiration maps from Meteosat imagery to reveal soil moisture patterns which may also assist with locust habitat monitoring (Rosema, 1990).

### Australian Plague Locust Chortoicetes terminifera

*Chortoicetes terminifera*, like *S. gregaria*, requires both moist soil and vegetation growth for population development. In a study by McCulloch (1979), Landsat imagery was obtained for the Channel Country, a very arid breeding area for *C. terminifera* in southwest Queensland, to produce false-colour composites of Landsat bands 4, 5 and 7. These gave a magenta colour where green vegetation was present. For three study areas, an absence of magenta in the imagery was consistently associated with an absence of *C. terminifera* and, where green vegetation was indicated in the imagery, there was a reasonable spatial coincidence with the occurrence of locusts.

A more detailed study (McCulloch & Hunter, 1983) in the same area between January and March 1981 was less promising. Landsat false-colour composites appeared to identify only ephemeral vegetation growth and not the perennial Mitchell grass (*Astrebla* spp.) which, due to its greater persistence following a single rainfall event, is favoured by *C. terminifera*.

More recently, work by Bryceson (1990) indicated that the use of digital Landsat data enabled areas of green vegetation to be discriminated from the soil background in this region, making it possible to detect potential outbreak areas. A study on the 1987 invasion of *C. terminifera* (Bryceson, 1989) showed that the NDVI indicated vegetated areas where locusts might be expected to aggregate after long-distance migration, and a maximum likelihood classification indicated preferred oviposition sites within these areas. More than 95% of all nymphal band targets were within areas defined by an NDVI threshold.

Bryceson (1991) used Landsat MSS data to characterise habitat suitability for locust infestations on three criteria:

1 Habitat type (maximum likelihood classification);
2 Habitat condition (NDVI);
3 Soil type (soil brightness transformation).

Her analysis showed 70%, 75% and 88% of locust bands to be contained within the areas defined by the ranges of selected parameters of each estimated variable, respectively. Though these statistics show great operational potential, Landsat data are prohibitively expensive for routine monitoring. Low-cost meteorological satellite data, such as NOAA and GOES, are being investigated as an operational tool with which to monitor the weather systems that produce rain in these areas (Bryceson & Cannon, 1990).

These studies indicate the potential for highly integrated GIS incorporating environmental variables, such as habitat type, vegetation activity, rainfall, soil type, temperature and wind run, some of them estimated from satellite data, as a framework for simulation models of population development in *C. terminifera* (see e.g. Wright, 1987).

### *African Armyworm* Spodoptera exempta

Severe infestations of *Spodoptera exempta* occur periodically in eastern, central and southern Africa and occasionally in western Africa and southwestern Arabia (Haggis, 1986). The larvae feed almost exclusively on Graminae and Cyperacae and exhibit a density-dependent phase polyphenism (Faure, 1943). High-density outbreaks comprising larvae of the gregaria phase occur during the rainy season, while low-density populations of solitaria-phase larvae are thought to persist throughout the year over much of the range of the species (Rose, 1979). The two phases differ radically in their appearance, behaviour and physiology.

Early in the rains in East Africa, high-density outbreaks originate with the concentration of flying moths, most frequently by convergent airflow associated with rainstorms, as they migrate from low-density source areas (Pedgley, 1990). Concentration of moths where rain has fallen is followed by egg-laying *en masse* and a high level of larval survival results in 'primary' outbreak populations, some of which are described as 'critical' (Rose *et al.*, 1987). Moths emerging (usually over approximately one week) from these critical outbreaks and migrating downwind cause subsequent 'secondary' outbreaks at generation intervals of approximately one month.

Interest is currently being shown in a strategic approach to the armyworm problem, involving control of critical primary outbreaks in order to interrupt the progression of secondary infestations. The success of strategic control depends on the prompt identification of the critical outbreaks.

The first attempts to analyse armyworm outbreaks in relation to remotely sensed data were by Garland (1985) and led to the installation of a Meteosat receiving facility at the headquarters of the Regional Armyworm Project in East Africa. These satellite data are used to derive CCD estimates to locate deep convective storm clouds with which are associated the gust fronts capable of concentrating flying moths. Although the images produced by this system are used regularly by the Regional Armyworm Forecaster, a quantitative analysis has never been conducted to assess their value.

The potential of GIS and remote-sensing techniques for modelling within- and between-season variations in the distribution of habitats suitable for armyworm population development has been examined in a collaborative project between NRI and Reading University (Robinson, 1991). Food availability was estimated from 7.6 km AVHRR-NDVI data (ARTEMIS) and long-term average-monthly-temperature images of East Africa were generated by interpolation of climatic statistics, by using a digital elevation model (DEM). Records of outbreaks in Kenya and Tanzania between 1982 and 1987 were then analysed with respect to both the NDVI and the temperature data, and a model was developed that predicts the seasonal distribution of habitat suitability for armyworm population development, in terms of the presence of larval host plants and favourable environmental temperatures. The model delimited areas within empirically derived lower NDVI ($<0.2$), and upper ($>27$ °C) and lower ($<17$ °C) average temperature thresholds.

The model can also be used to distinguish between areas that may be more suitable for low-density, source-population development and those that may be more suitable for outbreak-population development. This distinction depends on an estimate of habitat durational stability (HDS)

(Southwood, 1981), areas with long HDS being considered more suitable for low-density population development and those with short HDS more likely to contain outbreak populations. Robinson (1991) discusses how HDS may provide a direct measure of environmental selection pressure for migratory potential (Chapters 10 and 11, this volume).

The two main constraints on Robinson's (1991) analysis were the absence of reliable data on the effects of environmental conditions on *S. exempta* development rates and problems with the spatial and temporal integrity of the available satellite data. However, the study demonstrates how rasterised GIS can provide an ideal vehicle for the extension of biological models through space and time.

## Discussion and conclusions

In all three cases discussed above, there has been a lack of integration in the approach adopted, although this is clearly changing as the benefits of GIS are being realised. The interdisciplinary nature of the work requires cooperation between biologists, modellers, meteorologists, environmental physicists, geographers and computing experts, and the models developed need to consider both habitat suitability and pest movement. Efficient planning and coordination of research, in conjunction with control operations, is therefore a primary consideration.

Of the environmental variables, rainfall is probably the most important in determining habitat suitability, as well as being the most variable in space and time. The reliability of the CCD method of estimating rainfall depends on the nature of the rain-producing systems in an area. The method works reasonably well in the western Sahel, a region of relatively simple topography and climatic zonation, where the majority of the rain comes in heavy falls from deep convective cloud cells. It is likely to be less effective in areas where topography, weather and climate are more complex. An important source of error is that the estimates do not distinguish convective from layer, specifically cirrus, cloud. Furthermore, the soil in some areas may be moistened by water travelling by runoff arising from rainfall many kilometres away. Mapping of soil moisture by use of these techniques might, therefore, be more accurate if they could be used in conjunction with more general hydrological models incorporating features like elevation, slope and drainage, to account for movement of surface water.

Much work remains to be done on the calibration of satellite data and products such as the NDVI. NDVI has been correlated with vegetation features such as leaf-area index and biomass (see e.g. Tucker *et al.*, 1981; Asrar *et al.*, 1984), and photosynthetic rates and stomatal resistance to water-vapour transfer (Sellers, 1985; Tucker & Sellers, 1986). Its time integral relates well to primary production (Prince, 1991). A difficulty with the index is that different satellites record rather different sections of the red and NIR portions of the spectrum, so that NDVI estimates from different sensors may not be comparable. Other problems with NDVI as a measure of vegetation activity are reviewed by Robinson (1991).

Many studies have been conducted with the use of NDVI data from NOAA's AVHRR sensor to monitor vegetation, particularly in the Sahelian environment (see e.g. Justice, 1986; Prince & Justice, 1991). With the use of multi-temporal composited NDVI images from the same sensor, the continents of Africa, Asia and South America have been classified on the basis of their NDVI responses over time, or 'phenologies' (Tucker, Townshend & Goff, 1985*b*; Townshend, Justice & Kalb, 1987; Saull & Millington, 1991). While such vegetation classifications can rarely distinguish particular plant species, they can provide information on vegetation structure, and the methods need to be investigated further if their potential for estimating habitat suitability for insect population development is to be realised.

Understanding and overcoming atmospheric influences on satellite data are also important tasks. The launch of the ATSR-2 (along track scanning radiometer) instrument will provide an opportunity to improve our understanding of these influences by generating pairs of images at different viewing angles. This will make it possible to compare spatially and temporally contemporary imagery sensed through different thicknesses of the atmosphere.

In addition to rainfall and vegetation, other environmental variables, such as temperature, saturation deficit and possibly sunshine amount, are important in determining habitat suitability for insect population development. Methods to map these variables and incorporate them into the models need to be developed. Some variables may be estimated entirely from remotely sensed data (Chen *et al.*, 1983; Henderson-Sellers *et al.*, 1987; Mosalam Shaltout & Hassen, 1990), some may be mapped entirely by interpolation of ground measurements (Hutchinson, 1989; Robinson, 1991) and others may require the use of satellite data to assist interpolation of ground measurements (Arp *et al.*, 1976).

The appropriate spatial and temporal scales for these studies must be identified. Ecological patterns and processes occur on various scales of space and time and Wiens (1989) discusses how different patterns emerge at different scales of investigation. In many terrestrial environments, for example, broad-scale climatological and geological influences may be obscured by finer-scale micrometeorological (microhabitat) effects and biological interactions, such as competition. Because local heterogeneity is averaged out in coarser-scale studies, patterns often appear to be more predictable at these scales.

The choice of scale depends on:

1 The operational requirements of each study; for example, control operations may be launched at the level of a small farm or over very large areas of grassland;
2 The heterogeneity of the landscape in question, determining the resolution required;
3 The mobility of the organism in question, determining the geographical extent of the study.

For migrants, the extent of the study must be large enough to accommodate migration distances and, given the constraints imposed by data storage and processing, it is desirable to use a relatively coarse resolution. It is critical, therefore, to identify the coarsest resolution consistent with operational requirements and with habitat heterogeneity.

A number of authors have considered the problems of appropriate scale for satellite data (see e.g. Robinson, 1991; Justice *et al.*, 1989). Townshend & Justice (1988) describe some of the constraints imposed by the spatial, spectral and temporal properties of sensors on identifying patterns of land-cover change.

The way in which satellite data are degraded can limit their usefulness. The most widely available vegetation index data are the 4-km NOAA-AVHRR (GAC) data and the various derivative products. These data are produced by resampling the original 1.1 km (LAC) data by selecting only every third scan line and then averaging the channel values for a row of four pixels along the line, ignoring the fifth (Kidwell, 1990). Justice *et al.* (1989) have identified the deficiencies of this sampling procedure and Robinson (1991) discusses the spatial and temporal accuracy of a further resampled 10-d composite data set (7.6 km ARTEMIS-NDVI). It is apparent that inaccuracies introduced through the sampling procedures used in their generation severely limit the usefulness of these data for forecasting *S exempta* outbreaks.

As well as relating to different scales, spatial data from a variety of sources are often mapped to different projections. Before these data can be integrated, they must be transformed to a common scale and projection. Although this should be a fundamental component of any GIS, the number of projections that can be dealt with by commercial packages is limited and the accuracy with which the transformation can be achieved is sometimes questionable. This is especially true of unusual projections, such as the Hammer–Aitoff projection used in the ARTEMIS data system, and the Meteosat projection.

In the past, constraints have been imposed on the achievable spatial and temporal resolution of studies by delays involved in the processing and acquisition of the satellite data. However, this is now less of a problem as PC-based satellite receivers are becoming commercially available (e.g. for Meteosat and NOAA-AVHRR data), enabling users to obtain data at the original resolution and in real-time.

Attempts to use GIS and remote sensing to relate the distribution of migrant pests to the environments in which they thrive must address four key questions.

1  Which population variables most influence changes in numbers?
2  Which environmental variables most influence these population variables?
3  How, and with what accuracy, can computer maps of the relevant environmental variables be produced?
4  How can these computer maps be integrated to reveal the distribution of environmental suitability for population development?

To answer the first two questions, models relating population variables (e.g. fecundity, mortality, development rate, immigration and emigration) to environmental variables (e.g. vegetation, rainfall, temperature, saturation deficit, sunshine, airflow and topography) must be developed. Broadly speaking, two analytical approaches can be adopted: statistical and biological. The statistical approach compares the known distribution of a species over time and space with values for environmental variables. If a number of environmental variables is used, this approach can be used to define the presence or absence of a species in a multidimensional environment space and discriminant criteria can be sought to categorise unknown sites according to the probability of occurrence of the species. The biological approach, on the other hand, involves the derivation of models for population growth that take into account known or assumed effects of environmental variables on population variables. The biological approach is clearly preferable, but it requires detailed biological data and robust theoretical models. The statistical approach requires minimal information about the study species, i.e. presence or absence at different locations at different times, which will often be all that is available.

Once key environmental variables have been identified, computer maps can be generated from remotely sensed data and by interpolation of meteorological data. These data planes can then be incorporated into spatial models that predict changes in pest numbers using GIS. The accuracy of these predictions must then be tested in the field. A criticism of the *Schistocerca gregaria* and *Spodoptera exempta* work is the lack of quantitative field verification of model predictions, but the closer liaison between the research and operational components of the *Chortoicetes terminifera* work appears to have overcome this problem.

The potential of these techniques depends on the quality of the biological and physical models that they extend through space and time. The precision and robustness of the models is, in turn, a function of the selection and availability of the appropriate biological and environmental data from which they are derived. It is only through the conscientious collection and careful interpretation of these data, together with rigorous and quantitative field testing of the predictions made by the models, that the full practical potential of these systems will be realised.

In a discussion of the paper by Skaf *et al.* (1990), J. Hewitt referring to the control of locust upsurges commented that the 'real problem facing operations of any kind is that governments, while responding to crises, never anticipate them'. The techniques described in this chapter have the potential to help to predict the development and movement of populations of migrant pests, so that warnings of imminent crises can be issued in time for action to be taken to avert them.

**List of acronyms**

| | |
|---|---|
| ARTEMIS | African Real Time Environmental Monitoring Using Imaging Satellites |
| ATSR-2 | Along track scanning radiometer |
| AVHRR | Advanced very high resolution radiometer |
| CCD | Cold cloud duration |
| ERTS-1 | Earth Resources Technology Satellite – 1 |
| ESA | European Space Agency |
| FAO | Food and Agriculture Organisation |
| GAC | Global area coverage |
| GIS | Geographic information systems |
| GOES | Geo-Stationary Operational Environmental Satellites |
| HDS | Habitat durational stability |
| LAC | Local area coverage |
| MSS | Multi-spectral scanner |
| NASA | National Aeronautic and Space Administration |
| NDVI | Normalised difference vegetation index |
| NLR | National Aerospace Laboratory of the Netherlands |
| NOAA | National Oceanic and Atmospheric Administration |
| NIR | Near-infrared |
| NRI | UK Natural Resources Institute |
| PBAF | Potential breeding activity factor |
| SPOT | Système Probatoire de l'Observation de la Terre |
| TM | Thematic mapper |

Milford, J.R. & Dugdale, G. (1987). *Rainfall Mapping over West Africa in 1986, 1987*. Consultants Report GCP/INT/432/NET. Rome: FAO.

Milford, J.R. & Dugdale, G. (1990). Monitoring of rainfall in relation to the control of migrant pests. *Philosophical Transactions of the Royal Society of London B*, **328**, 689–704.

Mosalam Shaltout, M.A. & Hassen, A.H. (1990). Solar energy distribution over Egypt using cloudiness from Meteosat photos. *Solar Energy*, **45**, 345–51.

Nageswara Rao, P.P. (1988). Remote sensing in plant protection. *Workshop on Agrometeorological Information for Planning and Operation in Agriculture, Calcutta*, 22–26 August. WMO/FAO/BCKV/IMD pp. 129–53.

Pedgley, D.E. (1973). *Testing the Feasibility of Detecting Locust Breeding Sites by Satellite*. Final report to NASA in ERTS-1 Experiment. London: Centre for Overseas Pest Research.

Pedgley, D.E. (1990). Concentration of flying insects by the wind. *Philosophical Transactions of the Royal Society of London B*, **328**, 631–53.

Perry, B.D., Lessard, P., Norval, R., Kundert, K. & Kruska, R. (1990). Climate, vegetation and the distribution of *Rhipicephalus appendiculatus* in Africa. *Parasitology Today*, **6**, 100–4.

Prince, S.D. (1991). Satellite remote sensing of primary production: comparison of results for Sahelian grasslands. *International Journal of Remote Sensing*, **12**, 1301–11.

Prince, S.D. & Justice, C.O. (eds.) (1991). Coarse resolution remote sensing of the Sahelian environment. *International Journal of Remote Sensing*, **12** Special Issue, 1133–421.

Riley, J.R. (1989). Remote sensing in entomology. *Annual Review of Entomology*, **34**, 247–71.

Robinson, T.P. (1991). Modelling the seasonal distribution of habitat suitability for armyworm population development in East Africa using GIS and remote-sensing techniques. PhD thesis, University of Reading.

Roffey, J. (1975). *Programme Plan for Implementing Desert Locust Survey and Control by Satellite Remote Sensing*. Consultant's report. Rome: FAO.

Rogers, D.J. & Randolph, S.E. (1991). Mortality rates and population density of tsetse flies correlated with satellite imagery. *Nature*, **351**, 739–41.

Rose, D.J.W. (1979). The significance of low-density populations of the African armyworm, *Spodoptera exempta* (Walk.). *Philosophical Transactions of the Royal Society of London B*, **287**, 393–402.

Rose, D.J.W., Dewhurst, C.F., Page, W.W. & Fishpool, L.D.C. (1987). The role of migration in the life system of the African armyworm *Spodoptera exempta*. In *Recent Advances in Research on Tropical Entomology*, ed. M.F.B. Chaudhury, pp. 561–9. *Insect Science and its Application*, vol. 8, Special Issue. Nairobi: ICIPE Science Press.

Rosema, A. (1990). Comparison of Meteosat-based rainfall and evapotranspiration mapping in the Sahel region. *International Journal of Remote Sensing*, **11**, 2299–309.

Rosenberg, L. J. and Magor, J. I. (1986). Modelling the effects of changing windfields on migratory flights of the brown planthopper, *Nilaparvata lugens* Stål. In *Plant Virus Epidemics: Monitoring, Modelling and Predicting Outbreaks*, ed. G.D. McLean, R.G. Garrett & W.G. Ruesink, pp. 345–56. Sydney: Academic Press.

Saull, R.J. & Millington, A.C. (1991). *Pakistan Land Cover Zonation*. Final Report, World Bank/UNDP ESMAP, Household Energy Strategy Survey of Pakistan (PAK/88/036). New York: World Bank/UNDP.

Scott, R.W. & Achtemier, G.L. (1987). Estimating pathways of migrating insects carried in atmospheric winds. *Environmental Entomology*, **16**, 1244–54.

Sellers, P.J. (1985). Canopy reflectance, photosynthesis and transpiration. *International Journal of Remote Sensing*, **6**, 1355–72.

Skaf, R., Popov, G.B. & Roffey, J. (1990). The desert locust: an international challenge. *Philosophical Transactions of the Royal Society of London B*, **328**, 525–8.

Southwood, T.R.E. (1981). Ecological aspects of insect migration. In *Animal Migration*, ed. D.J. Aidley, pp. 197–208. Society for Experimental Biology Seminars Series 13. Cambridge: Cambridge University Press.

Townshend, J.R.G. & Justice, C.O. (1988). Selecting the spatial resolution of satellite sensors required for global monitoring of land transformations. *International Journal of Remote Sensing*, **9**, 187–236.

Townshend, J.R.G., Justice, C.O. & Kalb, V. (1987). Characterization and classification of South American land cover types using satellite data. *International Journal of Remote Sensing*, **8**, 1189–207.

Tucker, C.J., Hielkema, J.U. & Roffey, J. (1985*a*). The potential of satellite remote sensing of ecological conditions for survey and forecasting desert-locust activity. *International Journal of Remote Sensing*, **6**, 127–38.

Tucker, C.J., Holben, B.R., Elgin, J.H. Jr & McMurtrey, J.E. III (1981). Remote sensing of total dry matter accumulation in winter wheat. *Remote Sensing of the Environment*, **11**, 171–89.

Tucker, C.J. & Sellers, P.J. (1986). Satellite remote sensing of primary production. *International Journal of Remote Sensing*, **7**, 1395–416.

Tucker, C.J., Townshend, J.R.G. & Goff, T.E. (1985*b*). African land-cover classification using satellite data. *Science*, **227**, 369–75.

Tucker, M.R. (1984). Possible sources of outbreaks of the armyworm, *Spodoptera exempta* (Walk.) (Lepidoptera: Noctuidae), in East Africa at the beginning of the season. *Bulletin of Entomological Research*, **74**, 599–607.

Wiens, J.A. (1989). Spatial scaling in ecology. *Functional Ecology*, **3**, 385–97.

Wood, B.L., Beck, L.R., Washino, R.K. & Palchick, S.M. (1991). Spectral and spatial characterization of rice field mosquito habitats. *International Journal of Remote Sensing*, **12**, 621–6.

Woolley, J.T. (1971). Reflectance and transmittance of light by leaves. *Plant Physiology*, **47**, 656–62.

Wright, D.E. (1987). Analysis of the development of major plagues of the Australian Plague Locust *Chortoicetes terminifera* (Walker) using a simulation model. *Australian Journal of Ecology*, **123**, 423–37.

# 17

# Forecasting systems for migrant pests. I. The Brown Planthopper *Nilaparvata lugens* in China

B.-H. ZHOU, H.-K. WANG AND X.-N. CHENG

## Introduction

Rice is the staple food crop in China, which is the biggest rice producer and consumer in the world. In 1990, 33 million hectares of rice produced approximately 180 million tonnes of grain, representing 45% of the country's total grain output and 37% of world rice output. However, pre-harvest losses caused by insect pests are estimated at several million tonnes each year and a substantial proportion of these losses is attributable to the Brown Planthopper *Nilaparvata lugens*.

*N. lugens* has become an increasingly serious problem since the 1970s. Outbreaks have increased in frequency and the area regularly infested has extended into the Jianghuai region (between the Yangtze and Huaihe Rivers) and north of the Huaihe River. On average, some 13.3 million hectares of the crop are likely to be affected, with an annual loss of some half a million tonnes of grain (Table 17.1). In 1991, the worst year on record, severe damage extended over the whole rice-growing region and 17.3 million hectares of paddy fields were infested by *N. lugens*. In spite of great efforts at prevention and control, which were estimated to have saved approximately 9.8 million tonnes of grain, more than 2 million tonnes were still lost to these infestations. These efforts represented a substantial investment of manpower and material resources and the total loss sustained in that year was estimated at 2000 million Yuan (about US$ 400 million).

This chapter describes the development and operation of a monitoring and forecasting system for this important pest of rice in China.

Table 17.1. *The area of rice infested by the Brown (*Nilaparvata lugens)
*and White-backed (*Sogatella furcifera) *Planthoppers in China and estimated
yield losses for 1987–91 (from statistics of the National General Station
for Plant Protection). Plus signs indicate high, moderate and low levels
of infestation in affected areas*

| Year | Infested area (hectares $\times$ 10$^6$) | Population level | Yield loss (tonnes $\times$ 10$^6$) |
|------|------|------|------|
| 1987 | 18.3 | +++ | 1.02 |
| 1988 | 16.0 | ++ | 0.70 |
| 1989 | 14.9 | + | 0.40 |
| 1990 | 16.4 | ++ | 0.54 |
| 1991 | >20.0 | +++ | >2.00 |

## General features of *N. lugens* occurrence and the current monitoring and forecasting system

In the mid-1970s, the Ministry of Agriculture of China established a cooperative Brown Planthopper Research Project involving 134 research units in twenty provinces. Several aspects of the life cycle and migration of *N. lugens* were investigated by means of:

1 A survey of overwintering populations throughout the main rice-growing areas from Hainan Island to north of the Yangtze River;
2 Trap catches from aeroplanes at various altitudes, at sea, in mountain-top nets and in light traps;
3 Release and recapture of marked adults;
4 Studies of population dynamics in different localities;
5 Studies of migratory flight in relation to meteorological conditions.

The results showed that, in most years, the northern limit of over-wintering by *N. lugens* is the Tropic of Cancer and winter survival north of 25° N is unknown. All infestations at higher latitudes therefore occur as a result of immigration. The northward movement appears to occur in 4–5 successive waves each year, emigration being synchronised with the maturation of the rice plants in the source areas (Fig. 17.1). The migrants move northwards on the southwesterly winds, which prevail during spring and summer in eastern China. Analysis of the data collected in this programme resulted in the identification of eight *N. lugens* outbreak regions linked by these seasonal migrations (Cheng *et al.* 1979, 1983).

As a result of these studies, a joint monitoring and forecasting network was established in 1977. This comprises 2000 monitoring stations, of

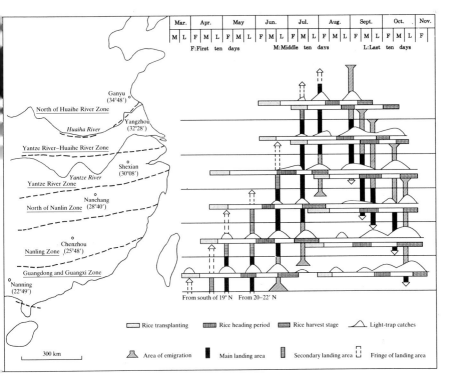

Fig. 17.1. Diagram illustrating the occurrence of the Brown Planthopper in China (see text). *Nilaparvata lugens* occurs between latitudes 18° and 40° N, but outbreaks are generally confined to regions south of 34° N. There are major differences in climate and systems of rice cropping within this latitudinal range. The number of generations of *N. lugens* varies with latitude, periods of immigration and emigration and the timing of transplanting the rice. The insect produces 12 generations annually in the southern part of Hainan Island but achieves only two generations north of the Huaihe River, a reduction of approximately one generation for every 2° increase in latitude. The seasonal occurrence of *N. lugens* can thus be classified into eight zones, of which six, centred on Nanning, Chenzhou, Nanchang, Shexian, Yangzhou and Ganyu, are shown in the figure (the two on Hainan Island are omitted). Names given to the outbreak zones are underlined. The right-hand part of the figure shows major migrations of *N. lugens* in relation to the phenology of the rice crop at these centres.

which 150 are designated as key District Monitoring Stations. At each station, a light trap is used to monitor numbers of immigrants. Field populations are monitored systematically by crop searches. One hundred rice hills in a check plot (>1500 m²) are sampled at four-day intervals, the planthoppers counted and life stages identified. These monitoring data are collected by the District Monitoring Stations and transmitted, by telegram every five days and by post, to the national General Plant Protection Station (GPPS) in Beijing.

The monitoring and forecasting system comprises three levels.

1  GPPS analyses the incoming data from areas of emigration to produce a long-term forecast for the current season. When peaks of emigration are detected, GPPS predicts migratory pathways and identifies the main sink areas for the migration, on the basis of the results of the co-operative research project (see above; Cheng *et al.*, 1979) and of current weather data. These results are sent out to the provincial Plant Protection Stations and to the monitoring-station network. When severe outbreaks are anticipated, as in 1982, 1984, 1987 and 1991, additional warnings are issued and released through television and radio broadcasts and the monitoring effort at provincial and county level is intensified.

2  After immigration has occurred, the county stations make a general estimate of the timing and population size of the generation expected to cause the major damage, on the basis of light-trap data and the densities of immigrants monitored in the field. They also predict the timing and population size of the next generation from the field-monitoring data and current weather reports. This information is issued as a medium-term forecast.

3  The county stations then issue control recommendations to farmers in the form of a short-term forecast, based on the population levels and life stages of the insects monitored locally, and the stages of development of the crop. These recommendations, which are tailored to local conditions, include advice on timing of insecticide application, choice of chemicals and dosages, and proper application methods. In recent years, farmers have been recommended to synchronise planthopper control at the village level and to adopt a preventive approach by attacking the generation preceding the one predicted to cause maximum damage.

Because of the large areas involved and variation in pest occurrence and in local conditions and requirements, there are no universally applied rules governing the preparation and issue of forecasts and recommendations at the county level. Staff of the pest-forecasting units use procedures and practices that have evolved in response to local experience and needs.

### New techniques used in migration and forecasting research
#### *Radar observation of migration*

Although there is incontrovertible evidence for the annual occurrence of northward migrations, the evidence for southward migrations of

*N. lugens* to lower latitudes in the late summer and autumn has been sparse and circumstantial. The resulting controversy over the extent and importance of these southward migrations has made it difficult to assess the relative contributions of immigration and local breeding to late-season infestations of the pest.

In recent years, this southward migration has been the subject of a collaborative project between scientists from the Natural Resources Institute (NRI) of the UK's Overseas Development Administration and Nanjing Agricultural University (NAU), using high-frequency (8-mm wavelength) radar and a net suspended from a kytoon. Radar has the great advantage that it can, except in the presence of precipitation, monitor migrations while they are actually in progress. The studies conducted in Jiangpu, Jiangsu province in 1988–1990 (Riley *et al.*, 1991) confirm that long-distance southward migrations do occur in the area of Nanjing in mid- and late September, *N. lugens* being carried on north-easterly winds towards autumn infestation and overwintering areas. The radar observations show that, after mass takeoff in the late afternoon or at dusk, the migrants fly for from several to more than 10 h during the night, often forming a dense layer at altitudes of between 400 and 1000 m. These dense layer concentrations generally have well-defined ceilings corresponding to an air temperature of about 16 °C. They occur above the top of the near-surface temperature inversion and so above the altitude of maximum air temperature.

Likely source areas for these migrations of *N. lugens* overflying the radar were identified by back-trajectories (Riley *et al.*, 1991). The analyses indicated that the planthoppers had probably emigrated from outbreaks in central Jiangsu province, up to 200 km away to the north-east. Forward-tracking analyses suggested that *N. lugens* leaving the vicinity of the radar site at Jiangpu at dusk would reach areas in south Anhui or north Jiangxi province, about 300 km to the southwest, if they continued to fly for 12 h. The progeny of these migrants, therefore, probably contribute to the infestations that occur on late rice in October and early November in these areas.

Radar observations also allow estimates to be made of *N. lugens* aerial densities and these were in general agreement with those based on the kytoon net catches (Riley, 1992).

### Computer simulation of aerial movements

A trajectory-analysis model to simulate *N. lugens* migration using meteorological charts was developed at the UK's Tropical Development and Research Institute (now NRI) (Magor, 1981; Rosenberg & Magor, 1983*a,b*). The model was derived from one developed in the USA (Heffter, 1980) to evaluate long-term pollution problems and to investigate individual pollution episodes. With the assistance of NRI, NAU applied this model to investigate *N. lugens* migration in China (H.-K. Wang, unpublished data). Values for the parameters used are given by Magor (1981) and Rosenberg & Magor (1983*a,b*, 1986). They include a maximum flight duration of 30 h and a minimum temperature threshold for flight of 17 °C.

The model reads a data file that contains the locations of meteorological stations and daily files that contain wind directions and speeds, and temperatures at specific altitudes for each station. After entering the trajectory start date, time and duration, and the name and location (longitude and latitude) of the study point, the model outputs the trajectory endpoint (latitude and longitude), distance, wind direction and speed, and temperature at up to 3–h intervals for the duration of the trajectory. The results can be shown on screen or plotted to charts.

The results of these trajectory analyses, when integrated with information on rice growth-stages and synopses of direct observations of *N. lugens* infestations in southern China, indicate the following potential sources of migrant *N. lugens* (Fig. 17.1).

1  Late June to the beginning of July: south of latitude 24.5° N, longitude 109–118° E – southern Guangdong and Guangxi.
2  Early to mid-July: latitude 24–28° N, longitude 109–119° E – northern Guangdong and Guangxi, southern Hunan and Jiangxi, and western Fujian. The most important source is likely to be in the region of Nanling.
3  Late–July: latitude 24–31° N, longitude 109–121° E – the area between northern Guangdong and Guangxi, and Hubei, Anhui and Zhejiang. The most important source is likely to be to the north of Nanling.

The trajectories produced also correspond, in general, with those suggested by catches on ships in the East China Sea and by recaptures of marked insects (H.-K. Wang, unpublished data), and they support the migration pathways of *N. lugens* in east China proposed by Cheng *et al.* (1979). The method is, therefore, useful for studying insect migration in the lower atmosphere, supplementing radar studies (Riley *et al.*, 1991), and the data generated can be used in mid- to long-term forecasting models of the trans-regional spread of *N. lugens* infestations.

## Simulation models and the decision-making system

Considerable progress has also been made in the development of short-, medium- and long-term forecasting and management models (Qi *et al.*, 1988, 1991; Cheng *et al.*, 1989; Cheng, Chen & Zhu, 1990*a*; Cheng & Holt, 1990; Cheng, Norton & Holt, 1990*b*; Holt, Cheng & Norton, 1990). These include population simulations, management and economic threshold models, and expert systems. They are based on systematic studies of selected biological parameters and environmental conditions involved in the dynamics of experimental and field populations.

The first population-simulation model for *N. lugens* in China was developed by Qi *et al.* (1988). It was based on a Leslie population matrix and used biological and ecological parameters derived from the results of Chen *et al.* (1986) and Ding *et al.* (1987). After the entering of field-population data (including age structure), stage of development of the rice plants and temperature, the model simulates subsequent population development and dynamics. Its performance was tested successfully against field data from Taihu Lake District, Jiangsu province.

Cheng *et al.* (1990*a*) proposed a computer-aided decision-making system that comprised the following four subsystems:

1 A database that stores and accesses information on immigration, temperature, population dynamics and pesticide application from Jiaxin and neighbouring counties in Zhejiang province;
2 A simulation model (Cheng *et al.*, 1989, 1990*a*; Cheng & Holt, 1990) for predicting timing and size of peak populations, population dynamics and yield losses;
3 A decision analysis subsystem composed of an expert system (Holt *et al.*, 1990), which was developed to improve recommendations for planthopper control;
4 A help component.

The system promises to provide a useful tool to support and improve decision-making in planthopper management.

These models have been developed for the regions mentioned and relate specifically to local conditions. Further research now in progress at Nanjing Agricultural University can draw upon computerised databases of pest incidence, cropping practices and environmental conditions in more than 200 counties in 14 provinces. Qi *et al.* (1991) described an *N. lugens* management system, utilising this database, that consisted of a simulation model and a management model. The structure of the simulation model was that of the boxcar-train model developed by Goudriaan (1973). The management model produces optimal control strategies determined by a dynamic programming method. By the choice of different parameter values from the database, this management system can be adapted to conditions in different locations.

The results of this multidisciplinary research provide a foundation for further improvements in current forecasting and pest-management systems. Proposals are being prepared for a collaborative programme between NAU and NRI, to establish a computer-based information, forecasting and decision-making system for eastern China utilising the techniques and information discussed above. This will improve the precision and reliability of forecasts and management recommendations and reduce the time required for their preparation and dissemination.

## Further research relevant to *N. lugens* migration and forecasting
### *Regional and local occurrence*

In addition to the large-scale projects on *N. lugens* migration described above, there have been a number of local studies focussed on areas with specific topographical features. Patterns of immigration, landing and distribution of *N. lugens* in coastal areas, lake margins and mountain valleys frequently differ from those observed on the plains, because of the characteristic atmospheric circulations associated with them. In these areas, peaks of immigration typically occur earlier and more frequently and population densities are often higher and the damage caused more serious (M. Hu, personal communication). For example, in 1985, the first wave of immigration at Jiuli Station (Ganyu county, Jiangsu province) 2.5 km from the Yellow Sea coast, was 17 days earlier than at Ganyu County Station 5.5 km from the coast, and total light-trap catches before the end of July at the former station were seven times larger than at the latter. Similar differences have been observed at stations in mountain valleys and on lake margins in China (M. Hu, unpublished data), and between mountain valleys and lowland paddy fields in Japan (Noda & Kiritani, 1989).

The earlier occurrence of immigration at these locations can provide warnings of wider-scale invasions and so contribute to forecasting. Such studies are also likely to make an important contribution to our understanding of the dynamics of planthopper migration.

## Biotypes

*N. lugens* can adapt to particular rice cultivars or resistant varieties rapidly, often within ten generations, this adaptation involving a change in biotype. For example, Wu *et al.* (1983) reared *N. lugens*, collected from the field and predominantly of biotype 1, on var. Mudgo (with the *Bph*1 resistance gene) for eleven generations in the laboratory. This resulted in a strain, identified as biotype 2, adapted to this resistant variety. Similar shifts in biotype have been observed with the extension of cultivation of resistant rice varieties.

From 1973, cultivation of IR26, an *N. lugens*-resistant variety, became widespread in southeast Asian countries, including the Philippines, Indonesia and Vietnam. After only three years, biotype 2, adapted to this variety, became dominant in the region (Feuer, 1976; Huynh, 1977; Mochida *et al.*, 1977). Because *N. lugens* cannot overwinter in most rice-growing areas in China, the biotypes of field populations in all areas except Hainan Island and southern Guangxi province would be expected to reflect those of the seasonal immigrants from sources in southeast Asia. However, samples of *N. lugens* subsequently collected from 118 counties and cities of ten provinces in China were all found to be biotype 1 (Wu, 1988). The reasons for this are not known, although it has been suggested that biotype 1 is the most vigorous among *N. lugens* biotypes and is thus likely to predominate in sink areas following long-distance migrations (Wu *et al.*, 1990). Even in Nanning, Yulin and Guilin, in Guangxi province, where *N. lugens* can occasionally overwinter, the proportions of biotype 2 in field populations showed no increase until 1987 (Li *et al.*, 1991). This delayed transition to the other biotypes in the major rice-growing regions of China has been a substantial but un-expected bonus to rice breeding.

### Insecticide resistance

Cao, Shou & Liao (1987) surveyed the susceptibility of *N. lugens* in the Chenzhou area, Hunan province, to 11 insecticides, obtaining $LD_{50}$ values similar to those reported for *N. lugens* in Japan in 1967 (Nagata, 1982). In laboratory experiments, however, malathion and carbofuran-selected strains developed 7.2- and 8-fold resistance, respectively, within eight generations (Wang, Ku & Chu, 1988*a*). This lack of insecticide resistance in the field, in spite of widespread chemical control throughout eastern Asia, can be attributed to the origin of different waves of immigration, within and between seasons, in source areas in which the extent of chemical control and the insecticides used are likely to differ widely. The absence of consistent selection for specific resistance mechanisms over several generations, as well as regular intermixing with susceptible immigrants, must impede the evolution of resistance. The observation that local differences in susceptibility to insecticides are apparent in autumn but not in spring in Taiwan supports this interpretation (Wang, Ku & Chu, 1988*b*).

## Conclusions

Although there has been significant progress in the development of forecasting and pest-management systems for *N. lugens* in China, the intermittent occurrence of outbreaks of this migratory pest continues to present farmers and plant-protection organisations with formidable management problems. In view of the continuing pressure of immigration from southeast Asia and within China, *N. lugens* can be expected to remain a major threat to the rice industry for the foreseeable future. Further progress in meeting the challenge posed by this pest depends, therefore, on continued research and development in the following areas:

1  The influence of atmospheric circulation and synoptic weather patterns on insect migration;
2  The characteristics of regional and local patterns of immigration and their implications for integrated pest management;
3  The establishment of radar observation stations and the deployment of alpine-net traps at critical sites to improve monitoring of migrations;
4  Biotype monitoring;
5  The ecology, physiology and genetics of insect migration;
6  The development of a computerised forecasting system.

## References

Cao, C.-R., Shou, Z.-B. & Liao, M.-H. (1987). Monitoring of insecticide resistance of *Sogatella furcifera* (Horváth) and *Nilaparvata lugens* (Stål). *Journal of Nanjing Agricultural University*, **4**, 135–8. In Chinese, English summary.

Chen, R.-C., Qi, L.-Z., Cheng, X.-N., Ding, Z.-Z. & Wu, Z.-L. (1986). Studies on the population dynamics of brown planthopper *Nilaparvata lugens* (Stål) I. Effects of temperature and diet conditions on the growth of experimental population. *Journal of Nanjing Agricultural University*, **3**, 23–33. In Chinese, English summary.

Cheng, J.-A., Chen, M.-G. & Zhu, Z.-R. (1990a). A computer system for decision making process in BPH management. *Acta Agriculturae Universitatis Zhejiangensis*, **16**, 129–33. In Chinese, English summary.

Cheng, J.-A. & Holt, J. (1990). A systems-analysis approach to brown planthopper control on rice in Zhejiang province, China. I. Simulation of outbreaks. *Journal of Applied Ecology*, **27**, 85–99.

Cheng, J.-A., Norton, G. A. & Holt, J. (1990b). A systems-analysis approach to brown planthopper control on rice in Zhejiang province, China. II. Investigation of control strategies. *Journal of Applied Ecology*, **27**, 100–12.

Cheng, J.-A., Zhang, L.-G., He, S.-Z., Fan, Q.-G., Holt, J. & Norton, G. A. (1989). A simulation model of population dynamics of rice brown planthopper and its validation. *Acta Agriculturae Universitatis Zhejiangensis*, **15**, 131–6. In Chinese, English summary.

Cheng, X.-N., Chen, R.-C., Hu, J.-Z. & Zhu, S.-S. (1983). Forecasting the migrations of Brown Planthopper (*Nilaparvata lugens* (Stål)) in China. In *First International Workshop on Leafhoppers and Planthoppers of Economic Importance*, ed. W.J. Knight, N.C. Pant, T.S. Robertson & M.R. Wilson, pp.359–64. London: CAB International.

Cheng, X.-N., Chen, R.-C., Xi, X., Yang, L.-M., Zhu, Z.-L., Wu, J.-C., Qian, G.-R. & Yang, J.-S. (1979). Studies on the migrations of Brown Planthopper *Nilaparvata lugens* (Stål). *Acta Entomologica Sinica*, **22**, 1–21. In Chinese, English summary.

Ding, Z.-Z., Wu, Z.-L., Qi, L.-Z., Chen, R.-C., Cheng, X.-N. & Li, R.-D. (1987). A study of the population dynamics of brown planthopper II. Some biological parameters affecting field population dynamics. *Journal of Nanjing Agricultural University*, **4**, 42–7. In Chinese, English summary.

Feuer, R. (1976). Biotype 2 brown planthopper in the Philippines. *International Rice Research Newsletter*, **1**, 15.

Goudriaan, J. (1973). Dispersion in simulation models of population growth and salt movement in the soil. *Netherlands Journal of Agricultural Science*, **21**, 269–81.

Heffter, J.L. (1980). *Atmospheric Transport and Dispersion Model (ARL-ATAD)*. Technical Memorandum ERL ARL-81. Silver Spring, Maryland: National Oceanic and Atmospheric Administration.

Holt, J., Cheng, J.-A. & Norton, G.A. (1990). A systems-analysis approach to brown planthopper control on rice in Zhejiang province, China. III. An expert system for making recommendations. *Journal of Applied Ecology*, **27**, 113–22.

Huynh, N.V. (1977). A new biotype of the brown planthopper in the Mekong Delta of Vietnam. *International Rice Research Newsletter*, **2**, 10.

Li, Q., Luo, S.-Y., Wei, S.-M., Huang, R.-Q., Shi, A.-X. & Huang, H.-Y. (1991). [Preliminary studies on biotypes of brown planthopper, *Nilaparvata lugens* (Stål), in Guangxi.] *Guangxi Nongye Kexue [Agricultural Science in Guangxi]* **1**, 29–32. In Chinese.

Magor, J.I. (1981). *COPR/HFC Collaborative Project on the Migration of the Brown Planthopper* Nilaparvata lugens *(Stål) in Northeastern India. Report of Preliminary Studies 1981*. London: Tropical Development Research Institute.

Mochida, O., Oka, I.N., Dandi, S., Harahap, Z., Sutjipto, P. & Beachell, H.M. (1977). IR26 found susceptible to the brown planthopper in North Sumatra, Indonesia. *International Rice Research Newsletter*, **2**, 10.

Nagata, T. (1982). Insecticide resistance and chemical control of the Brown Planthopper, *Nilaparvata lugens* (Stål). *Bulletin of the Kyushu National Agricultural Experiment Station*, **22**, 49–164.

Noda, T. & Kiritani, K. (1989). Landing places of migratory planthoppers, *Nilaparvata lugens* (Stål) and *Sogatella furcifera* (Hováth) (Homoptera: Delphacidae) in Japan. *Applied Entomology and Zoology*, **24**, 59–65.

Qi, L.-Z., Chen, R.-C., Cheng, M.-Y. & Ding, Z.-Z.. (1988). [A prediction model for Brown Planthopper, *Nilaparvata lugens* (Stål), population dynamics: BASIC programme and its usage.] *Kunchong Zhishi [Entomological Knowledge]*, **25**, 257–61. In Chinese.

Qi, L.-Z., Huang, F.-N., Huang, J.-Y. & Li, R.-D. (1991). Studies on a management model of the population life system of brown planthopper. *Chinese Journal of Applied Ecology*, **2**, 214–20. In Chinese, English summary.

Riley, J.R. (1992). A millimetric radar to study the flight of small insects. *Electronics & Communication Engineering Journal*, **4**, 43–8.

Riley, J.R., Cheng, X.-N., Zhang, X.-X., Reynolds, D.R., Xu, G.-M., Smith, A.D., Cheng, J.-Y., Bao, A.-D. & Zhai, B.-P. (1991). The long-distance migration of *Nilaparvata lugens* (Stål) (Delphacidae) in China: radar observations of mass return flight in the autumn. *Ecological Entomology*, **16**, 471–89.

Rosenberg, L.J. & Magor, J.I. (1983*a*). Flight duration of the brown planthopper, *Nilaparvata lugens* (Homoptera: Delphacidae). *Ecological Entomology*, **8**, 341–50.

Rosenberg, L.J. & Magor, J.I. (1983*b*). A technique for examining the long-distance spread of plant virus diseases transmitted by the brown plant-hopper, *Nilaparvata lugens* (Homoptera: Delphacidae), and other wind-borne insect vectors. In *Plant Virus Epidemiology*, ed. R.T. Plumb & J.M. Thresh, pp. 229–38. Oxford: Blackwell Scientific Publications.

Rosenberg, L.J. & Magor, J.I. (1986). Modelling the effects of changing windfields on migratory flights of the brown planthopper, *Nilaparvata lugens* Stål. In *Plant Virus Epidemics: Monitoring, Modelling and Predicting Outbreaks*, ed. G.D. McLean, R.G. Garrett & W.G. Ruesink, pp. 345–56. Sydney: Academic Press.

Wang, S.-C., Ku, T.-Y. & Chu, Y.-I. (1988*a*). Resistance patterns in the brown planthopper *Nilaparvata lugens* (Homoptera: Delphacidae) after selection with four insecticides and their combinations. *Plant Protection Bulletin (Taiwan)*, **30**, 59–67. In Chinese, English summary.

Wang, S.-C., Ku, T.-Y. & Chu, Y.-I. (1988*b*). Local difference of insecticide resistance of Brown Planthopper (*Nilaparvata lugens*) in Central Taiwan. *Plant Protection Bulletin (Taiwan)*, **30**, 52–8. In Chinese, English summary.

Wu, G.-R., Chen, F.-L., Tao, L.-Y., Huang, C.-W. & Feng, B.-C. (1983). Studies on biotype of Brown Planthopper (*Nilaparvata lugens* (Stål)). *Acta Entomologica Sinica*, **26**, 154–9. In Chinese, English summary.

Wu, G.-R., Tao, L.-Y., Yu, X.-P. & Saxena, R.C. (1990). [Present situation of biotypes of Brown Planthopper, *Nilaparvata lugens* (Stål).] *Kunchong Zhishi [Entomological Knowledge]*, **27**, 47–51. In Chinese.

Wu, R.-Z. (1988). Advances in rice breeding for resistance to insect pests in China. In *The Developments in Integrated Control of Rice Diseases and Insect Pests in China*, ed. Zeng Z.-H., pp.62–76. Zhejiang: Science and Technology Press of Zhejiang. In Chinese, English summary.

# 18

# Forecasting systems for migrant pests. II. The rice planthoppers *Nilaparvata lugens* and *Sogatella furcifera* in Japan

T. WATANABE

## Introduction

The long-range migrations of the the Brown Planthopper *Nilaparvata lugens* and the White-backed Planthopper *Sogatella furcifera* extend over the whole area of rice cultivation in East Asia and are closely associated with the wind systems of the monsoon climate of the region (Chapter 3, this volume). Because they do not overwinter in Japan, field populations of these planthoppers originate with overseas immigration in the *Bai-u* (rainy) season in June and July.

Kisimoto (1976) found that immigration of rice planthoppers into Japan was associated with the activity of frontal depressions during the *Bai-u* season and concluded that mass immigration was correlated with depressions that had passed through eastern central China. The south-west winds in these depressions carry the migrating planthoppers and Rosenberg & Magor (1983) used trajectory analysis to estimate the source of *N. lugens* arriving in Japan or caught on board ship in the East China Sea. They showed that trajectories based on winds at 1.5 km above sea level originated in potential emigration source areas more frequently than trajectories at 10 m.

Seino *et al.* (1987) obtained highly significant correlations between the immigration of planthoppers into Japan and the development of low-level jet (LLJ) airflows (see below) that coincide with the occurrence of heavy *Bai-u* rainfall. This chapter discusses the use of analysis of LLJs in monitoring systems to determine the location and timing of rice-planthopper immigration into Japan.

## Monitoring systems

### *Definition of LLJ*

The *Bai-u* front often incorporates a fast-moving stream of air with a windspeed maximum at a height of between 1 and 3 km: the *Bai-u* LLJ (Ninomiya, 1979). The LLJs referred to in this chapter are strong, southwesterly air currents in the warm sector of a depression moving along the *Bai-u* front (Akiyama, 1973). They are most intense 200–300 km south of the front and occur over the East China Sea and Japan. An LLJ is tentatively identified as being present in this region if windspeeds at the 850–hPa standard pressure level (approximately 1500 m above sea level) exceed 20 knots (approximately 10 m s$^{-1}$) (Seino *et al.*, 1987).

### *A computer program for monitoring LLJs*

A computer program has been developed to identify LLJs objectively (Watanabe *et al.*, 1988, 1990). Windspeed and directions at 74 sites on the 850-hPa weather charts covering East China and Japan were used for the analysis (Fig. 18.1). The atmospheric observations were carried out at 0900 h and 2100 h Japan Standard Time (JST) every day and 850-hPa weather charts were obtained by radio facsimile from the Japan Meteorological Agency or via computer network from the Japan Meteorological Association. Meteorological stations are not uniformly distributed, so grid values of the windfield were estimated as follows. The area for analysis was divided into a 21 × 17 grid with a spacing of approximately 150 km east–west and 125 km north–south, and windspeed and direction for each grid point were interpolated with the use of data from the three closest meteorological stations. An example of the observational data and the analysed output is given in Fig. 18.2.

### *Correlation between LLJs and planthopper immigration*

It is well established that major peaks of immigration usually occur between late June and early July in Kyushu in southern Japan (Sogawa & Watanabe, 1989). Kisimoto (1991) compared the efficiency of various traps for monitoring the windborne immigration of rice planthoppers and concluded that net traps were the most effective. Two net traps (1 m in diameter and 1.5 m in depth), deployed 10 m above the ground, were

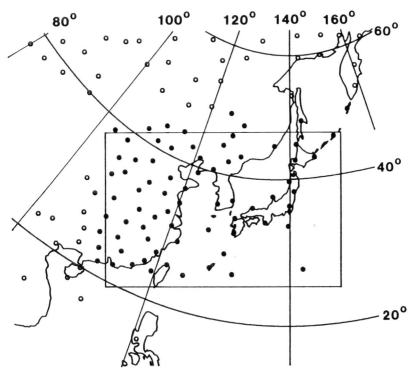

Fig. 18.1. Map showing the area (rectangular frame) over which low-level jet streams (LLJs) are monitored. Each circle indicates the location of a meteorological station, the filled circles showing those providing data for the analysis. From Watanabe *et al.* (1988) with permission.

used routinely for monitoring planthopper immigration at Kyushu National Agricultural Experiment Station (KNAES) in Chikugo, Fukuoka Prefecture, Kyushu. Analysis of LLJs during the *Bai-u* season commenced at the Laboratory of Pest Management Systems in KNAES in 1987.

Seino *et al.* (1987) identified 37 possible immigration periods on the basis of their analysis of the development of LLJs during the *Bai-u* seasons of 1980–86 and found that 35 of these coincided with actual immigration detected by net traps at KNAES. For example, the correlation between immigration of rice planthoppers and the development of LLJs in 1987 (Watanabe *et al.*, 1991) is shown in Fig. 18.3. In early June, minor immigration of *S. furcifera* was observed when an LLJ became established over the South China Sea across the Ryukyu Islands and Kyushu in 1–10 June (e.g. Fig. 18.4a). However, massive immigration of

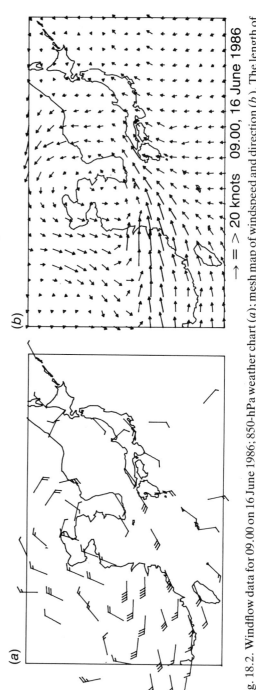

→ = > 20 knots    09.00, 16 June 1986

Fig. 18.2. Windflow data for 09.00 on 16 June 1986; 850-hPa weather chart (*a*); mesh map of windspeed and direction (*b*). The length of each arrow indicates the relative windspeed. From Watanabe *et al.* (1988) with permission.

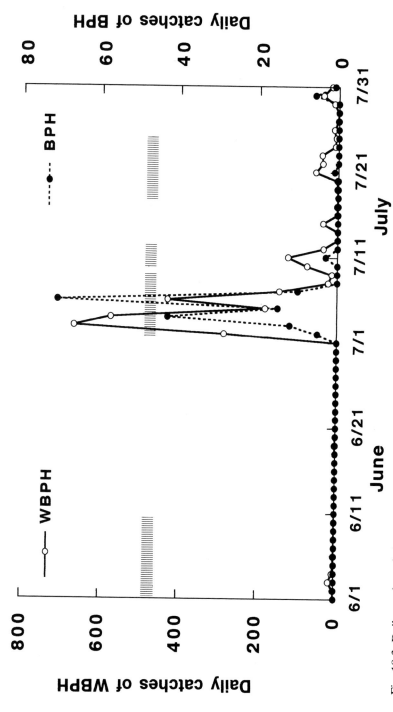

Fig. 18.3. Daily catches of planthoppers by the two net traps at Chikugo, Fukuoka, Japan, in June and July, 1987. WBPH: *Sogatella furcifera*; BPH: *Nilaparvata lugens*. Horizontal bar indicates the periods during which an LLJ was established. From Watanabe *et al.* (1991) with permission.

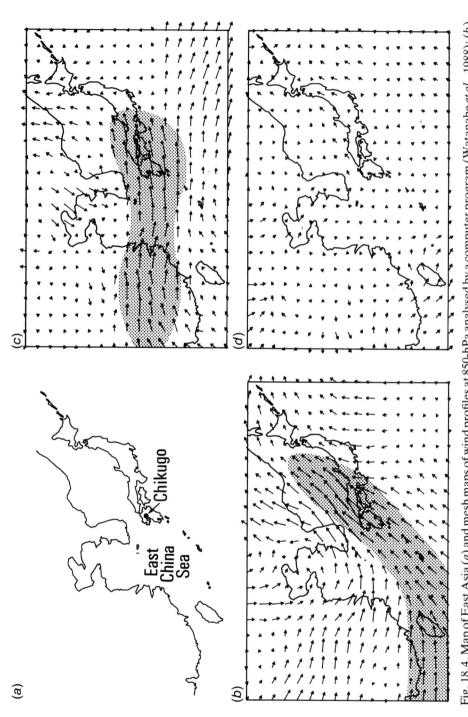

Fig. 18.4. Map of East Asia (*a*) and mesh maps of wind profiles at 850-hPa analysed by a computer program (Watanabe *et al.*, 1988): (*b*)

both *S. furcifera* and *N. lugens* into Kyushu took place in 2–9 July. During this period, a typical LLJ developed, extending from south China to Kyushu, via the Yangtze River delta and the East China Sea (Fig. 18.4*b*). In the absence of LLJs connecting mainland China and southern Japan, planthopper immigration did not occur in Kyushu (e.g. Fig. 18.4*c*).

Akiyama (1973) showed that the area of Japan covered by LLJs shifted gradually northward as the *Bai-u* season progressed. In order to verify the expected change in the pattern of immigration over Japan correlated with this seasonal movement of the LLJs, a database of light-trap catches at 69 locations during the immigration periods in 1980–85 was utilised (Watanabe & Seino, 1991). Although the efficiency of light traps depends on weather conditions, they have been widely used to monitor the seasonal abundance of insect pests in Japan for about 50 years. Thirty-two periods of major immigration of *S. furcifera* were detected from 1980 to 1985 and their geographical distribution reflected the seasonal shift in the pattern of LLJs.

Early in the *Bai-u* season (early to mid-June), when LLJs blow from the coastal regions of south China to coastal areas of western Japan, over the southwestern part of the East China Sea (e.g. Fig. 18.5*a*), immigration was restricted to southwestern coastal areas in Japan (Fig. 18.5*b*).

The most frequent pattern of association between the distribution of planthopper immigration and the path of LLJs is illustrated in Fig. 18.5*c* and 18.5*d*, showing an LLJ stretching from the Yangtze River delta to cover the greater part of southwestern Japan. Immigration in these conditions was widely distributed in the western part of Japan and was particularly concentrated in the western coastal regions of Kyushu, decreasing towards the northeast. This pattern, which can be observed during the whole of the *Bai-u* season, but occurs mainly from late June to mid-July, accounted for approximately half the instances of major immigration into Japan over the five-year period.

When the path of an LLJ turned northeastwards into the Sea of Japan (e.g. Fig. 18.5*e*), the main areas of immigration were located in the western coastal region, with some immigrants reaching the northeast of the country (Fig. 18.5*f*). This pattern was observed seven times late in the *Bai-u* season, from mid- to late July, in the years 1980–85.

### *Forecasting service for rice planthoppers*

A forecasting service, utilising a computer network and covering a range of pests and plant diseases, has been provided by the Plant Protection

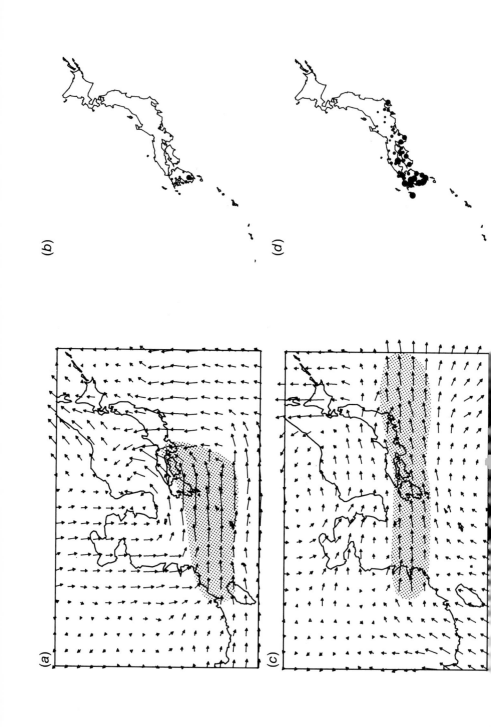

(a)

(b)

(c)

(d)

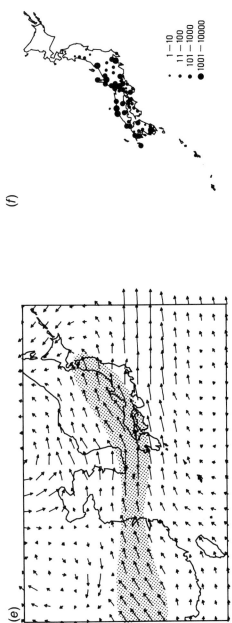

(e)

(f)

```
·    1—10
·    11—100
●    101—1000
●    1001—10000
```

Fig. 18.5. Wind profiles at 850-hPa with shaded areas showing the location and path of LLJs (left) and the distribution of *Sogatella furcifera* immigration (right) in Japan. (*a,b*) 21.00, 12 June 1983; (*c,d*) 21.00, 17 July 1982; (*e,f*) 21.00, 21 July 1983 (Japan Standard Time). From Watanabe & Seino (1991) with permission.

Division of the Ministry of Agriculture, Forestry and Fisheries since 1987. The Pest Forecasting Service System (Yokota, 1987) is based on daily catches of pests, including rice planthoppers, in traps at more than 200 sites, operated by staff of the Pest Control Offices of the 47 prefectures in Japan. Traps are deployed at several sites within each prefecture and selected 'observation fields' are routinely monitored for pests and diseases using standardised procedures. These data are transmitted through a computer network, to be collated on a host computer. Information on pest and disease incidence for the whole country can then be accessed through terminal computers in the prefectural Pest Control Offices. The data are also used to compile forecasts issued by the Plant Protection Division in a national bulletin every two weeks.

In addition to the results of these local surveys, synoptic-scale meteorological information is required to predict the onset and magnitude of outbreaks of migratory pests, such as the rice planthoppers, that reach Japan from overseas. A unit at the Kagoshima Prefectural Agricultural Experiment Station, located in a region consistently reporting the highest densities of immigrant planthoppers, utilises the computer programme described above to monitor the development of LLJs and to determine their paths. The resulting forecasts of immigration are disseminated by facsimile to the Pest Control Offices in each prefecture, where they are integrated with the survey data.

## Conclusions

It is apparent that these analyses of the development and paths of LLJs, based on 850-hPa weather charts, make a useful contribution to forecasting and monitoring of the timing and location of the immigration of rice planthoppers into Japan. However, information on LLJs relates only to the processes transporting the migrant planthoppers and tells us nothing about the extent and timing of take-off in the source areas or of arrival at the destinations of these migrations. The only information relating to the embarkation of the migratory planthoppers that is available in Japan consists of limited reports of the occurrence of these pests and of the cropping pattern in the source areas in mainland China.

Both the magnitude of immigrations of *S. furcifera* and the annual fluctuations in the scale of immigration of both planthopper species have been increasing since the late 1970s (Watanabe, Sogawa & Suzuki, 1994). Recently, it has been observed that these changes in the numbers of immigrants have been associated with qualitative changes in the resulting populations, including changes in wing-form responses to density (Iwanaga, Tojo & Nagata, 1985; Iwanaga, Nakasuji & Tojo, 1988) and in the biotypes present (Sogawa, 1992). International cooperation is clearly essential if the population and immigration dynamics of rice planthoppers in East Asia are to be understood.

## References

Akiyama, T. (1973). A geostrophic low-level jet stream in the *Bai-u* season associated with heavy rainfalls over the sea area. *Journal of the Meteorological Society of Japan*, **51**, 205–8.

Iwanaga, K., Nakasuji, F. & Tojo, S. (1988). Wing polymorphism in Japanese and foreign strains of the Brown Planthopper, *Nilaparvata lugens*. *Entomologia Experimentalis et Applicata*, **43**, 3–10.

Iwanaga, K., Tojo, S. & Nagata, T. (1985). Immigration of the Brown Planthopper, *Nilaparvata lugens*, exhibiting various responses to density in relation to wing morphism. *Entomologia Experimentalis et Applicata*, **38**, 101–8.

Kisimoto, R. (1976). Synoptic weather conditions inducing long-distance immigration of planthoppers, *Sogatella furcifera* (Horváth) and *Nilaparvata lugens* (Stål). *Ecological Entomology*, **1**, 92–109.

Kisimoto, R. (1991). Long-distance migration of rice insects. In *Rice Insects: Management Strategies*, ed. E.A. Heinrichs & T.A. Miller, pp. 167–95. New York: Springer-Verlag.

Ninomiya, K. (1979). Low-level jet stream and rain band on the *Bai-u* front. *Kisho Kenkyu Note*, **138**, 118–41. In Japanese.

Rosenberg, L.J. & Magor, J.I. (1983). Flight duration of the Brown Planthopper, *Nilaparvata lugens* (Homoptera: Delphacidae). *Ecological Entomology*, **8**, 341–50.

Seino, H., Shiotsuki, Y., Oya, S. & Hirai, Y. (1987). Prediction of long distance migration of rice planthoppers to northern Kyushu considering low-level jet stream. *Journal of Agricultural Meteorology*, **43**, 203–8.

Sogawa, K. (1992). Rice Brown Planthopper (BPH) immigrants in Japan change biotype. *International Rice Research Newsletter*, **17**, 26–7.

Sogawa, K. & Watanabe, T. (1989). Outline of long-term yearly fluctuations of the rice planthopper occurrence from light-trap data at Kyushu National Agricultural Experiment Station. *Proceedings of the Association for Plant Protection of Kyushu*, **35**, 65–8. In Japanese, English summary.

Watanabe, T. & Seino, H. (1991). Correlation between the immigration area of rice planthoppers and the low-level jet stream in Japan. *Applied Entomology and Zoology*, **26**, 457–62.

Watanabe, T., Seino, H., Kitamura, C. & Hirai, Y. (1988). Computer program for forecasting the overseas immigration of long-distance migratory rice planthoppers. *Japanese Journal of Applied Entomology and Zoology*, **32**, 82–5. In Japanese, English summary.

Watanabe, T., Seino, H., Kitamura, C. & Hirai, Y. (1990). A computer program, LLJET, utilising an 850-hPa weather chart to forecast long-distance rice planthopper migration. *Bulletin of the Kyushu National Agricultural Experiment Station*, **26**, 233–60. In Japanese, English abstract.

Watanabe, T., Sogawa, K., Hirai, Y., Tsurumachi, M., Fukamachi, S. & Ogawa, Y. (1991). Correlation between migratory flight of rice planthoppers and low-level jet stream in Kyushu, south-western part of Japan. *Applied Entomology and Zoology*, **26**, 215–22.

Watanabe, T., Sogawa, K. & Suzuki, Y. (1994). Analysis of yearly fluctuations in the occurrence of migratory rice planthoppers, *Nilaparvata lugens* Stål and *Sogatella furcifera* Horváth, based on light-trap data in northern Kyushu. *Japanese Journal of Applied Entomology and Zoology*, **38**, 7–15. In Japanese, English summary.

Yokota, T. (1987). New on-line system for forecast information. *Shokubutu-boeki [Plant Protection (Japan)]*, **41**, 454–6. In Japanese.

# 19

# Forecasting systems for migrant pests. III. Locusts and grasshoppers in West Africa and Madagascar

M. LECOQ

## Introduction

Many acridoid species are known to undertake long-distance migratory flights (Farrow, 1990). The migrations of gregarious-phase locusts are a well-known phenomenon (Rainey, 1951, 1963; Uvarov, 1977; Pedgley, 1981) but those of solitary locusts and grasshoppers have been observed only relatively recently, over the past 30 to 40 years (Golding, 1948; Davey, Descamps & Demange, 1959; Waloff, 1963; Têtefort, Dechappe & Rakotoharison, 1966; Botha & Jansen, 1969; Launois, 1974*b*; Lecoq, 1975, 1978*a,b*). For the last 20 years, French scientists now at CIRAD-PRIFAS in Montpellier have studied the factors influencing the migrations of solitary locusts and grasshoppers and their role in the events leading to outbreaks of pest species. The work has led to the development of survey and warning systems for three of these, the Migratory Locust, the Senegalese Grasshopper and, more recently, the Desert Locust (Lecoq, 1991*a*). This chapter describes the work on the Migratory Locust *Locusta migratoria capito* in Madagascar (with a brief reference to *L. m. migratorioides* in West Africa) and the Senegalese Grasshopper *Oedaleus senegalensis* in West Africa.

## Flight movements and a warning system for the Malagasy Migratory Locust

### *Occurrence and importance of flight movements*

In Madagascar, gregarisation of *L. m. capito* often begins in an extensive outbreak area in the southwest of the country and swarms originating here may subsequently invade the entire island (Fig. 19.1). The pioneering work of Zolotarevsky (1929) provided the impetus for research which

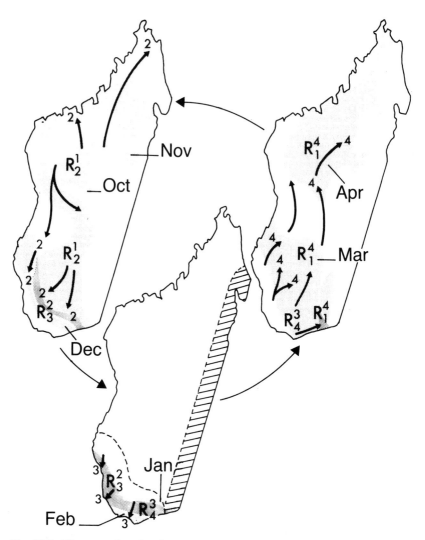

Fig. 19.1. The annual cycle of movements of solitary-phase *Locusta migratoria capito* in Madagascar (modified after Launois, 1974*b* and Duranton *et al.*, 1982)

The shaded areas represent the zones favourable for the development of the Migratory Locust in the months specified. Arrows indicate migratory movements by which the insects follow the seasonal displacements of these zones. The areas occupied by each generation are indicated by $R_b^a$, with *a*, the number of that generation in the seasonal cycle and *b*, the number of the subsequent generation. The number associated with each arrow indicates the generation to which the migrants belong.

In the middle of the rainy season, in January and February, the populations become concentrated in the southeastern coastal region of the island. This is the culminated, in the 1970s, in an FAO/UNDP project. Its aim was to develop a preventive control strategy based on improved knowledge of the dynamics of solitarious populations and the ecological conditions leading to locust outbreaks (FAO, 1973).

The annual cycle of solitarious *L. m. capito*, which normally complete 4–5 generations per year, includes a series of nocturnal migrations. These movements occur throughout the year and are influenced by wind direction. Detailed studies of the dynamics of local populations indicate an association between influxes of migrants and winds from the direction of potential sources (Lecoq, 1975). Further evidence comes from the association of observed departures and arrivals of migrants, at locations in the south of the island, with wind direction (Launois, 1974b), and from large-scale mark–recapture studies (Têtefort *et al.*, 1966). The results of the mark–recapture studies, which involved several million individuals, also demonstrate that displacements frequently exceeded several tens of kilometres; some marked individuals were recovered several hundred kilometres from their point of departure.

Favourable habitats for *L. m. capito* depend on rainfall, which is, therefore, an essential factor regulating these migrations (Launois, 1974b; Lecoq, 1975). Archive data at the Malagasy Anti-acridian Service (MAS) have been used to plot the distribution of locust populations, and the occurrence of major migrations and upsurges, in relation to monthly rainfall. The results provide evidence of a redistribution of populations from areas of very low or high rainfall, to those where it has been moderate (Darnhofer & Launois, 1974). It appears, therefore, that both inadequate and excessive rainfall induce emigration.

Field studies of population dynamics also provide evidence of the influence of rainfall on both migration and oviposition. Under unfavourable conditions, locomotory activity and, in particular, flight appear to be enhanced, whereas, in favourable conditions, field studies show that migration is inhibited and populations become sedentary and reproduce (Lecoq, 1975). In outbreak areas, rainfall may be followed by the arrival of locusts. If no further rain falls, these insects may stay only for a short

Caption for fig. 19.1 (*cont.*).
only zone in which favourable conditions for the species persist and it is also where gregarisation takes place. For much of the dry season, from May to September, the whole of the island is too dry for reproduction and the first generation ($R^1$) adults remain dispersed.

The dotted line indicates the approximate limit of the area of gregarisation. The rest of the island, except for the afforested, humid, eastern coastal region (hatched), makes up the invasion area.

time (about five days in the hot and rainy season) before resuming their migration, but, if there is sufficient rain, migration is terminated. Specifically, monthly rainfall in the range 50–100 mm has been shown to provide optimal conditions for the insects to aggregate and reproduce (Darnhofer & Launois, 1974; Launois, 1974*b*).

The precise microclimatic conditions to which the insects respond have not been defined, although humidity is likely to be critical. Operationally, only the correlation between monthly rainfall (a parameter that is readily accessible over the whole island in near real-time) and the distribution of the locust populations is important. Thus, populations of *L. m. capito* are displaced with the winds, migrating across unfavourable areas, where monthly mean population densities therefore remain low, and stopping for varying periods where conditions are more favourable.

Because migrating individuals are carried downwind, maximum densities of immigrants are observed in areas where there has been strong wind convergence. In a field study (Lecoq, 1975), locust movements were monitored by almost daily sampling of populations at approximately 30 stations in the zone of gregarisation in the extreme south of the island. Immigration from distant sources (several tens of kilometres, as indicated by the sampling data) was generally associated with the presence of a zone of convergence over or in the vicinity of the location concerned. Major influxes were observed in conditions of intense convergence associated with depressions, corresponding to the weather types 7c and 8 of Duvergé (1949).

The extent and location of immigration is, therefore, a function of the location and intensity of zones of wind convergence, the distribution and intensity of rainfall over the whole southwestern part of the island, and the resulting flight potential and densities of these solitarious populations (Lecoq, 1975).

A correlation is apparent over the whole island between changes in the distribution of rainfall and the displacements and distribution of *L. m. capito* populations (Launois, 1974*b*; Lecoq, 1975). Except in the north of the island, most migrations are from the north-northeast to south-southwest early in the rainy season and in the opposite direction towards the beginning of the cool dry season (Fig. 19.1). The direction of migratory flight is also influenced by topographical features (rivers, valleys, etc.) or swaths of deforestation through forested regions of the country (Lecoq, 1975). Although the mechanisms involved are unknown, immigration was often observed at sampling stations associated with them. When winds are in the appropriate direction, i.e. along the axes of such topographical features, they appear to guide the flying locusts, channelling them into areas where 'densation' (an increase in density due to immigration into a restricted area; Pasquier, 1950) can occur suddenly, with a high probability of subsequent gregarisation (Duranton *et al.*, 1979*a*).

Migratory flight by solitarious individuals plays a crucial part in the onset of gregarisation in *L. m. capito*. Net reproductive rates of sedentary populations, estimated using population analysis techniques, e.g. LEXIS diagrams (Lexis, 1875 – a technique originally developed for human demographic studies and applied here, with modifications, to acridoid population dynamics) and life tables, are usually low and the critical gregarisation threshold (about 2000 solitarious individuals ha$^{-1}$; Launois, 1974*b*) is never reached in less than 2–3 generations, even under the most favourable conditions. When this does happen, it is usually at the beginning of the dry season, when, in any case, deteriorating conditions prevent gregarisation (Launois, 1974*b*). Suitable local conditions are, therefore, only one factor leading to phase transformation and there must also be a prior influx of solitarious migrants leading to highly aggregated oviposition (Lecoq, 1975). In these circumstances, the density threshold for gregarisation may be surpassed very rapidly, within one to a few days.

However, even these concentration phenomena may not, in themselves, be enough to initiate outbreaks. Conditions favourable for gregarisation must also extend over very large areas (of the order of several hundred square kilometres) so that many swarms are formed. When these encounter each other during their displacements, they coalesce to form the large aggregations that constitute outbreaks. When gregarisation occurs in isolated pockets, the resulting swarms in which the change in phase is initially only weakly established, are likely to disperse rapidly (Launois, 1974*b*). The sequence of events leading to gregarisation and outbreaks is summarised in Table 19.1.

### Outbreak warning system for Locusta migratoria capito

The research programme in the 1970s led to the development of a system to prevent *L. m. capito* outbreaks in Madagascar, based on the 'optimum rainfall concept' and recognition of the importance of flight movements in achieving gregarisation and the consequent outbreaks.

The warning system involves monitoring the distribution of optimal rainfall zones on a monthly basis, in relation to the solitarious hopper and

Table 19.1. *Successive stages in gregarisation from solitary populations of the Migratory Locust in Madagascar; favourable conditions and relevant information/data sources for a pest control service (after Lecoq, 1975)*

| Stages leading to gregarisation | Favourable conditions | Information/data sources |
| --- | --- | --- |
| Takeoff | 50 > MR > 100 mm | Rainfall maps |
| Displacement | Wind direction | Synoptic meteorological maps |
| Concentration | Weather categories 7c, 8 | Synoptic meterological maps |
| | Restriction by migration routes | Topographical and vegetation maps |
| Sedentarisation | Rainfall | Rainfall maps |
| Rapid reproduction and | | |
| swarming | 50 < MR < 100 mm | Rainfall maps |
| Invasion | Favourable conditions extending over vast areas | Rainfall maps |

MR, monthly rainfall; weather categories 7c and 8 according to the classification of Duvergé (1949).

adult populations in the main areas of habitat in the south and southwest of the island (Launois, 1974*a*). The survey infrastructure includes 22 sites at which intensive surveys, involving regular assessment (once or twice a week) of the density and structure of locust populations, are centred. The sites are representative of the main areas of habitat and are also located along known migration routes of the solitarious adults. These data are complemented by more extensive surveys carried out along transects chosen to sample an entire ecological zone.

As regards rainfall, three simple warning criteria were defined following analysis of extensive archive data accumulated by the MAS.

1 Occurrence of optimal rainfall conditions early in the season (November–December) in the outbreak zone in the southwest of the island.

2 Persistence of these conditions over an adequate area (several hundred square kilometres) for at least three consecutive months. Where optimal conditions occur in one month, immigration can be expected, accompanied by reproduction; when they persist in the same area for 2–3 consecutive months, there is a high probability of the appearance of hopper bands followed by swarming.

3 Excessive rainfall in other zones on the island. The extent of immigration and concentration in the southern outbreak zone is related to the area in the north affected by unfavourably heavy rainfall.

These criteria do not provide the basis for detecting every individual upsurge, but they do account for about 80% of major locust outbreaks recorded over 30 years. More important, they identify the conditions that have preceded all outbreaks leading to plagues over the whole island (Launois, 1974*a,b*).

At the end of each month, rainfall and locust distribution maps for that month are superimposed on those for the preceding two months; transmission of rainfall data to the MAS by radio allows the maps to be prepared within a few days. Population density data are then analysed in relation to the gregarisation threshold and the extent of zones receiving optimum rainfall for one, two or three consecutive months, to identify areas at serious risk. Surveillance and control teams can then be dispatched promptly to verify the situation on the ground and to take any necessary action. This system, which came into operation in 1973, was claimed by MAS personnel to have performed satisfactorily for several years, before being disrupted following political and administrative reorganisation on the island. It is, however, now being revived following the deterioration in the locust situation since the beginning of 1992; Madagascar experienced the most serious upsurge for 30 years in 1992–93.

This optimal rainfall concept has, therefore, contributed to understanding the seasonal distribution and gregarisation potential of *L. m. capito* and has provided the basis for an operational warning system.

**Locusta migratoria migratorioides *in West Africa***

Following gregarisation, the Migratory Locust *L. m. migratorioides* is capable of invading most of sub-Saharan Africa. There are two main outbreak areas (Batten, 1966, 1967): one in the flood plain of the central Niger delta in Mali (Davey, 1959; Farrow, 1975), and the other in the Lake Chad Basin (Descamps, 1953; Davey & Johnston, 1956). However, the latter has, until now, been known to give rise only to local upsurges.

The life cycle of *L. m. migratorioides* is complex and migration by solitarious adults, as in Madagascar, is an essential feature of the sequence of events leading to gregarisation and outbreaks. However, the pattern of flooding in these extensive seasonally inundated areas, as well as seasonal rainfall associated with the advance and retreat of the inter-tropical convergence (ITC), underlie movements between the complementary breeding areas (Farrow, 1975, 1990). Generally, two generations are produced during the dry season in zones that become favourable after the floodwaters subside (Fig. 19.2 – D1 and D2 or S1 and S2), with a further 2–3 during the rainy season (Fig. 19.2 – P1 to P3).

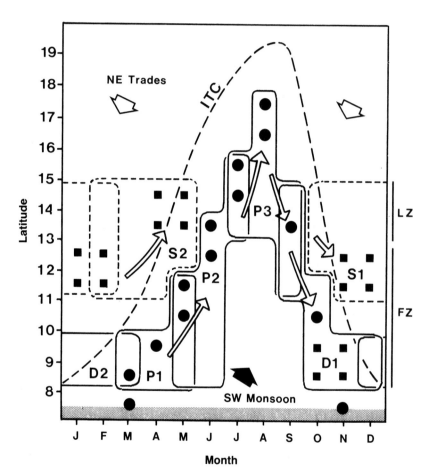

Fig. 19.2. Life cycle and movements of solitarious *Locusta migratoria migratorioides* in the Lake Chad Basin outbreak area (modified after Lecoq, 1974). ●, zones with suitable rainfall; ■, zones where subsiding floodwaters have created suitable habitats; LZ, lacustrine and rain flood zone; FZ, fluvial flood zone; ITC, inter-tropical convergence; P1, P2, P3, rainy season generations; S1, S2, dry season generations in the north; D1, D2, dry season generations in the south. Thin arrows indicate main population movements.

The migrations and population dynamics of solitarious *L. m. migratori-oides* were studied, using a similar approach to that used for *L. m. capito*, for the outbreak areas in Mali (Launois, 1975) and the Lake Chad Basin (Lecoq, 1974; Duranton, 1980; Fig. 19.2). For the Lake Chad outbreak area, application of the ecological optimum method has provided a new approach to the interpretation of recent local upsurges, based on rainfall and flood-level data. The method shows promise as the basis for an operational surveillance system, but one has never been established, due, primarily, to difficulties in real-time monitoring of favourable breeding sites that appear as the floodwaters subside (Darnhofer, 1976; Duranton, 1980). Although the data required are now accessible with remote-sensing technology, the usefulness of such a surveillance system for *L. m. migratorioides* in these regions is debatable. It has been argued that the effects of climatic change and of human activity have reduced to neglig-ible proportions the probability of an upsurge leading to major emigra-tion (Lecoq, 1991*b*). However, continued vigilance is likely to be necessary to monitor local swarms and upsurges.

## Flight movements and a warning system for West African grasshoppers

### Occurrence and importance of flight movements

In West Africa, long-distance movements of grasshoppers take the form of seasonal migrations between complementary breeding areas as they become suitable successively. These migrations are a major factor in the population dynamics of many species of the Sahelian and Sudanese zones. They have been observed directly with radar (Schaefer, 1976; Riley & Reynolds, 1983; Reynolds & Riley, 1988), as well as by means of light-trap catches (Davey *et al.*, 1959; Lecoq, 1978*b*), population counts (Golding, 1934; Joyce, 1952; Descamps, 1961; Popov, 1988) and population-dynamics studies (Lecoq, 1978*a,b*, 1980). The combined results of these investigations have revealed the extent of migration in West African grasshoppers, the populations of which appear to be characterised, at any one site, by a rapid succession of waves of immi-gration and emigration. The population studies, in particular, have involved detailed analysis of migratory movements to assess their role in population dynamics and to investigate their association with ecological factors on a local and regional scale (Lecoq, 1978*a,b*).

Regionally, these migrations are linked to seasonal variations in rain-fall which, in turn, are affected by the movements of the ITC. Wide-spread migrations of the acridoid fauna accompany the slow, progressive northward shift of the ITC at the beginning of the rainy season. They are induced as rainfall associated with the front causes the habitats the insects occupy to become progressively too wet. The southwesterly winds normally carry the migrants northeastwards to regions where the rainfall is lighter and there is a higher probability of finding suitable ecological conditions for reproduction. For as long as conditions remain favourable, the species remain sedentary and breed. At the end of the rainy season, the grasshoppers gradually leave the drying biotopes in the northern part of their range and are generally displaced to the south by the northerly winds which set in as the ITC retreats in that direction (Popov, 1976; Lecoq, 1978*a*,*b*; Riley & Reynolds, 1983).

This general pattern of migration is characteristic of many grass-hoppers, although its details vary according to individual species' life cycles and ecological requirements, especially their degree of hygrophily (Dahdouh, Duranton & Lecoq, 1978; Lecoq, 1978*b*, 1980, 1984; Duranton *et al.*, 1979*b*; Duranton & Lecoq, 1980; Fishpool & Popov, 1984). Together with diapause in the embryonic or adult state (in some species), migration enables them to exploit seasonally complementary habitats in the ephemeral and harsh ecological conditions of Sahelian and Sudanese zones. Thus, at any one site, their population densities are determined by both long-distance migrations, resulting in the often very rapid influx and emigration of insects, and local breeding (Lecoq, 1978*b*).

### Outbreak warning system for Oedaleus senegalensis

Long-distance migration has been shown to play an important part in the development of outbreaks of the Senegalese Grasshopper *Oedaleus senegalensis* and an ecological model has been developed to identify areas at risk.

This grasshopper is one of the main crop pests of the Sahel zone in the southern Sahara (Batten, 1969: Lecoq, 1978*c*; Cheke, Fishpool & Forrest, 1980). It undergoes an embryonic diapause in the dry season, with three generations in the rainy season, migrating to follow the movement of favourable habitats as the ITC shifts progressively north-ward and then southward (Launois, 1978; Lecoq, 1978*b*).

*O. senegalensis* tends to aggregate and breed where rainfall is at an optimal level, within the range 25–50 mm per month (Launois, 1978). A first generation (G1) hatches at the beginning of the rainy season in the breeding area in the south of the species' range (Fig. 19.3). The young G1 adults migrate, breed and give rise to a second generation (G2) in the

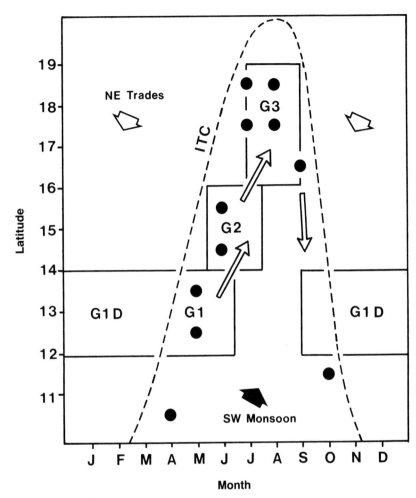

Fig. 19.3. Life cycle of *Oedaleus senegalensis* in West Africa and the main migrations in relation to the shifting optimal ecological zones (modified after Launois, 1979). G1, G2, G3, first, second, third generations; G1D, diapausing eggs during the dry season; ●, optimal breeding zones for *Oedaleus senegalensis*. Thin arrows indicate main population movements.

transitory breeding area located further north. The G2 adults then migrate and give rise to a G3 in the northern breeding area. All of these migratory movements are determined by the northward shift of the monsoon front and the progressively increasing rainfall, the migrants being carried on the southwesterly monsoon winds. At the end of the rainy season, G3 adults emerge into habitats that are beginning to dry out and migrate southward to their southern breeding range where they eventually lay diapause eggs (Launois, 1978). Clearly, this outline is schematic and cannot take account of the complexities of the real situation. For example, diapause eggs are not all deposited in the southern part of the range, as indicated in Fig. 19.3. There is a north–south gradient of oviposition and the locations of reproduction shown are those where, on average, most oviposition takes place. In fact, the majority of diapause eggs can be deposited at very different latitudes from one year to the next, depending on the position of the ITC at the end of the rains (Launois, 1978; Popov, 1988). Furthermore, the different generations are not, in fact, distinct from one another, as suggested by the figure, and complex and multiple overlapping of generations frequently occurs (Launois, 1978; Popov, 1988; Cheke, 1990; Steedman, 1990). The figure therefore presents a simplified outline of a complex cycle of events, which differs from year to year according to the spatial distribution of rainfall and the movements of the ITC.

The spatial and temporal pattern of rainfall determines the extent and timing of migration and the distribution and success of breeding in a particular year. Of particular importance in determining the breeding success of each generation, and therefore the probability of outbreaks and the threat to crops, is the degree of synchrony between the distribution of favourable habitats in space and time and the phenology of the grasshopper populations.

Temporal and spatial monitoring of favourable habitats is, therefore, of prime importance for crop protection. At PRIFAS, a model has been developed to support analysis of the population dynamics and migratory movements of *O. senegalensis*, taking into account rainfall, day-length, temperature and moisture content of the soil (Launois, 1978, 1979, 1984; Arnaud, Forest & Launois, 1982).

This synoptic model is based on the principle that spatiotemporal variation in environmental conditions determines the spatial and temporal dynamics of *O. senegalensis* populations. Development of the model has entailed the following steps:

1 Identification of the critical environmental variables;
2 Subdivision of the range of values of each variable into a number of discrete classes;
3 Combination of these classes of the different variables to characterise different categories of environmental conditions the grasshoppers might encounter;

4 Accumulation of data on biological performance (development rate, survival and reproduction) for each of these categories.

These data, relating biological performance to environmental conditions in a matrix of relationships, form the core of the model. The model takes account of three environmental variables:

1 Photoperiod, which determines the production of diapausing or non-diapausing eggs;
2 Mean temperature ($T$);
3 A coefficient of water-balance, $K$ = AET/PET (AET, actual evapotranspiration; PET, potential evapotranspiration).

Twenty-four categories of environmental conditions were defined on the basis of these parameters. The season is divided into 10-day periods and $T$ is entered into the program as the class representing the mean temperature, across years, for each period. $K$ is calculated from 10-day rainfall data and an estimate of the water retention capacity of soils in potential habitats (the selected arbitrary values are 150 mm for field capacity and 40 mm for permanent wilting point). $T$ and $K$ are input for each degree square (approximately 100 × 100 km) and the program calculates, for each life stage, a development period index ($V$) and a survival index ($S$), whose product is the index of success, $R$. The program then calculates an overall index of environmental favourability to estimate the probability of swarming in a degree-square. This final index integrates indices of success for each life stage over the two preceding 10-day periods, and includes further indices representing the probabilities of immigration and emigration.

For each location for which data are entered, it is possible to use the model to make qualitative daily assessments of the potential for development of grasshopper populations. However, the principal product of the model consists of series of maps, issued every 10 days, which indicate the risk of swarms developing in each degree square. Risk is expressed at five levels – negligible, very low, low, average, and high – and the maps cover the whole Sahel, including all the Sahelian countries, from Mauritania to Chad.

It is essential to emphasise that the model attributes a level of risk to a degree square solely on the basis of the suitability of the prevailing environmental conditions for aggregation and development of grasshopper populations. There is no implication that those populations are actually present. For this reason, the maps provide only the starting point for a manual process of monitoring the development of the grasshopper situation through the rainy season.

In practice, the maps are used in several ways. At the beginning of the season, they can be superimposed on maps of the distribution of egg-laying by the last generation of the previous season. These are compiled from field survey data and the model's output covering the period of oviposition at the end of the previous season (see below). This procedure improves the accuracy with which areas at risk from large-scale emergence of the G1, and consequent serious damage to newly germinated crops, can be identified.

During the season, the maps provide a framework for monitoring the principal migratory movements of the grasshoppers, by following the spatiotemporal development of favourable habitats, and can be used to evaluate, at 10-day intervals, the potential severity of the current *O. senegalensis* season. If necessary, warnings can then be issued of the risk of serious damage to maturing crops by the G3 migrating southwards, from the northern pastures of the Sahel to the southern agricultural zones.

Finally, at the end of the rains, the maps can be used to identify areas favourable for oviposition of diapausing eggs by the G3. This information is useful for planning dry-season field surveys to map the location and extent of these ovipositions.

At present, the model's output is in a semi-quantitative form and the reliability of its predictions is subject to several sources of error. Additional experimental data are required on the effects of environmental variables on the dynamics of *O. senegalensis* populations. However, the major problem is availability of dependable meteorological data in real-time. The ground-based meteorological network is inadequate; currently, 10-day rainfall data are regularly available from only 120 stations and the large number of interpolations necessary to complete the grid must introduce a significant bias. Only satellite remote-sensing can provide estimates of rainfall and other relevant environmental data, e.g. soil moisture, in real-time and with the required resolution for the whole Sahel. Research is now in progress on the use of satellite data, particularly from NOAA polar-orbiting satellites and the geostationary METEOSAT, and on the development of a geographic information system to support the model (Tappan, Moore & Knausenberger, 1991; Berges, Chiffaud & Mestre, 1992).

In its present form, therefore, the model can only be used to identify areas that may be at risk of outbreaks, to direct and focus field surveillance, not replace it. Its usefulness is heavily dependent on the users' common sense and experience. In making the best use of the limited ecological data available in real-time, it can, however, contribute to the effective deployment of the limited resources available to the Sahelian countries, for monitoring and controlling *O. senegalensis* populations.

The model is currently being used as an operational tool by the crop protection services of the Sahelian countries, to support the organisation of field surveys and decisions on where to deploy control equipment (Launois & Lecoq, 1990). Several years of observations in the Sahel (Launois, 1978) suggest that the model's results provide a sounder basis for decision-making than the often incomplete and unreliable data available from the field. A seminar on modelling techniques held in May 1988 in Niamey provided an opportunity for plant protection service personnel to assess the results of the *O. senegalensis* model in relation to their experience of events on the ground. All participants concluded that it afforded sufficiently reliable indications of field situations on which to base decisions on appropriate responses (M. Launois, A. Monard, J. Gigault & M.F. Grimaux, unpublished data). However, high priority needs to be given to evaluating the model's performance, with tests of the sensitivity of its internal parameters and by systematic validation of its output and predictions against reliable field data (CIRAD, 1992).

## Conclusions

Seasonal changes in ecological conditions cause many acridoid species to migrate over great distances. These migratory movements allow locusts and grasshoppers to survive in arid and semi-arid environments and enable them to exploit seasonally complementary habitats that are often far apart. Migration thus plays a crucial role in their population dynamics. It often results in massive aggregation of acridoids (as a result of accumulation in zones of wind convergence, channelling by routes of migration, and concentration in favourable areas), leading to local outbreaks. In locusts, such aggregation of solitarious adults leads to gregarisation and upsurges.

These migratory movements are closely associated with changes in environmental conditions and, for some species, synoptic models have been developed to link their occurrence to a small number of important environmental factors, especially rainfall. These models are the basis of operational systems to forecast outbreaks of *L. m. capito* in Madagascar and *O. senegalensis* in West Africa.

However, the reliability and operational potential of these forecasting systems is currently limited by the quality of the available real-time meteorological and ecological data. Their replacement with remotely sensed satellite data, integrated with geographic information systems, offers an opportunity to circumvent this constraint. Research to this end is in progress.

## References

Arnaud, M., Forest, F. & Launois, M. (1982). Automatisation d'un modèle écologique original propre à *Oedaleus senegalensis* (Krauss, 1877) (Orthoptera: Acrididae). *L'Agronomie Tropicale*, **37**, 159–71.

Batten, A. (1966). The course of the last major plague of the African migratory locust, 1928 to 1941. *FAO Plant Protection Bulletin*, **14**, 1–16.

Batten, A. (1967). Seasonal movements of swarms of *Locusta migratoria migratorioides* R & F in Western Africa in 1928 to 1931. *Bulletin of Entomological Research*, **57**, 357–80.

Batten, A. (1969). The Senegalese grasshopper, *Oedaleus senegalensis* (Krauss). *Journal of Applied Ecology*, **6**, 27–45.

Berges, J.-C., Chiffaud, J. & Mestre, J. (1992). Contribution of spatial remote sensing to the identification of zones favourable to the Senegalese Locust in the Sahel. *Veille Climatique Satellitaire*, **40**, 25–33. In English and French.

Botha, D. H. & Jansen, A. (1969). Night flying by brown locusts, *Locustana pardalina* (Walker). *Phytophylactica*, **1**, 79–92.

Cheke, R. A. (1990). A migrant pest in the Sahel: the Senegalese grasshopper *Oedaleus senegalensis*. *Philosophical Transactions of the Royal Society of London B*, **328**, 539–53.

Cheke, R. A., Fishpool, L.D.C. & Forrest, G.A. (1980). *Oedaleus senegalensis* (Krauss)(Orthoptera: Acrididae: Oedipodinae): an account of the 1977 outbreak in West Africa, and notes on eclosion under laboratory conditions. *Acrida*, **9**, 107–32.

CIRAD (1992). *Rapport de la première revue externe du PRIFAS*. Montpellier: Centre de Coopération Internationale en Recherche Agronomique pour le Développement. In French, English summary.

Dahdouh, B., Duranton, J.F. & Lecoq, M. (1978). Analyse de données sur l'écologie des acridiens d'Afrique de l'Ouest. *Cahiers de l'Analyse des Données*, **3**, 459–82.

Darnhofer, T. (1976). *Rapport final d'expert*. Project PNUD(FS)/FAOReport AML/MET/8. Rome: FAO.

Darnhofer, T. & Launois, M. (1974). *L'optimum pluviométrique du Criquet migrateur malgache: principe et applications*. FAO Project PNUD(FS)/ FAO AGP:DP/MAG/70/523 Report MML/MET/9. Rome: FAO.

Davey, J.T. (1959). The African migratory locust (*Locusta migratoria* Rch. and Frm., Orth.) in the Central Niger Delta. Part Two. The ecology of *Locusta* in the semi-arid lands and seasonal movements of populations. *Locusta*, **7**, 1–180.

Davey, J.T., Descamps, M. & Demange, R. (1959). Notes on the Acrididae of the French Sudan with special reference to the Central Niger Delta. *Bulletin de l'Institut Francais d'Afrique Noire (A)*, **21**, 60–112 and 565–600.

Davey, J.T. & Johnston, H. B. (1956). *The African Migratory Locust* (Locusta migratoria migratorioides *R. & F.) in Nigeria.* Anti-Locust Bulletin, 22. London: Anti-Locust Research Centre.

Descamps, M. (1953). Observations relatives au criquet migrateur africain et à quelques autres espèces d'Acrididae du Nord Cameroun. *L'Agronomie Tropicale*, **8**, 567–613.

Descamps, M. (1961). Le cycle biologique de *Gastrimargus nigericus* Uv. (Orthoptera: Acrididae) dans la vallee du Bani (Mali). *Revue de Pathologie Vegetale et d'Entomologie Agricole de France*, **40**, 187–99. In French, English summary.

Duranton, J.-F. (1980). *Etude sur le Criquet migrateur africain dans son aire grégarigène du Bassin tchadien. (Assistance à l'OICMA). Mars-avril 1980.* FAO AGP:DP/RAF/77/049 Report 1980/1. Rome: FAO.

Duranton, J.-F., Launois, M., Launois-Luong, M.H. & Lecoq, M. (1979*a*). Les voies privilégiées de déplacement du Criquet migrateur malgache en phase solitaire. *Bulletin d'Écologie*, **10**, 107–23. In French, English summary.

Duranton, J.-F., Launois, M., Launois-Luong, M.H. & Lecoq, M. (1979*b*). Biologie et écologie de *Catantops haemorrhoidalis* (Krauss) en Afrique de l'Ouest. *Annales de la Société Entomologique de France (N.S.)*, **15**, 319–43. In French, English summary.

Duranton, J.-F., Launois, M., Launois-Luong, M.H. & Lecoq, M. (1982). *Manuel de Prospection Acridienne en Zone Tropicale Sèche*, 2 vols. Paris: Ministère des Relations Extérieures – Coopération et Développement – et GERDAT.

Duranton, J.-F. & Lecoq, M. (1980). Ecology of locusts and grasshoppers (Orthoptera, Acrididae) in Sudanese West Africa. 1. Discriminant factors and ecological requirements of acridian species. *Acta Oecologica/ Oecologia Generalis*, **1**, 151–64.

Duvergé, M. (1949). *Principes de météorologie dynamique et types de temps à Madagascar.* Publication no. 13. Tananarive: Service Météorologique de Madagascar.

FAO (1973). *Rapports sur les résultats, conclusions et recommandations du Projet relatif aux recherches sur le Criquet migrateur malgache.* FAO PNUD/FAO AGP:DP/MAG/70/523 Terminal report. Rome: FAO.

Farrow, R.A. (1975). The African migratory locust in its main outbreak area of the Middle Niger: quantitative studies of solitary populations in relation to environmental factors. *Locusta*, **11**, 1–198.

Farrow, R. A. (1990). Flight and migration in Acridoids. In *The Biology of Grasshoppers*, ed. R.F. Chapman & A. Joern, pp. 227–314. London: John Wiley.

Fishpool, L.D.C. & Popov, G.B. (1984). The grasshopper faunas of the savannas of Mali, Niger, Benin and Togo. *Bulletin de l'Institut Fondamental d'Afrique Noire Série A*, **43**, 275–410.

Golding, F. D. (1934). On the ecology of Acrididae near Lake Chad. *Bulletin of Entomological Research*, **25**, 263–303.

Golding, F. D. (1948). The Acrididae (Orthoptera) of Nigeria. *Philosophical Transactions of the Royal Society of London B*, **99**, 517–87.

Joyce, R.J.V. (1952). *The ecology of grasshoppers in East Central Sudan.* Anti-Locust Bulletin 11. London: Anti-Locust Research Centre.

Launois, M. (1974*a*). *Le service d'avertissement antiacridien à Madagascar. Conception et réalisation. Projet relatif aux recherches sur le Criquet* *migrateur malgache*. PNUD/FAO AGP:DP/MAG/70/523. Report PNUD(FS)MML/BIO/9. Rome: FAO.

Launois, M. (1974*b*). *Influence du Facteur Pluviométrique sur l'Évolution Saisonnière du Criquet Migrateur* Locusta migratoria capito *(Saussure) en Phase Solitaire et sur sa Grégarisation à Madagascar*. Paris: Ministère de la Coopération.

Launois, M. (1975). *Report of the visit of the expert-consultant to Mali 21 July–21 August 1974. African Migratory Locust Project AGP:DP/RAF/69/146. Biological studies*. Report UNDP(SF)AML/BIO/4. Rome: FAO.

Launois, M. (1978). *Modélisation Écologique et Simulation Opérationnelle en Acridologie. Application à* Oedaleus senegalensis *(Krauss, 1877)*. Paris: Ministère de la Coopération et GERDAT.

Launois, M. (1979). An ecological model for the study of the grasshopper *Oedaleus senegalensis* in West Africa. *Philosophical Transactions of the Royal Society of London B*, **287**, 345–55.

Launois, M. (1984). Les biomodèles à géométrie variable appliqués à la surveillance des criquets ravageurs. *L'Agronomie Tropicale*, **39**, 269–74.

Launois, M. & Lecoq, M. (1990). Biomodélisation et stratégies de lutte anti-acridienne en Afrique et à Madagascar. *Medelingen van der Faculteit Landbouwwetenschappen Rijksuniversiteit Gent*, **55**, 225–34. In French, English summary.

Lecoq, M. (1974). *Report of expert on a visit to Mali and the Lake Chad basin. African Migratory Locust Project AGP:DP/RAF/69/146. Biological studies*. Report UNDP(SF)AML/BIO/5. Rome: FAO.

Lecoq, M. (1975). *Les Déplacements par Vol du Criquet Migrateur Malgache en Phase Solitaire: leur Importance sur la Dynamique des Populations et la Grégarisation*. Paris: Ministère de la Coopération.

Lecoq, M. (1978*a*). Les déplacements par vol à grande distance chez les acridiens des zones sahélienne et soudanienne en Afrique de l'Ouest. *Comptes Rendus de l'Académie des Sciences Paris D*, **286**, 419–22. In French, English summary.

Lecoq, M. (1978*b*). Biologie et dynamique d'un peuplement acridien de zone soudanienne en Afrique de l'Ouest (Orthoptera, Acrididae). *Annales de la Société Entomologique de France (N.S.)*, **14**, 603–81. In French, English summary.

Lecoq, M. (1978*c*). Le problème sauteriaux en Afrique soudano-sahélienne. *L'Agronomie Tropicale*, **33**, 241-58.

Lecoq, M. (1980). Biologie et dynamique d'un peuplement acridien de zone soudanienne en Afrique de l'Ouest (Orthopt., Acrididae). Note complémentaire. *Annales de la Société Entomologique de France (N.S.)*, **16**, 49–73. In French, English summary.

Lecoq, M. (1984). Ecology of locusts and grasshoppers (Orthoptera, Acrididae) in Sudanese West Africa. 2. Ecological niches. *Acta Oecologica/Oecologia Generalis*, **5**, 229–42.

Lecoq, M. (1991*a*). Le Criquet pélerin. Enseignements de la dernière invasion et perspectives offertes par la biomodélisation. In *La Lutte Anti-Acridienne*, ed. A. Essaid, pp. 71–98. Paris: John Libbey Eurotext.

Lecoq, M. (1991*b*). *The Migratory Locust,* Locusta migratoria, *in Africa and Malagasy*. The Orthopterists' Society Field Guide Series, C2E. Québec: Orthopterists' Society.

Lexis, W. (1875). *Einleitung in die Theorie der Bevolkerungsstatistik*. Strassburg, Trubner. (In German. Pages 5–7 translated into English by

N. Keyfitz, 1977 and printed, with Figure 1, in *Mathematical Demography*, ed. D. Smith & N. Keyfitz. Berlin: Springer-Verlag)

Pasquier, R. (1950). Sur une des causes de la grégarisation chez les acridiens: la densation. *Annales de l'Institut Agronomique et des Services de Recherche et d'Expérimentation Agricole de l'Algérie*, **5**, 1–9.

Pedgley, D.E. (1981). *Desert Locust Forecasting Manual*, 2 vols. London: Centre for Overseas Pest Research.

Popov, G.B. (1976). The 1974 outbreak of grasshoppers in western Africa. In *Report of the Sahel Crop Pest Management Conference*, held at the invitation of the Agency for International Development, United States Department of State, Washington, DC, USA December 11–12, 1974, ed. R.F. Smith & D.E. Schlegel, pp. 35–43. Berkeley: University of California.

Popov, G.B. (1988). Sahelian grasshoppers. *Overseas Development Natural Resources Institute Bulletin*, **5**, 1–87.

Rainey, R.C. (1951). Weather and the movements of locust swarms: a new hypothesis. *Nature*, **168**, 1057–60.

Rainey, R.C. (1963). *Meteorology and the migration of desert locusts. Applications of synoptic meteorology in locust control*. World Meteorological Organisation Technical Note 54. Geneva: World Meteorological Organisation.

Reynolds, D.R. & Riley, J.R. (1988). A migration of grasshoppers, particularly *Diabolocatantops axillaris* (Thunberg) (Orthoptera, Acrididae), in the West African Sahel. *Bulletin of Entomological Research*, **78**, 251–71.

Riley, J.R. & Reynolds, D.R. (1983). A long-range migration of grasshoppers observed in the Sahelian zone of Mali by two radars. *Journal of Animal Ecology*, **52**, 167–83.

Schaefer, G.W. (1976). Radar observations of insect flight. In *Insect Flight*, ed. R.C. Rainey, pp. 157–97. *Symposia of the Royal Entomological Society of London* 7. Oxford: Blackwell Scientific Publications.

Steedman, A. (1990). *Locust Handbook*, 3rd edn. Chatham: Natural Resources Institute.

Tappan, G.G., Moore, D.G. & Knausenberger, W.J. (1991). Monitoring grasshopper and locust habitats in Sahelian Africa using GIS and remote sensing technology. *International Journal of Geographical Information Systems*, **5**, 123–35.

Têtefort, J., Dechappe, P. & Rakotoharison, J.M. (1966). Etude des migrations du Criquet migrateur malgache *Locusta migratoria capito* (Sauss.) dans sa phase solitaire. *L'Agronomie Tropicale*, **12**, 1389–97. In French, English and Spanish summaries.

Uvarov, B. P. (1977). *Grasshoppers and Locusts*, vol. 2. London: Centre for Overseas Pest Research.

Waloff, Z. (1963). *Field studies on solitary and transiens Desert Locusts in the Red Sea area*. Anti-Locust Bulletin 40. London: Anti-Locust Research Centre.

Zolotarevsky, B. (1929). Le Criquet migrateur (*Locusta migratoria capito* Sauss.) à Madagascar. *Annales de l'Institut National de Recherche Agronomique. Serie C: Annales des Epiphyties*, **15**, 185–236.

# Part four

Overview and synthesis

# 20

# Forecasting migrant insect pests[1]

## J. I. MAGOR

### Introduction

An important objective of studying insect migration is to develop fore-casting systems to predict damaging pest events in time for preventive action to be taken. Operational forecasts are prepared for a variety of decision makers to strict deadlines and the forecasters have to work within the limitations imposed by current knowledge, hypotheses, communications and budgets. They are dependent on pest monitoring for evidence of the current situation and these datasets are inevitably incomplete. Additionally, in many if not all cases, the population dy-namics of the pest are imperfectly understood, so that the basis of forecasting is flawed. It follows that forecasters should seek results from population studies that would improve their predictions.

Migrant-pest forecasters have concentrated on predicting the temporal and spatial distribution of pests, as well as changes in severity of attack, as discussed below. They have paid less attention to understanding how users (finance officers, agrochemical manufacturers and retailers, plant protection officers, farmers and pastoralists) seek to achieve better control. There is, however, a growing awareness that forecasts need to be written and evaluated with the users' requirements in mind, if they are to influence decisions.

This chapter shows how entomological studies are used in forecasting of the Desert Locust *Schistocerca gregaria*, and gives examples of advances arising from recent research and technological changes. It outlines current attempts to improve the accuracy and utility of these operational forecasts and examines approaches adopted for other species.

---

[1] Crown copyright 1992.

## Forecasting *Schistocerca gregaria*

### *Scientific hypotheses and data*

All forecasters of insect pests need to receive data that allow them to integrate interactions between the pest, its habitat and the weather, in time to transmit timely warnings. For migrant pests, the data must relate to migration patterns and take account of the geographical scale of the control strategy.

*Schistocerca gregaria* forecasting began in 1941. The first step was taken ten years earlier, when the practical implications of the theory of phase transformation (Uvarov, 1921; Faure, 1923) suggested a feasible control strategy for locusts. This involved locating permanent outbreak areas and intervening to prevent the solitary locusts in them forming swarms and escaping to initiate a plague in the surrounding invasion area, with consequent damage to crops and grazing.

In 1929, the countries within the very extensive *S. gregaria* distribution area (Fig. 20.1*a*) agreed to report regularly to a central location, by responding to a questionnaire. This was structured to obtain information on the timing, behaviour, habitat and weather associated with current and past events. Questions were arranged to elucidate the life-cycle stages separately: immigration, egg-laying, hatching, hoppers and, finally, emigration. Mapping and analysis of the reports and parallel field studies were initiated simultaneously to locate the outbreak areas and to provide information for directing control campaigns (Uvarov, 1930).

Mapping data between 1931 and 1939 established temporal and spatial links between widely separated breeding areas of *S. gregaria* (Fig. 20.1*b–d*), but no permanent outbreak areas were found (Uvarov, 1933–1935; Uvarov & Milnthorpe, 1937–1939). Breeding coincided with, and was regulated by, seasonal rains (Waloff, 1946*a*; Donnelly, 1947; Davies, 1952; Fortescue-Foulkes, 1953) and forecasting began as soon as these findings could be used predictively (Uvarov, 1951). The need for joint investigations by entomologists and meteorologists was stressed from the beginning. Rainey (1951) summarised studies on the downwind displacement of swarms and established its adaptive significance. Subsequently, forecasters used daily weather maps to predict short-term migration (Rainey, 1969; Roffey, 1970). These predictions were aided by many retrospective case studies that used improved knowledge of flight physiology, behaviour and downwind displacement, to establish links between long-distance movements of up to 4000 km and the concurrent weather (Rainey, 1963; Pedgley, 1981, 1989).

*Schistocerca gregaria* plagues are intermittent and have lasted 1–22 years. Their occurrence has varied from four out of every five years from 1860 (when reasonable records began) to the end of the 1949–63 plague, to only one year in six subsequently (Waloff, 1976; Symmons, 1992). It is uncertain whether preventive control (FAO, 1985), drier weather or a combination of the two caused this change (Magor, 1989). Since forecasting began, therefore, the emphasis has changed from collecting data to predict the movement of swarms during plagues, to identifying sequences of rain and associated vegetation that could lead to population growth and plague initiation. The latter involves evaluating survey reports on locusts, weather and habitat changes in the ephemeral outbreak areas (Waloff, 1966; Roffey & Popov, 1968; Bennett, 1976; Skaf, 1978; Hemming *et al.*, 1979; Hielkema, Roffey & Tucker, 1986). The continuing iterative process of developing remote-sensing imagery to monitor rainfall and habitat suitability for this task is described by Robinson (Chapter 16, this volume).

Assessing the long-term impact of intermittent migrant pests has proved a perennially difficult problem. The *S. gregaria* control strategy assumes that destroying initial gregarising populations will minimise damage, by preventing the spread of locusts from the desert to major agroeconomic zones (Singh, 1967; Bennett, 1976). Unfortunately, the relative roles of swarming and non-swarming populations at the beginning of plagues are not fully understood. Consequently, the environmentally and economically desirable moment to initiate control is disputed (Skaf, Popov & Roffey, 1990; Symmons, 1992). Furthermore, evaluation of the control strategy needs to include estimates of the damage prevented in those agricultural zones that would have been invaded but for successful control. Damage caused by *S. gregaria* is poorly reported; questionnaires have been used periodically to collect information from all affected countries simultaneously (Fourth International Locust Conference, 1937; FAO, 1953; FAO, 1958; Lean, 1965) but cost–benefit studies similar to that of Wright (1986) for the Australian Plague Locust *Chortoicetes terminifera*, which included extrapolated downwind loss avoidance, have yet to be carried out.

### *Data collection, transmission and interpretation*

National survey teams monitor seasonal breeding areas and record standard observations on locusts and their habitat. They also note recent rainfall along their route and collect information from nomads and from

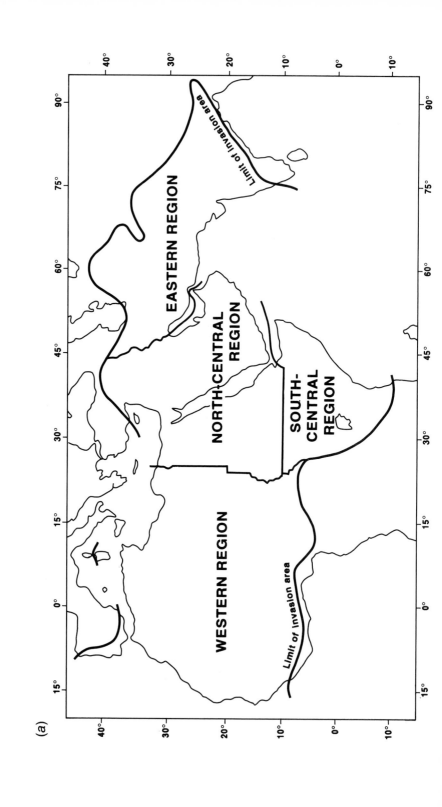

(a)

EASTERN REGION

NORTH-CENTRAL REGION

SOUTH-CENTRAL REGION

WESTERN REGION

Limit of invasion area

Limit of invasion area

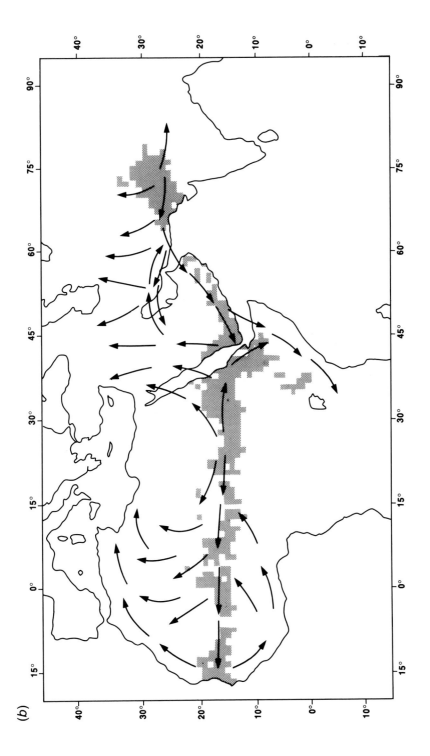

(b)

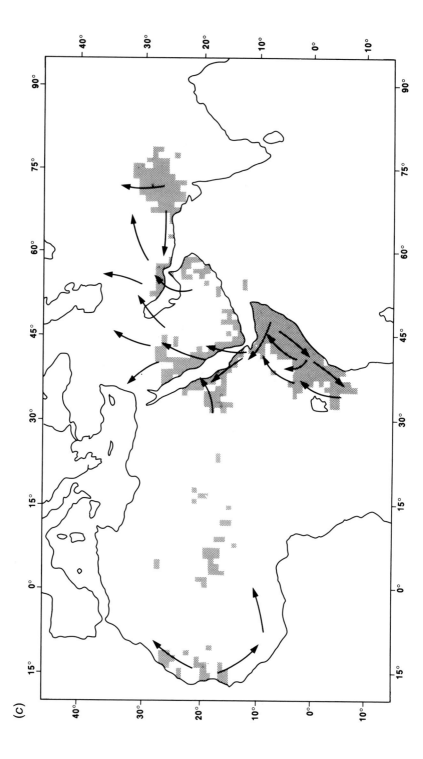

(c)

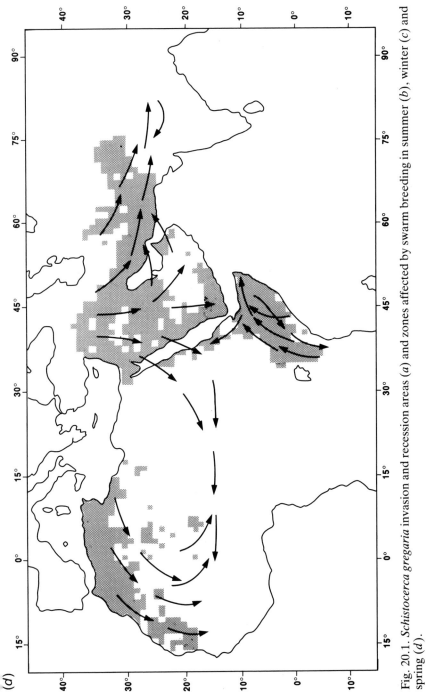

Fig. 20.1. *Schistocerca gregaria* invasion and recession areas (*a*) and zones affected by swarm breeding in summer (*b*), winter (*c*) and spring (*d*).

other travellers. These data are usually transmitted daily, by radio, to the team's headquarters, where they are used to plan and conduct control. Information signifying important changes is transmitted immediately, by fax or telex, to the forecasters at the Food and Agriculture Organisation (FAO), whereas routine survey data are summarised and sent at the end of the month. Forecasters receive additional weather observations from meteorological services, and information on the distribution of vegetation and rainfall derived from satellite images (Chapter 16, this volume).

*Schistocerca gregaria* recession populations may exist at very low densities. They fly at night (Roffey, 1963) and their desert habitat is inhospitable and partially inaccessible (Skaf *et al.*, 1990). The logistics of data collection, therefore, present difficulties that affect the quality and completeness of information available to forecasters and researchers (Symmons, 1972). Assessments of the risk posed by different infestations remain subjective and the introduction and regular evaluation of tiered sampling, to estimate numbers of non-plague populations, would greatly improve these judgements. This may become more feasible once affordable remote sensing can be used to identify habitat-suitability classes (FAO, 1993). Remote-sensing products, however, need calibration by expensive and logistically difficult field validation in many more regions, before they can be confidently used throughout the *S. gregaria* area (Cherlet & Di Gregorio, 1991). In addition, the range of locust reponses to defined habitat-suitability classes has yet to be fully established.

The speed and reliability of data transmission is one factor that determines the decisions that should be guided by locally, regionally or centrally prepared forecasts. Local forecasts and decision making may be necessary if communications are slow and only a short delay is practicable between observation and control. Computer and satellite communication links promise faster, and reliable, data transmission but it must be borne in mind that more data arriving quickly are not necessarily more useful than a smaller number of accurate and well-ordered reports that arrive later, though soon enough for timely predictions to be issued.

Pedgley (1981) described maps of survey data used by forecasters to visualise the geographical relationship of near-contemporary reports and their change over time. The initial plotting is usually on a large scale (1:500 000 to 1:4 million) and for one month. Maps at this scale allow the distribution, date and life-stages of all sightings to be shown and indicate the scale of local infestations. They are too large, however, for the whole situation to be visualised easily and so data are transferred to medium-scale maps (1:11 million). Symbols denoting similar population types are combined and their spatial accuracy is kept to within 50 km. Smaller A4-sized maps, showing the main population type present in a month, accompany the Desert Locust Bulletin issued by FAO and are used subsequently as quick guides to between-month and between-year changes.

The current situation is assessed from these maps and reports, plus all available information on habitat quality and weather in the region during the month. Survey teams will not have visited all previously reported populations and forecasters interpolate these missing data before evaluating likely developments (Pedgley, 1981). Historical data are used in this process: archived monthly locust maps, case studies of past events, frequency maps indicating the probability of locust populations infesting an area, and dates of annual invasions (Rainey, 1963; Pedgley, 1981). Life-stage duration (Reus & Symmons, 1992) and trajectory (windborne-displacement) models (Waloff, 1946*b*; Taylor, 1979; Ritchie & Pedgley, 1989) are available for interpreting incoming data (Table 20.1).

### *Types of forecast*

*Schistocerca gregaria* forecasts, which were originally produced at the Anti-Locust Research Centre in London, are now prepared by FAO in Rome. Three types of forecast may be issued: long-term predictions to help officials allocate budgets to regional and national locust organisations, and medium- or short-term forecasts to guide those responsible for the seasonal and, ultimately, daily deployment of survey and control teams (Table 20.2). Pedgley (1981) described the data, office procedures and methods for preparing forecasts. Only medium-term forecasts are produced regularly. Each month's locust, weather and habitat data are described in the FAO Desert Locust Bulletin, which contains a forecast for the following six to eight weeks. The same forecast goes to all decision makers and it assumes a sound background in locust events, which recipients such as aid donors may not possess.

Forecasters use different information when preparing long-, medium- and short-term forecasts (Table 20.3). Long-term forecasts are prepared up to a year in advance. Consequently, they have to rely on statistical data: climate normals and frequency of pest incidence. The accuracy of such forecasts is inevitably low, as locust numbers are highly responsive to the area's extremely variable rainfall and their destinations are determined by windspeed and direction during migration.

Weather can be forecast reliably at most for a week or ten days ahead.

Table 20.1. *Products used in* Schistocerca gregaria *forecasting*

| Current inputs | Archived data/research results/models |
| --- | --- |
| *Desert Locust* Schistocerca gregaria | |
| Ground and aerial survey reports | Monthly distribution maps – hoppers and adults |
| | Monthly frequency of incidence – hoppers and adults during plagues and recessions |
| | First and last dates of invasions and breeding seasons |
| | Case studies |
| |   movements and associated weather |
| |   breeding success/failure |
| |   plague onset and decline sequences |
| |   predicating forecasting rules |
| | Development period model |
| | Trajectory model |
| *Weather and habitat* | |
| Rainfall: daily, decadal, monthly estimates, cloud temperatures | Monthly means |
| Remotely sensed vegetation; vegetation seen on locust surveys | Habitat distribution maps |
| Winds, temperature, weather systems – meterological service charts | Climate means |

Table 20.2. *Decisions supported by long-, medium- and short-term* Schistocerca gregaria *forecasts*

| Forecast type | Decision/option | Decision maker(s) |
|---|---|---|
| *Long-term forecasts* next 1–2 seasons; 2–12 months | Budgets, bids for material, staff | Central governments/regional organisations/ donors/pesticide manufacturers/suppliers |
| *Medium-term forecasts* Next 1–2 generations; 1–2 months | Deployment of survey and control teams in region | Regional campaign managers/farmers/pesticide suppliers |
| *Short-term forecasts* Current generation; 1–2 days | Finding control targets | Local control teams/farmers |

Table 20.3. *Types of information used in long-, medium- and short-term*
Schistocerca gregaria *forecasts*

| Information | Forecast lead-in time | | |
|---|---|---|---|
| | Long | Medium | Short |
| Historical frequency | ☐ | | |
| Case studies of immigration | ☐ | ☐ | |
| Climate normals | ☐ | ☐ | |
| Summer fledging areas | | ☐ | |
| Current adult sightings | | ☐ | ☐ |
| Current emigration areas | | | ☐ |
| Current weather | | | ☐ |
| Local habitat suitability | | | ☐ |

It follows that the spatial resolution of forecasts improves as the duration of the forecast period shortens. Long-term and medium-term predictions can indicate only the statistical associations, as yet only partially established by case studies, of occasions when locusts migrated between one or more parts of complementary seasonal breeding areas. For example, part of the progeny of spring breeding in Saudi Arabia often moves through Egypt to the summer-breeding areas of the Sudan, occasionally spreading to Chad and as far west as Niger, Mali and southern Algeria (Ballard, Mistikawi & El Zoheiry, 1932; Rainey, 1963; Omar, 1965; Hosni, 1966; Bennett, 1976; Pedgley,1981). The other frequent movement in spring takes swarms eastwards to the Indo-Pakistan frontier, occasionally, as in 1954, spreading as far east as Assam (Rainey, 1963; Rainey, Betts & Lumley, 1979; Chandra, Sinha & Singh, 1988). Less frequently, they move southwards from Saudi Arabia to Ethiopia, Yemen and northern Somalia (Pedgley & Symmons 1968; Pedgley 1981). The initial forecast, based on past analogues and the most likely weather, cannot narrow down the area to be infested. This can only be done for the few days covered by short-term forecasts, when weather forecasts are reliable enough to project swarm tracks forward from known positions down the forecast windfields.

Information about the whole distribution area is needed for long-term forecasts, but the area that has to be considered declines as the forecast period shortens. A forecaster asked, in spring, to predict the probability of *S. gregaria* invading a district in Pakistan during the following winter, needs data from the whole distribution area, since established seasonal displacements from as far away as northwest Africa in spring could, over successive generations, reach Pakistan by the next winter. For medium-term forecasts issued during the autumn, only populations in the central and eastern regions are relevant. Finally, short-term warnings are issued only in the few days when a weather system known to transport locusts to Pakistan is present or is reliably forecast. By this time, the relevant migrating populations will be in Pakistan or its neighbouring countries.

Long-term forecasts should, therefore, offer decision makers a series of dates at which the worst scenario envisaged will be confirmed or downgraded by subsequent developments. Medium-term forecasts should clearly indicate the weather systems associated with predicted movements so that recipients and their local weather forecasters are aware of the potential importance of these systems.

## Improving forecasts

The need to improve predictions of the timing and seriousness of migrant-pest attack remains an exciting challenge to researchers. The main areas of current interest are developing computer-aided information systems to manage the large quantities of data used in forecasting, and producing principles and guidelines around conceptual or mathematical models (Fisher, 1985). Some multidisciplinary groups are also undertaking key field experiments on dispersal, to improve forecasting and the formulation of control strategies (Tu, 1980; National Coordinated Research Group for White-backed Planthoppers, 1981; Hendrie *et al.*, 1985).

Migrant pests can be considered as falling into two groups: those like locusts and the African armyworm *Spodoptera exempta*, which are accidental feeders on crops (Rose *et al.*, 1987), and those pests such as aphids and the Brown Planthopper *Nilaparvata lugens*, for which all or part of the life cycle is dependent upon particular crops. The former pests, unlike the latter, also spread from permanent or temporary outbreak areas where their numbers are too low to cause damage. In addition, current control strategies aim to limit their spread onto and between crops. This is seldom the chosen control strategy for crop pests where crop variety and agronomic practices affect population size after immigration, and control is an on-farm decision. These differences affect data collection in three ways: first, locust and *S. exempta* forecasting systems are international or inter-state activities; second, data are collected in non-crop as well as in crop areas; and third, information on farming practices is not sought.

The distinction, however, is not absolute. Many crop pests overwinter in sub-tropical sites and cause damage in much larger areas, after migrating in spring and summer to temperate latitudes on southerly winds in the warm sectors of depressions. Johnson (1969) cited many examples in North America: the leafhoppers *Empoasca fabae* and *Macrosteles fascifrons*, and the aphid *Toxoptera graminum*. Notable examples in China are the Brown Planthopper *Nilaparvata lugens* (Cheng *et al.*, 1979), the White-backed Planthopper *Sogatella furcifera* (National Coordinated Research Group for White-backed Planthoppers, 1981), the Rice Leafroller *Cnaphalocrocis medinalis* (Chang *et al.*, 1980) and the Oriental Armyworm *Mythimna separata* (Chen & Bao, 1987). Knipling (1979) advocated regional management by suppressing pests, including the *Heliothis* complex, Fall Armyworm *Spodoptera frugiperda* and Pink Bollworm *Pectinophora gossypiella*, which cause heavy damage following their northward spread in the USA, as they overwinter. An important aspect in developing such a control strategy is establishing the boundaries of the regions concerned. Hendrie *et al.* (1985) described a multidisciplinary group of experiments for establishing the redistribution of aphids in the USA. Hutchins, Smelser & Pedigo (1988) showed how atmospheric trajectory analysis was used to predict the multiple origins of Black Cutworms *Agrotis ipsilon* reaching the mid-western United States from overwintering sites in the Gulf States. In addition, they examined a potential practical application of this migration model, namely, planning the shipment, at minimum cost, of insecticides to match the predicted severity of infestations within the area.

Most if not all forecasters use historical datasets of pest incidence and case studies of migration events to provide analogue forecasts. Tatchell (1991), reviewing aphid forecasting, noted the importance of establishing continuous, standard, historical datasets for comparison with current populations. He described the use of the Rothamsted system of suction trapping to achieve this objective, but concluded that quantitative values on the abundance and timing of aerial populations needed to be supplemented by other population attributes, such as resistance to pesticides and virulence to crop varieties.

For many migrants, fixed-site trap catches identify the timing and intensity of immigration well, subsequent forecasts being based on relationships established between trap catches, weather and outbreaks nearby. However, few forecasting methods use suction-trapping data and so lack the comparable, quantitative data afforded by this method. Net- and light-trap networks exist for *N. lugens* and other migrant pests in many Asian countries (Ooi & Heong, 1988; Chapters 17 and 18, this volume) and light trapping has been largely replaced by pheromone trapping to monitor *S. exempta* distribution (C.F. Dewhurst, personal communication). It follows that relationships between catches and outbreaks will be more tentative than in the Rothamsted system, since catches by light and pheromone traps are more difficult to interpret, as they vary with moonlight and windspeed. Light trapping plays a minor role in establishing locust immigration, although catches at light are used during monitoring surveys to identify movement of night-flying solitaries.

### Data management, interpretation and integration

The use of computers as aids in the management and integration of data from diverse sources is an area of current major research that promises improved control and ability to manipulate data. Certainly, its implementation will lead to significant changes in operational procedures and, it is hoped, to more accurate and timely forecasts. Robinson (Chapter 16, this volume) and Lecoq (Chapter 19, this volume) summarise approaches to estimating and presenting weather and habitat data which drive the population dynamics of *S. exempta* and of locusts and grasshoppers, and both advocate using models to simulate these processes over space and time. Robinson also discusses their inclusion in a geographic information system (GIS).

The recent *S. gregaria* plague (1985–89) prompted research to replace the current generation of pesticides. Further priorities are improving forecasts and reviewing control strategies, tactics and application methods, to maximise the effectiveness of both current and new pesticides (FAO, 1989). A GIS is being designed to help forecasters and researchers achieve these objectives, by improving their access to current and historical data, and their ability to manipulate them.

The prime function of the GIS is to support the administration, plotting and analysis of data for operational forecasting, initially at FAO but, in principle, anywhere. A simplified structure of the system is shown in Fig. 20.2. The database design is based on tables of 'events'. These describe the life stages, density and size of locust populations, vegetation type and cover, and weather. The database tables are built around the location and date of events. This facilitates building up full descriptions of the events from a multiplicity of partial reports and the correlation and display of synchronous events. All information is added in a similar manner with the use of forms, menus and maps. The form for an event-type is opened and the attributes of a particular event are specified.

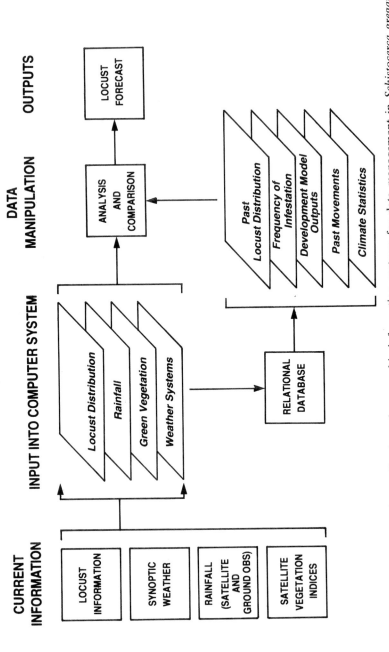

Fig. 20.2. Representation of the application of geographic information systems for data management in *Schistocerca gregaria* forecasting. OBS, observations.

Simply selecting the location on the map by pointing and clicking the mouse adds the information to the database. Similar forms and menus are used to select and display information already in the database.

The strength of using GIS for *S. gregaria* forecasting and research is that spatial data from different sources can be displayed simultaneously (overlaid) on a range of background maps, including remote-sensing products of habitat quality. In addition, map zoom and pan operations are available and several maps can be seen simultaneously and at different scales. This saves the time previously used for manually compiling and replotting locust data at different scales, both to produce a single overview of the area and to provide overlays for the habitat maps and weather charts supplied by external agencies. The system automatically prepares and updates frequency-of-incidence maps.

The GIS design team has completed two preliminary phases: an initial design exercise, which discussed system content and display options, and the provision of a prototype for users to test in an operational setting. The operational system will incorporate modifications to the initial specifications requested by forecasters, as the scope of the system becomes apparent. The prototype uses readily accessible, low-cost, machine-readable reference datasets for background maps, historical climatic records and gazetteers and covers the *S. gregaria* area (25° W to 100° E and 15° S to 45° N).

All data from current locust reports will be stored, but capturing the 60–year historical locust dataset for such a large area is not a trivial task. Consequently, only an agreed subset has been abstracted from archived maps dating from 1930. This information replaces the monthly distribution maps used by forecasters which record the presence or absence, during each month, of a defined set of fifteen locust population types in a degree square (1° latitude × 1° longitude) grid. It will not be adequate for all studies and researchers will be able to add point location, date and other attributes, for each locust, habitat or weather event in archived reports. Data collecting and mapping practices have evolved since 1930 as awareness of key steps in gregarisation of recession populations has increased (Waloff, 1976). Consequently, user manuals will be provided to describe how the availability and quality of data has varied over space and time.

The initial stage of prototype development concentrated on database design and the user interface for the data processes performed by forecasters. Two other tasks are in progress, aimed at giving forecasters rapid access to the results of case studies on plague upsurges and swarm invasions, and at incorporating externally prepared digital products: rainfall estimates and vegetation indices processed by the Remote Sensing Group at FAO, and weather charts and rainfall records provided by meteorological services. An important issue is to maintain compatibility with the GIS, as the preparation of these products improves and changes over time.

Day and Knight (Chapter 15, this volume) mention two other operational data-management systems, WORMBASE (African Armyworm) and FLYPAST (aphids). Both give improved access to current and past distributions of population counts at fixed trapping sites; WORMBASE also includes past outbreaks. Both allow output to maps but lack the full functions of a GIS. Unlike the *S. gregaria* system, both have related expert systems to aid information interpretation. FLYPAST has two functions: the preparation of a weekly bulletin for advisers, summarising the counts of 21 agriculturally important aphids, and the development of decision support systems. The latter will use information on on-farm practices, supplied by the user, to drive expert systems: one to predict outbreaks in field beans, and the other to assess the levels of potato virus Y in seed potatoes (Knight *et al.*, 1992). The development of comparable simulation models and support systems for rice pests in China is described by Zhou *et al.* (Chapter 17, this volume). Odiyo (1990) described an experimental expert system designed to run alongside WORMBASE, which he structured to capture the knowledge of an armyworm forecaster.

### Field research and modelling

Forecasters are seldom in a position to commission research directly or to highlight constraints that warrant further investigation. Usually, researchers and forecasters are in different institutes that are often in separate ministries; control teams may be in a third command. It is interesting to note that major improvements in forecasting the Australian Plague Locust *Chortoicetes terminifera* resulted when operational research and control were brought together in a single organisation. The annual research supplements and research reports of the Australian Plague Locust Commission (1980–1991) reveal a tightly focussed but multifaceted programme, driven by the need to prevent damage in eastern Australia and benefiting from the results of long-term research from other groups. An independent but similarly practically oriented system has also evolved in Western Australia (Chapter 9, this volume).

A practical problem in Australia was to control those adult *C. terminifera* populations that would emigrate and invade crops. The forecasting process is described in the context of a retrospective study by Symmons and Wright (1981). It became feasible once the group could quantify the rainfall and vegetation associated with three distinct types of population: potential migrants, adults that would breed locally, and those unable to migrate or to survive. Botanists had already established the role of climate on grass growth in Australia. These components were integrated, with sub-models of egg and nymphal development periods, to predict the probability of adults migrating, laying nearby or perishing. Field observations established that mass takeoff by migrants occurs within a narrow time-window after fledging, and only during a single weather type. Simple projections downwind predicted the path of migrating swarms adequately, so that true trajectory modelling was unnecessary. An operational system for remote sensing of habitats was being evaluated in 1991–92. The choice of control targets became regulated, ultimately, by the relative costs and efficacy of locating and spraying diffuse pre-migration and concentrated post-migration targets.

No single, compact grouping of staff could cover the much more extensive *S. gregaria* distribution area, but the current task of capturing and displaying data in a GIS for the central forecasters at FAO offers opportunities to consider user objectives and to identify soluble constraints on forecasting. Modelling and validating the spread of infestations over space and time in relation to concurrent changes in weather and vegetation was given high priority (FAO, 1989). A biomodel that simulates the development and movement of populations simultaneously in the whole distribution area is described by Launois *et al.* (1992). Consequently, an objective of the GIS project is to define the methodological and technical developments necessary to provide a satisfactory interface with population dynamics, migration and habitat-suitability models, as well as to enable researchers and forecasters to access and compare past and present events in their bid to improve forecasts.

These developments will be fully exploited only when field studies establish the population dynamics of the early stages of *S. gregaria* plagues. Weather and vegetation parameters still have to be adequately defined to confirm qualitative judgements by forecasters that plague-inducing rains are distinguishable, by their quantity, frequency and geographical extent, from those that cause localised but transient upsurges. Determining the movements of *S. gregaria* recession populations is another essential part of establishing their population dynamics since Waloff (1966) established that plagues followed multiplication through three to four generations in two complementary seasonal areas, with adults migrating between them. Trajectory models use windfields, which themselves continually change over space and time, to simulate the downwind displacement of locusts. It follows that an understanding of the initial emigration behaviour from breeding sites is essential if these models are to forecast, rather than to explain, well-documented recent or historical locust movements. Finally, a fundamental controversy remains to be resolved. The current control strategy is based on the assumption that controlling the initial gregarious locusts will prevent plagues, whereas research suggests that non-swarming populations, which are uneconomic spray targets, play a continuing role in plague initiation (Bennett, 1976; Symmons, 1992).

*Schistocerca gregaria* forecasters are acutely conscious of the imperfections in the data and knowledge of the pest that impede data interpretation and its integration in models. At present, therefore, they wish to retain the ability to access and compare separate sources of information. Examples include measured and satellite-derived estimates of the distribution and quantity of rainfall and vegetation which contribute to habitat suitability. Forecasters also wish to make their own assessments and to compare them with model simulations. This accords with Fisher's (1985) view that end users may find it difficult to adopt solutions provided by conceptual or mathematical models, because these either are, or are seen as, too complicated. He was considering on-farm control decisions. Here, the end users are forecasters who are stressing their need to assess the quality of data-entering models.

### Evaluating forecasts

Forecasts are written to alert people to the time, location and severity of migrant-pest populations. The recipients' task is to assess the importance of predicted events and the consequences of not taking action. Most forecasters, however, only attempt to evaluate the technical content of their predictions and fail to assess their impact on decision-makers (Betts, 1976; Odiyo, 1979). Consequently, information on how well people understand the forecasts, and how well the predictions aid their decisions, are poorly documented (Chapter 15, this volume).

There is clearly a need for a decision-tree analysis more comprehensive than that in Table 20.2 and for discussions with recipients, to act as a focus for tailoring *S. gregaria* forecasts to users' needs. No questionnaire has been distributed specifically to determine how recipients use locust and armyworm forecasts, comparable to that organised for the Rothamsted Aphid Bulletin (Bardner, French & Dupuch, 1980). This established that advisers wanted the data more rapidly and that there were three main uses for the bulletin data: advisory work (51%), research (42%) and other uses, mainly teaching (7%). Comment by recipients of locust and armyworm forecasts is mainly through institutional structures, such as formal committees, advisory groups and conferences (Uvarov, 1951; Lean, 1965; Odiyo, 1979).

All recipients want timely forecasts and, for migrant pests, a trade-off exists between the delay involved in sending field data to a distant forecasting centre and the improved prediction of long-distance invasions. An *S. gregaria* working party in 1972 placed a higher value on rapid exchange, within and between regions, of information and the accompanying forecasts from regional locust centres, which were being made 'without any qualifying terms', than on receiving an assessment for the whole area, accompanied by medium-term probabilistic forecasts issued by the central office in London. This review took place during a recession, when economically damaging populations were not moving between regions, and their recommendation to discontinue centrally issued forecasts was accepted (FAO, 1972). The decision was rescinded during the 1978–79 upsurge (FAO, 1978) when it was recognised that a central office should be re-established, but at FAO, Rome, rather than in London, to facilitate the prediction of inter-regional swarm movements.

Migrant-pest predictions are mainly based on analogy with past events (Symmons, 1972; Betts, 1976; Pedgley, 1981; Odiyo, 1990; Tatchell, 1991) and have an expected frequency of occurrence. Yet the FAO Working Party (FAO, 1972), whose members were all key users of *S. gregaria* forecasts, by recommending that such probabilistic forecasts should be discontinued, ignored this fundamental feature of the forecasting method. This suggests that a considerable lack of understanding existed between the forecasters in the central office, and surveyors and control managers in national and regional units. This misunderstanding should be remedied, since forecasts will not be used effectively until forecasting procedures and limitations are fully appreciated by key users.

Forecasters give three reasons for testing the accuracy of their predictions against subsequent events:

1  To reveal factors, not considered when making the forecast, that led to errors, so that such omissions are not repeated (Betts, 1976; Pedgley, 1981; Odiyo, 1990);

2  To increase the understanding of events, so that future predictions can be placed in the 'almost-certain' or 'almost-impossible' categories (Betts, 1976; Odiyo, 1990);

3  To guide recipients on the reliability of the service provided (Betts, 1976).

Evaluating the accuracy, timeliness and utility of forecasts is not easy. Pedgley (1981) noted two sources of difficulty: finding evidence to test a prediction, since populations are incompletely reported, and deciding how to treat those predictions for which only part (severity of infestation, geographical location or timing) occurred as forecast. Equally important is that some predictions are conditional on another forecast event or are possible only in a particular set of circumstances. This emphasises the importance of issuing follow-up warnings and the necessity for recipients to re-evaluate forecasts as soon as information that allows a more certain, and therefore useful, prediction becomes available.

Since events cannot be predicted with high probability, except in very short-term warnings, it is the correctness of the probability assigned to an event that must be evaluated. Each defined prediction can be noted as occurring or not occurring. The correctness of the probability assigned (say 'in three out of ten years') can then be ascertained from a run of similar events during a suitably long period. It follows that a single forecast event cannot be assessed in isolation and that different events cannot be pooled. In addition, the terms for severity of infestation, the geographical areas involved, and probability of each predicted event, should be defined in advance (Symmons, 1972). *S. gregaria* forecasters have always used terms denoting differing degrees of probability for the severity of infestations, geographical areas, and expected frequency of occurrence, but they were defined explicitly and in advance in only four of the 50 years for which forecasts have been prepared. How, then, have forecasts been verified?

Betts (1976), for *S. gregaria*, and Odiyo (1979), for *S. exempta*, both counted as correct, those events that occurred and as wrong, those that did not. They then pooled all events in the same probability class and derived the percentage of 'correct' forecasts. They were, therefore, assessing the frequency of outbreaks associated with the terms used to indicate high, medium and low probability of occurrence. This method gives guidance on the likelihood of pest outbreaks associated with these probability terms, which Betts defined as the 'reliability' of a service. The pooling of events may, however, have concealed differences in 'success rate' achieved for particular events of use to recipients.

Furthermore, a forecaster's test for correctness may not coincide with a user's view of a forecast's utility. The geographical units used in these forecasts are large and an occurrence just reaching an area would be counted as 'a success' during evaluation. However, a user may take a less positive view, since the administrative districts used in *S. exempta* forecasts in Tanzania average 14 500 km² (range 500–63 000 km²) and the average size of districts in southern, central and eastern Kenya is 8300 km² (range 1000–39 000 km²). Many geographical units used in *S. gregaria* forecasting are even larger.

No satisfactory method of testing pest forecasts has been developed to date. Clearly, forecasts have dubious value unless they influence user decisions and they can be used effectively only if key recipients have an appreciation of forecasting procedures and limitations. Evaluation based on user needs presupposes dialogue between the two groups and is the starting point for achieving these goals.

## Discussion

The *S. gregaria* forecasting system may be unique in gathering and processing information from the whole distribution area of a pest at a single forecasting office. Plague upsurge sequences often involve movements between regions and economically damaging swarms move between the four recognised subregions too rapidly and frequently for regional centres to predict events, although such centres work satisfactorily for *Chortoicetes terminifera* and *Spodoptera exempta*. In contrast, *Nilaparvata lugens* forecasting is both a national and a local responsibility (Chapters 17 and 18, this volume).

Locusts and African Armyworm both spread from natural vegetation or grazing areas, where they exist in low numbers, to crops. The control strategy and the supporting forecasting system are designed to prevent this spread. In contrast, the population dynamics of *N. lugens* is closely related to local farming practices. Management and forecasting of this species are initiated by monitoring immigration, and preventing the spread of the migrants to minimise downwind damage is not a component of current control strategies.

It is interesting to speculate whether these differences are due to a reasoned response to a pest's biology and impact on agriculture or to the spread of ideas between closely linked institutions. For example, *S. exempta* control has been much more influenced by locust organisations than has *N. lugens* control. Knipling's (1979) conclusion that the spread of many long-distance migrants in the USA should be prevented, suggests that delays in the spread and implementation of ideas on migrant-pest control may have predominated.

A challenge still facing forecasters and recipients of migrant-pest predictions is to agree how forecasts are presented and associated parameters defined, and for probability forecasting to be understood. Without this agreement, comparable forecast events cannot be identified and checked for accuracy, and improvements will not be made.

## References

Australian Plague Locust Commission (1980–1991). *Australian Plague Locust Commission Annual Report: Research Supplements July–June 1978 to 1981 and Research Reports July–June 1984 to 1990.* Canberra: Australian Government Publishing Service.

Ballard, E., Mistikawi, A.M. & El Zoheiry, M.S. (1932). *The Desert Locust,* Schistocerca gregaria *Forsk, in Egypt.* Bulletin no. 110. Cairo: Ministry of Agriculture, Egypt.

Bardner, R., French, R.A. & Dupuch, M.J. (1980). *Agricultural benefits of the Rothamsted Aphid Bulletin*, Report for 1980, Part 2, pp. 21–39. Harpenden: Rothamsted Experimental Station.

Bennett, L.V. (1976). The development and termination of the 1968 plague of the desert locust. *Bulletin of Entomological Research*, **66**, 511–52.

Betts, E. (1976). Forecasting infestations of tropical migrant pests: desert locust and African armyworm. In *Insect Flight*, ed. R.C. Rainey, pp. 113–34. *Symposia of the Royal Entomological Society of London* 7. Oxford: Blackwell Scientific Publications.

Chandra, S., Sinha, P.P. & Singh, R.P. (1988). The desert locust build-up in western Rajastan during the monsoon season of 1986. *Plant Protection Bulletin*, **40**, 21–8.

Chang, S.-S., Lo, Z.-C., Keng, C.-G., Li, G.-Z., Chen, X.-L. & Wu, X.-W. (1980). Studies on the migration of the rice leaf roller, *Cnaphalocrocis medinalis* Guenée. *Acta Entomologica Sinica*, **23**, 130–40. In Chinese, English summary.

Chen, R.-L. & Bao, X.-Z. (1987). Research on the migration of oriental armyworm in China and a discussion of management strategy. In *Recent Advances in Research on Tropical Entomology*, ed. M.F.B. Chaudhury, pp. 571–2. *Insect Science and its Application*, vol. 8, Special Issue. Nairobi: ICIPE Science Press.

Cheng, S.-N., Chen, J.-C., Si, H., Yan, L.-M., Chu, T.-L., Wu, C.-T., Chien, J.-K. & Yan, C.-S. (1979). Studies on the migrations of brown planthopper *Nilaparvata lugens* Stål. *Acta Entomologica Sinica*, **22**, 1–21. In Chinese, English summary.

Cherlet, M. & Di Gregorio, A. (1991). *Calibration and integrated modelling of remote sensing data for desert locust habitat monitoring.* Report on Projects ECLO/INT/004/BEL & GCP/INT/439/BEL. Rome: FAO.

Davies, D.E. (1952). *Seasonal Breeding and Migrations of the Desert Locust (Schistocerca gregaria Forskål) in North-eastern Africa and the Middle East.* Anti-Locust Memoir 4. London: Anti-Locust Research Centre.

Donnelly, U. (1947). *Seasonal Breeding and Migrations of the Desert Locust (Schistocerca gregaria Forskål) in Western and North-western Africa.* Anti-Locust Memoir 3. London: Anti-Locust Research Centre.

FAO (1953). *Technical Advisory Committee on Desert Locust Control. Report of the Third Meeting, Rome, Italy, 21–24 April 1953.* Report FAO/53/4/3224. Rome: FAO.

FAO (1958). *Desert Locust Control Committee. Report of the Fifth Session, Rome, Italy, 16–21 June 1958.* FAO Meeting Report 58/11. Rome: FAO.

FAO (1972). *Desert Locust Control Committee. Report of the Sixteenth Session, Rome, Italy, 23–27 October 1972.* FAO Meeting Report AGP:1972/M/7. Rome: FAO.

FAO (1978). *Desert Locust Control Committee. Report of the Extraordinary Emergency Session (Twenty-second Session), Rome, Italy, 26–28 July 1978.* FAO Meeting Report AGP:1978/M/9. Rome: FAO.

FAO (1985). *Report of Consultation on Desert Locust Plague Prevention in the Central Region. Rome, 28–30 May 1985.* Rome: FAO.

FAO (1989). *Desert Locust Research Priorities. Report of the FAO Research Advisory Panel, FAO Rome, 2–5 May 1989.* Rome: FAO.

FAO (1993). *The Desert Locust Guidelines. II. Surveying for Desert Locust.* Rome: FAO.

Faure, J.C. (1923). The life-history of the brown locust, *Locusta pardalina* (Walker). *Journal of the Department of Agriculture of the Union of South Africa*, **7**, 205–24.

Fisher J.P. (1985). Forecasting and decision making: an aid to crop protection. *SPAN*, **28**, 50–2.

Fortescue-Foulkes, J. (1953). *Seasonal Breeding and Migrations of the Desert Locust* (Schistocerca gregaria Forskål) *in South-western Asia.* Anti-Locust Memoir 5. London: Anti-Locust Research Centre.

Fourth International Locust Conference (1937). *Proceedings of the Fourth International Locust Conference*, Cairo, April 22, 1936. Cairo: Government Press, Bulâq.

Hemming, C.F., Popov, G.B., Roffey, J. & Waloff, Z. (1979). Characteristics of desert locust plague upsurges. *Philosophical Transactions of the Royal Society B*, **287**, 375–86.

Hendrie, L.K., Irwin, M.E., Liquido, N.J., Ruesink, W.J., Mueller, E.A., Voegtlin, D.J., Achtemeier, G.L., Steiner, W.W.M., Scott, R.W., Larkin, R.P. & Jensen, J.O. (1985). Conceptual approach to modeling aphid migration. In *The Movement and Dispersal of Agriculturally Important Biotic Agents,* ed. D.R. MacKenzie, C.S. Barfield, G.C. Kennedy, R.D. Berger & D.J. Taranto, pp. 541–82. Baton Rouge: Claitor's.

Hielkema, J.U., Roffey, J. & Tucker, C.J. (1986). Assessment of ecological conditions associated with the 1980/81 desert locust plague upsurge in West Africa using environmental satellite data. *International Journal of Remote Sensing*, **7**, 1609–22.

Hosni, M.H. (1966). *Meteorological Factors Affecting Invasions of the UAR by the Desert Locust (*Schistocerca gregaria *Forskål) during the period December 1960 to June 1961.* UNDP Desert Locust Project Progress Report FAO/UNSF/DLP/RSF/9. Rome: FAO.

Hutchins, S.H., Smelser, R.B. & Pedigo, L.P. (1988). Insect migration: atmospheric modeling and industrial application of an ecological phenomenon. *Bulletin of the Entomological Society of America*, **34**, 9–16.

Johnson, C.G. (1969). *Migration and Dispersal of Insects by Flight.* London: Methuen.

Knight, J.D., Tatchell, G.M., Norton, G.A. & Harrington, R. (1992). FLYPAST: an information management system for the Rothamsted aphid database to aid pest control research and advice. *Crop Protection*, **11**, 419–26.

Knipling, E.F. (1979). Strategic and tactical use of movement information in pest management. In *Radar, Insect Population Ecology and Pest Management*, ed. C.R. Vaughn, W. Wolf & W. Klassen, pp. 41–57. NASA Conference Publication 2070. Wallops Island, Virginia: NASA Wallops Flight Center.

Launois, M., Duranton, J.-F., Launois-Luong, M.-H., Lecoq, M., Popov, G.B., Gay, P.-E., Gigault, J., Balança, G., Coste, C., Foucart, A. & Monard-Jahiel, A. (1992). *La modélisation du criquet pèlerin* Schistocerca gregaria *(Forskål, 1775) sur l'ensemble de son aire d'habitat: le biomodèle SGR. Rapport final du project janvier 1989–mai 1992, D442.* Montpellier: Commission des Communautés Européennes; Bruxelles: Ministère de la Coopération; Paris CIRAD-GERDAT-PRIFAS. In French, English summary.

Lean, O.B. (1965). FAO's contribution to the evolution of international control of the desert locust, 1951–1963. *Desert Locust Newsletter*, Special Issue 1965.

Magor, J.I. (1989). Joining battle with the desert locust. *Shell Agriculture*, **3**, 12–5.

National Coordinated Research Group for White-backed Planthoppers (1981). Studies on the migration of white-backed planthoppers (*Sogatella furcifera,* Horváth). *Scientia Agricultura Sinica*, **5**, 25–31. In Chinese.

Odiyo, P.O. (1979). Forecasting infestations of a migrant pest: the African armyworm, *Spodoptera exempta* (Walk.). *Philosophical Transactions of the Royal Society B*, **287**: 403–13.

Odiyo, P.O. (1990). Progress and development in forecasting outbreaks of the African armyworm, a migrant moth. *Philosophical Transactions of the Royal Society B*, **328**, 555–69.

Omar, M.H. (1965). Review of previous work on the relation of two synoptic meteorological situations associated with locust incidence in spring and summer in the United Arab Republic. In *Meteorology and the Desert Locust. Proceedings of the WMO/FAO Seminar on meteorology and the Desert Locust*, Teheran, 25 November – 11 December 1963, pp. 205–13. WMO Technical Note 69 (WMO 171.TP.85). Geneva: World Meteorological Organization.

Ooi, A.C. & Heong, K.L. (1988). Operation of a brown planthopper surveillance system in the Tanjung Karang Irrigation Scheme in Malaysia. *Crop Protection*, **4**, 39–50.

Pedgley, D.E. (ed.) (1981). *Desert Locust Forecasting Manual*, vol. 1 text, vol. 2 maps. London: Centre for Overseas Pest Research.

Pedgley, D.E. (1989). Weather and the current desert locust plague. *Weather*, **44**, 168–71.

Pedgley, D.E. & Symmons, P.M. (1968). Weather and the desert locust upsurge. *Weather*, **23**, 484–92.

Rainey, R.C. (1951). Weather and the movement of locust swarms: a new hypothesis. *Nature*, **168**, 1057–60.

Rainey, R.C. (1963). *Meteorology and the Migration of Desert Locusts. Applications of Synoptic Meteorology in Locust Control.* Anti-Locust Memoir 10. London: Anti-Locust Research Centre. Also as *Meteorology and the Migration of Desert Locusts*, Technical Note No. 54 (WMO 138, TP 64). Geneva: World Meteorological Organization.

Rainey, R.C. (1969). Effects of atmospheric conditions on insect movement. *Quarterly Journal of the Royal Meteorological Society*, **95**, 424–34.

Rainey, R.C., Betts, E. & Lumley, A. (1979). The decline of the desert locust plague in the 1960s: control operations or natural causes. *Philosophical Transactions of the Royal Society B*, **287**, 315–44.

Reus, J.A.W.A. & Symmons, P.M. (1992). A model to predict the incubation and nymphal development periods of the desert locust, *Schistocerca gregaria* (Orthoptera: Acrididae). *Bulletin of Entomological Research*, **82**, 517–20.

Ritchie, J.M. & Pedgley, D.E. (1989). Desert locusts cross the Atlantic. *Antenna*, **13**, 10–12.

Roffey, J. (1963). *Observations on Night Flight in the Desert Locust* (Schistocerca gregaria *Forskål*). Anti-Locust Bulletin 39. London: Anti-Locust Research Centre.

Roffey, J. (1970). *The Anti-Locust Research Centre. A Concise History to 1970.* London: Anti-Locust Research Centre.

Roffey, J. & Popov, G.B. (1968). Environmental and behavioural processes in a desert locust outbreak. *Nature*, **219**, 414–15.

Rose, D.J.W., Dewhurst, C.F., Page, W.W. & Fishpool, L.D.C. (1987). The role of migration in the life system of the African armyworm *Spodoptera exempta*. In *Recent Advances in Research on Tropical Entomology*, ed. M.F.B. Chaudhury, pp. 561–9. *Insect Science and its Application*, vol. 8, Special Issue. Nairobi: ICIPE Science Press.

Singh, G. (1967). *The Current Desert Locust Recession and FAO's Policy of Control and Prevention.* FAO Report FAO/PL/DL/1–Rev.1. Rome: FAO.

Skaf, R. (1978). *Etude sur les cas de grégarisation du criquet pèlerin en 1974 dans le sud-ouest Mauritanien et du Tamesna Malien.* FAO Report FAO/AGP/DL/TS/17. Rome: FAO.

Skaf, R., Popov, G.B. & Roffey, J. (1990). The desert locust: an international challenge. *Philosphical Transactions of the Royal Society B*, **287**, 7–20.

Symmons, P.M. (1972). Assessing and forecasting in relation to the work of the Desert Locust Information Service. In *Proceedings of the International Study Conference on the Current and Future Problems of Acridology*, ed. C.F. Hemming and T.H.C. Taylor, pp. 397–402. London: Centre for Overseas Pest Research.

Symmons, P.M. (1992). Strategies to combat the desert locust. *Crop Protection*, **11**, 206–12.

Symmons, P.M. & Wright, D.E. (1981). The origins and course of the 1979 plague of the Australian plague locust, *Chortoicetes terminifera* (Walker)

(Orthoptera: Acrididae), including the effect of chemical control. *Acrida*, **10**, 159–90.

Tatchell, G.M. (1991). Monitoring and forecasting aphid problems. In *Aphid–Plant Interactions: Populations to Molecules*, ed. D.C. Peters, J.A. Webster & C.S. Chlouber, pp. 215–321. Oklahoma Agricultural Experiment Station, Miscellaneous Publication 132.

Taylor, R.A.J. (1979). A simulation model of locust migratory behaviour. *Journal of Animal Ecology*, **48**, 577–602.

Tu, C.-W. (1980). The brown planthopper and its control in China. In *Rice Improvement in China and other Asian Countries*, ed. T.R. Hargrove, pp. 149–56. Los Banos: International Rice Research Institute.

Uvarov, B.P. (1921). A revision of the genus *Locusta*, L. ( =*Pachytylus*, Fieb.) with a new theory as to the periodicity and migrations of locusts. *Bulletin of Entomological Research*, **12**, 135–63.

Uvarov, B.P. (1930). *Instructions for Observations on Locusts*. London: Imperial Bureau of Entomology.

Uvarov, B.P. (1933–1935). The locust outbreaks in Africa and Western Asia, 1925–34. *Economic Advisory Council Committee on Locust Control*, 63–80–1 to 4. London: HMSO.

Uvarov, B.P. (1951). *Locust Research and Control 1929–1950*. Colonial Research Publication, No. 10. London: HMSO.

Uvarov, B.P. & Milnthorpe, W. (1937–1939). The locust outbreaks in Africa and Western Asia, 1935–1937. *Economic Advisory Council Committee on Locust Control*, 63–80–5 to 7. London: HMSO.

Waloff, Z. (1946*a*). *Seasonal Breeding and Migrations of the Desert Locust* (Schistocerca gregaria *Forskål*) *in Eastern Africa*. Anti-Locust Memoir 1. London: Anti-Locust Research Centre.

Waloff, Z. (1946*b*). A long-range migration of the desert locust from southern Morocco to Portugal, with an analysis of concurrent weather conditions. *Proceedings of the Royal Entomological Society of London A*, **21**, 81–4.

Waloff, Z. (1966). *The Upsurges and Recessions of the Desert Locust Plague: an Historical Survey*. Anti-Locust Memoir 8. London: Anti-Locust Research Centre.

Waloff, Z. (1976). *Some Temporal Characteristics of Desert Locust Plagues*. Anti-Locust Memoir 13. London: Centre for Overseas Pest Research.

Wright, D.E. (1986). Economic assessment of actual and potential damage to crops caused by the 1984 locust plague in south-eastern Australia. *Journal of Environmental Management*, **23**, 293–308.

# 21

## Insect migration:
## a holistic conceptual model

V. A. DRAKE, A. G. GATEHOUSE AND
R. A. FARROW

### Introduction

The previous chapters of this volume review and provide a contemporary synthesis of current knowledge about several of the world's most significant insect migration systems, and of the adaptations that enable these systems to function and persist. They illustrate the wealth of information that has been accumulated, and the understanding that has been developed, during the last two to three decades: the period since the 'turning point in the study of insect migration' declared by Kennedy (1961), Southwood's (1962) association of migration with temporary habitats, the publication of the comprehensive review by Johnson (1969), and the development of 'entomological' radars by Schaefer (1976).

Along with previous general (e.g. Johnson, 1969; Dingle, 1980, 1984; Kennedy, 1985) and more specific (e.g. Drake & Farrow, 1988; Gatehouse, 1989; Farrow, 1990) reviews, the foregoing chapters clearly indicate that migration is a common and widespread phenomenon, both geographically and across taxa. They also show that, while the details may vary from species to species and between geographical or climatic regions, many features of insect migration are common to most species and localities. The recognition (often gradual and implicit) of these more general features over the last decade or so has provided a series of insights that together amount to a major conceptual advance. Among the most important of these insights are:

1 Migration is actively initiated, maintained and terminated; it is not a 'passive' process or in any sense 'accidental' (Kennedy, 1961).
2 Migration can be distinguished behaviourally from other types of movement: migratory behaviour is characterised by persistent, straightened-out movement and some temporary inhibition of

station-keeping responses (Kennedy, 1961, 1985; Gatehouse, 1987).
Its outcome is that the insect is relocated to some new habitat beyond
its previous foraging range (Southwood, 1981).

3 Migration is an adaptation to temporary habitats that vary in availability and quality both in space and in time (Southwood, 1962, 1977);
seasonal variations may be quite regular but others are often unpredictable.

4 By determining where and when reproduction occurs, migration has a
major influence on fitness; it is thus a fundamental component of the
life histories of migratory species and plays a central role in population processes (Dingle, 1984, 1986b, 1989).

5 Migration may be obligatory (e.g. Gatehouse, 1986) but more usually
occurs facultatively in response to environmental cues (Johnson,
1969; Dingle, 1985). Induction of migration may involve developmental as well as physiological and behavioural responses, most obviously
in wing dimorphic species where environmental cues initiate production of macropters as well as regulating their migratory behaviour
(e.g. Shaw, 1970a,b).

6 In females, migration usually occurs early in the adult stage, before
reproduction begins (Johnson, 1960, 1963, 1969; Kennedy, 1961);
males exhibit more varied patterns of integration of migration and
reproduction (Johnson, 1969). Migration by larval stages occurs less
commonly in insects (Kennedy, 1975).

7 Long-distance migrations usually involve flight above the 'flight
boundary layer' (the zone near the ground in which the wind is slow
enough for the insect to be able to make headway against it – Taylor,
1958), frequently at altitudes up to 1 km (and occasionally higher),
and often at night (Schaefer, 1976). Such migrations are achieved
almost entirely through transport on the wind (Drake & Farrow,
1988). However, some long-distance migrants fly exclusively or predominantly within their flight boundary layer (Schmidt-Koenig, 1985;
Walker, 1985; Gibo, 1986).

8 Some migrations occur in regions and at seasons where the wind blows
predominantly from one particular direction; others may be subject to
winds that are highly variable (Drake & Farrow, 1988). Correlations
between wind direction and other environmental cues that regulate
migration (e.g. temperature) may reduce the extent of variability in
migration outcomes (Chapter 1, this volume). Winds are, however,
never constant and there is always an element of unpredictability
about the destinations of wind-assisted migrations.

9  The migrant is initially unresponsive to stimuli associated with favourable habitats, but this initial phase of migration is terminated spontaneously and followed by flight (or flights) with the migrant in responsive mode (Kennedy, 1985; Gatehouse, 1995).

10 Migrants are characterised by a suite of co-adapted biochemical, physiological, behavioural and demographic traits, the 'migration syndrome' (Dingle, 1984).

11 The phenotypic variance observed in many of the migration-syndrome traits has been shown to have a substantial additive genetic component, indicative of polygenic inheritance (Harrison, 1980; Dingle, 1986*a*; Gatehouse, 1989). Some (but not all) of the traits are associated by a system of genetic correlations (Dingle, 1986*b*, 1991; Gunn & Gatehouse, 1993).

12 The observed values and variances of the syndrome traits are likely to be the outcome of contemporary natural selection (Chapter 11, this volume).

In this overview, we endeavour to provide a synthesis of these generalisations, i.e. a unified description of insect migration that incorporates these insights. The description we present is a general one; specific variants will need to be derived from it for each individual migration system. Its basis is inevitably evolutionary and its scope is limited to the types of migration described in the previous chapters, i.e. long-distance windborne migrations of insects; however, we imagine that it could be applied, with varying degrees of modification and extension, to other taxa and types of transport.

Our objective in this synthesis is to enhance our understanding of insect migration by presenting a holistic view of the phenomenon, one that recognises all the physical and biological components of a migration system and emphasises the interlinkages between them. By stating explicitly our understanding of the roles and functions of these components and relationships, we aim to provide an impetus for testing, modifying and extending them through discussion and research. We hope that the conceptual structure and defined (sometimes novel) terminology introduced here will ease this process of development and debate, and reduce the opportunities for confusion that have sometimes resulted from the prolonged debate over the definition of migration (Kennedy, 1985; Taylor, 1986; Gatehouse, 1987; Preface to this volume) and the failure to define terms such as 'migratory strategy' and 'migration system'.

## A conceptual model of insect migration

The description of insect migration that we present here takes the form of a conceptual model (Andrewartha, 1957). It is fundamentally a model of a population, extending over many generations and, like any comprehensive population model, it necessarily incorporates both a spatial and a genetic dimension. Its central concern is the maintenance of the population by the process of evolution on the contemporary time-scale.

We will take this population to be isolated (i.e. not exchanging members with neighbouring populations) and, at any one time, to be composed of a number of subpopulations at separate locations. A subpopulation is the basic unit of the model: it will be taken to be an interbreeding subset of the population which, between migration events, occupies a single, continuous area in which it experiences essentially the same environmental conditions. Except in the probably unusual circumstances in which an entire subpopulation migrates together to a single destination, subpopulations subdivide and diverge, or converge and may fuse (see below), at each migration event. The pattern of subpopulations is, therefore, generally disrupted and reassembled as a result of every migration. However, convergence and fusion of the component subpopulations must also occur at intervals so that, over long periods, the population is unitary.

The model consists of 'components' and 'processes'; the latter link the former and together they constitute a 'migration system'. Migration systems are characterised, and distinguished from population systems of non-migrants, by the frequency and extent of changes in the locations of subpopulations, and by the migratory behaviour that produces these changes.

### The components of a migration system

Our model of a migration system has four primary components (Fig. 21.1). These are:

1 The environment in which migration occurs – the 'migration arena';
2 The spatiotemporal population demography that results from migration – the 'population trajectory';
3 The traits that implement migration and determine the fitness of the migrants – the migration syndrome;
4 The genetic complex underlying the migration syndrome.

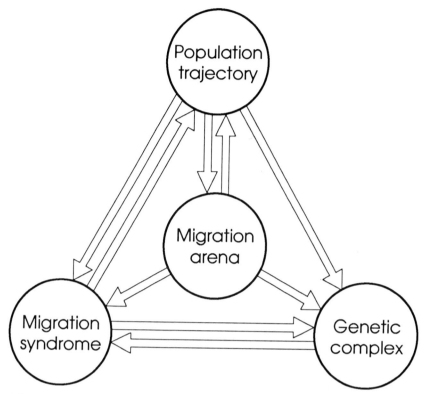

Fig. 21.1. Conceptual model of a migration system. The circles indicate the four primary components of the system, and the arrows the principal processes that connect them.

We will now consider each of these components in more detail, identifying those aspects that differentiate migration systems from population systems in general.

### The migration arena

A population's migration arena is defined to encompass the entire region within which the population's migrations take place, including regions that are reached only occasionally or accidentally. It incorporates both the terrestrial 'substrate' on which habitat patches are distributed and the atmospheric 'medium' through which migrants fly. Its boundaries in both the horizontal and vertical dimensions could, at least in principle, be determined by observation of the population, e.g. by a wide-ranging programme of field surveys and aerial sampling continued over many generations.

The arena incorporates the physical and biotic properties of the population's environment that influence a migrant's fitness, either through their effects on development, fecundity and survival, or through their influences on the occurrence, timing and outcome of migration. Three principle attributes can be recognised: invariant abiotic, variable abiotic and biotic components.

1 The more or less invariant abiotic components – land, water, air, climatic zones and geomagnetism. These include: geographical and topographical features (the disposition of land, sea, lakes, mountains, etc.), which limit the distribution of habitat patches and may act as barriers to migration; geology (underlying rock types and their effects on drainage and soils), which may have a major influence on habitat location and quality; the global circulation and the regional patterns of temperature and rainfall it produces; and the more permanent features of the atmosphere (the general decrease of temperature with altitude, and the characteristic day- and night-time patterns of wind and convection in the planetary boundary layer), which affect the ability of migrants to attain, and maintain themselves at, high altitudes. Geomagnetism may be important for orientation and navigation.

2 The variable abiotic components include weather, climate and climatic variability, the annual cycle of photoperiod and the lunar cycle. Weather and climate are the components of the physical environment that vary most and that have the greatest modulating effect on biological activity. 'Weather' is the chaotic (i.e. unpredictable far in advance) sequence of changing atmospheric conditions comprising phenomena with time-scales ranging from ~1 h to ~1 week: from thermals through thunderstorms and sea breezes to synoptic-scale depressions and anticyclones. The weather 'elements' of particular importance to migrant insects are temperature, rainfall and wind; they affect habitat quality, development, survival, natural enemies and the incidence, timing and outcome of migration (Drake, 1994). 'Climate' is the statistical integration of weather: the long-term averages, extreme values and probabilities (e.g. of rainfall), and their seasonal variations – the last being of particular significance for both migrant and non-migrant populations. 'Climatic variability' encompasses relatively unpredictable and irregular variations with time-scales of a few months to a few years; the best documented of these are the droughts and periods of above-average rainfall associated with the El Niño / Southern Oscillation phe-

nomenon (Nicholls, 1991). Finally, the annual cycle of photoperiod and the lunar cycle are both highly predictable; they are important primarily as cues that induce or suppress migration (and diapause).

3 The biotic environment includes host plants (for both food and shelter) or prey, and natural enemies (including disease). The abundance and quality of these factors vary in both space and time, modulating, with the abiotic attributes, the changing patterns of habitat favourabililty within the arena. They also influence the occurrence and timing of migration by their effects on development and by providing cues.

These arena attributes act on a population in combination, and they also frequently interact with each other. For example, weather and climate at a particular location are influenced by topographical factors and the proximity of the ocean, while vegetation produces its own microclimate, experienced by insects exploiting it. Perhaps most significantly, the third attribute is to a considerable extent driven by the first two: the condition of the biotic environment is determined primarily by physical factors such as soil type, temperature and rainfall, although density-dependent influences from the population itself, e.g. through the intensity of exploitation of resources or disease transmission, may sometimes also be important.

A migration arena differs from the environment of a non-migratory population in that the conditions and resources for survival and reproduction are relatively short-lived at many locations. New patches of favourable habitat that arise are so widely dispersed (e.g. where they depend on tropical rainfall), or at such distances (e.g. for insects migrating between climatic zones), that they can only be reached by migration. The arena will therefore often encompass extensive regions, within which conditions and resources are inadequate for reproduction or even survival, that must be traversed by migratory flight. Unlike the environment of a non-migratory species, it has a significant vertical dimension, extending to altitudes beyond the flight boundary layer; because of their importance for migratory transport, the winds in this zone are a particularly significant component of the arena environment.

### The population trajectory

The population trajectory is a representation of the spatiotemporal component of the population's demographic history – the 'march [of a population through its] habitats in space and time' (Southwood, 1977). It describes, for a series of generations, the spatial disposition of the population, the locations and times of year at which key life-history events

(a)

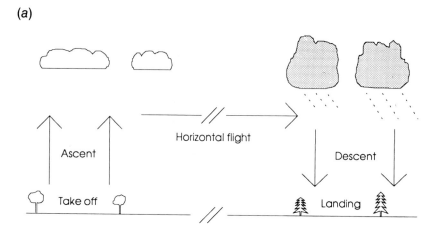

(b)

□　Birthplace
●　Breeding site
■　Place of death
────　Foraging area
- - - -　Parental foraging
　　　　area
──▷　Migration

Breeding range　　　　　　　　Non-breeding range

Fig. 21.2. Spatio-temporal representations of migration. (*a*) A single migration event, with its three phases of 'take-off' and ascent, horizontal translation, and descent and landing. After Kampmeier (1994). (*b*) A 'lifetime track', representing the spatio-temporal component of an individual's life history. Modified from Howard (1960). (*c*) The 'fern stele' representation of the time development of a spatially distributed population – what we term a 'population trajectory'. After Taylor & Taylor (1977). (*d*) The typical seasonal migrations (arrows) and breeding areas of a population (*Schistocerca gregaria* in central West Africa) – a form of 'population pathway'. After Farrow (1990).

(c)

(d)

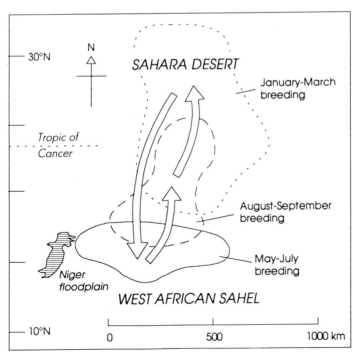

(reproduction, dormancy, mortality) occur, and the movements of sub-populations between these locations. It incorporates and extends earlier representations of migration as a spatiotemporal process (Fig. 21.2) and can be considered as the ensemble of 'lifetime tracks' (Baker, 1978) of all individual members of the population. It can be expressed schematically, in terms of areas rather than points and broad (and often diverging) arrows rather than simple trajectory lines (Fig. 21.2). A quantitative representation, in terms of densities and fluxes for each location and time, also appears possible (see Summary and discussion).

Population trajectories consist of one or more 'strands', of varying size and shape (Fig. 21.3). Each strand represents the location and development of a temporary subpopulation (or of the population as a whole if there is only one strand) through space and time. When migration occurs, strands may branch and diverge, or converge and fuse, as a new set of subpopulations is formed at new locations. In facultative migrants, strands representing individuals that fail to migrate remain at the same location. Branching (or fusion) of strands can also occur when part of a subpopulation enters (or emerges) from dormancy, or simply develops so much more slowly than the remainder that interbreeding becomes impossible; in these circumstances, the strands all remain at the same location. Along the trajectory, generations can be represented by dividing the strands into sectors, each of which can be further divided into stages representing periods of development, migration or dormancy, and reproduction (Fig. 21.3a).

It may sometimes be valuable to represent the information contained in the population trajectory in probabilistic terms: to indicate the probability of each location being occupied by a subpopulation of a particular size and life stage at a particular time, and the likelihood of migrations occurring to or from each other location. This alternative representation is termed the 'population pathway'. It incorporates all the year-to-year variability in the full population trajectory, and is therefore the distribution from which annual trajectory segments are drawn. It is essentially a projection of the trajectory reticulum (as depicted in Taylor & Taylor's (1976) 'fern stele' model, Fig. 21.2c) down onto a single annual cycle; it can be represented schematically by a two-dimensional network with labels to indicate the time of year (Fig. 21.2d). Note that the pathway, by definition, defines the boundaries of the arena, and that the arena in turn forms the 'domain' of the trajectory.

Trajectories of migrant populations differ from those of non-migratory species by the magnitude of the displacements they incorporate (and therefore their geographic extent), and often (though not necessarily) by their reticulate topology. Strands rarely, if ever, persist in one location for more than a few generations; rather, they form simple or complex patterns as the spatial disposition of the population within the arena changes, often radically, within and between generations. The large displacements result predominantly from migratory behaviour, rather than from foraging activity or adventitious displacement (Kennedy, 1985; Gatehouse, 1987). The trajectory patterns are to some extent predictable: they represent the outcomes of flights that are initiated either at every generation (for obligate migrants) or in response to environmental conditions (facultative migrants).

## The migration syndrome

Migratory insects exhibit a suite of co-adapted traits, absent in non-migratory species, which are termed the 'migration syndrome'. It comprises morphological, biochemical, physiological and behavioural traits that implement migration, and demographic traits that determine the migrants' schedules of reproduction and mortality. Among the most important of these traits are:

1 Mechanisms, in facultative migrants, that determine whether the insect migrates at all, e.g. induction of macroptery and accumulation of lipid reserves;
2 A flight apparatus (wings, musculature, fuel reserves and mobilisation mechanisms, etc.) and a flight-control system adapted for sustained flight;
3 Endocrine mechanisms controlling the differential development of the flight apparatus and the reproductive organs (and so implementing the 'oogenesis-flight syndrome' in females – Johnson, 1969); these and mechanisms controlling wing-muscle histolysis (after or between migrations – Dingle, 1985) and dealation determine the timing of migration in relation to other life-history events, especially reproduction and diapause;
4 Mechanisms regulating the insect's flight capacity (the duration of each migratory flight) and the migratory period (the number of successive days or nights that the insect remains capable of migratory flight); the migratory period may be determined by the endocrine mechanisms (number 3 above) controlling the length of the pre- (or inter-) reproductive period (Chapter 10, this volume);
5 Behaviour influencing the destination of a migration event (e.g. flight

initiation and termination behaviour, and behaviour controlling the migrant's altitude and orientation);
6 Responses to environmental cues (photoperiod, crowding, host quality, weather elements, illumination, etc.) that modulate expression of the traits in numbers 1, 4, and 5;

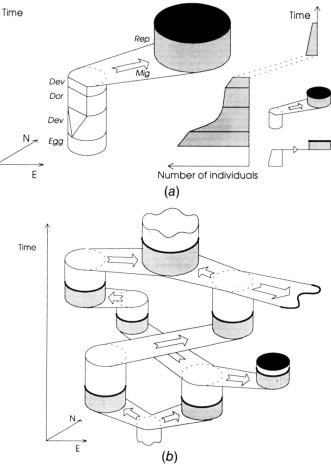

Fig. 21.3. Examples of population trajectories. (*a*) Schematic representation of one generation of a single subpopulation, from egg stage (*Egg*) through to death (shown black), with periods of development (*Dev*), dormancy (*Dor*), migration (*Mig*), and reproduction (*Rep*, shaded). Contraction of the area occupied by the subpopulation due to local mortality during the initial development phase, and divergence during migration (without branching), are illustrated in this example. Subpopulation size is indicated separately by the time series. Two simplified representations of the generation are shown at bottom right. (*b*) Schematic trajectory fragment showing strand branching, crossing, and fusing, and (at right)

7 Demographic traits, including development rates and schedules of reproduction and mortality, which determine the migrants' capacity to exploit the habitats they reach and so their fitness.

The migration syndrome for an individual is represented by the phenotypic character (e.g. macroptery) and values of these component traits,

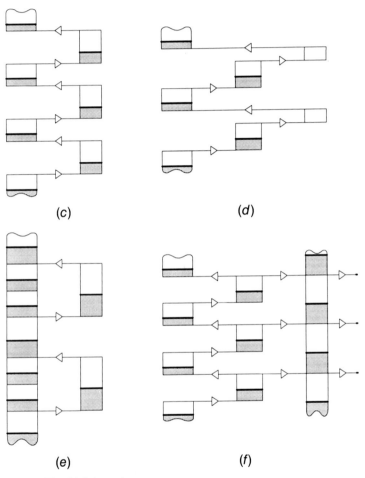

(c)                    (d)

(e)                    (f)

Caption for Fig. 21.3 (*cont.*)
two spurs. (*c–f*) Schematic trajectory fragments showing a to-and-fro movement between alternate breeding sites (*c*), a circuit incorporating two breeding sites and one dormancy site (*d*), migrant and sedentary subpopulations interbreeding in a permanent breeding area (*e*), and a subsidiary population periodically receiving immigrants from the continuing strand but never returning emigrants to it (*f*). In the simplified representations (*c–f*), only one spatial dimension (e.g. latitude) is shown explicitly; as in (*a*) and (*b*), time advances from bottom to top.

while a subpopulation can be characterised by the distributions of these values among its members. The model thus recognises explicitly that the members of a subpopulation are phenotypically diverse. The distributions of the trait values will vary between subpopulations, and with time as subpopulations advance along strands of the trajectory.

### *The genetic complex underlying the migration syndrome*

The genetic complex underlying the migration syndrome comprises the genes that code for the syndrome traits, their modes of inheritance, dominance and epistatic effects between them, and the existence or absence of positive and negative genetic correlations between loci due to linkage disequilibrium and pleiotropy[1]. It also encompasses mechanisms that influence the phenotypic expression of these genes, including genotype × environment interactions and maternal effects. For an individual, this genetic complex comprises all elements of the genome that code for traits contributing to the migration syndrome and any interactions between them; for subpopulations and populations, it is characterised by the frequencies of the alleles at these loci.

Most physiological, behavioural and demographic components of the migration syndrome are quantitative traits, which are inherited polygenically (i.e. they depend on the action of genes at several loci – Dingle, 1986*a*, 1991; Gatehouse, 1989) and show continuous variation. Variation in migration-syndrome traits is evident even in clonal migrants (e.g. *Aphis fabae* and *Rhopalosiphum padi* – Nottingham & Hardie, 1989; Nottingham, Hardie & Tatchell, 1991), possibly generated by some form of stochastic polyphenism (Walker, 1986). Inheritance of wing morph is sometimes determined at a single locus (e.g. in some Coleoptera and Hemiptera) but is more often polygenic (Roff, 1986*a*; Roff & Fairbairn, 1991), being controlled by a threshold mechanism that allows continuous variation in the proportions of the morphs (Roff, 1986*b*; Falconer, 1989).

Loci contributing to syndrome traits may be autosomal or sex-linked. X-linked traits may have a decreased rate of fixation under directional selection, compared with autosomal ones, which may help to maintain variation. However, many traits that have a major bearing on fitness appear to be sex-linked (Charlesworth, Coyne & Barton, 1987; Colvin & Gatehouse, 1993).

[1] The existence of genetic correlations between traits has been taken to be a necessary feature of a 'syndrome' in the sense of a 'complex adaptation' (Dingle 1986*a*); however, the migration syndrome, as we define the term, comprises all traits that contribute to the successful induction and outcome of migration.

Syndrome traits may be inherited independently, or genetically corre-lated; for example, in *Oncopeltus fasciatus*, flight capacity has been shown to be positively correlated with demographic traits that result in rapid early reproduction following migration and effective colonisation of the new habitat (Palmer & Dingle, 1986, 1989). Other syndrome traits, e.g. pre-reproductive period when it is a component of migratory poten-tial, may need to be independent of the constraints of correlated re-sponses to selection (Dingle, 1984, 1986*a*, 1991; Gatehouse, 1995; Chapter 10, this volume). These genetic correlations, and their sign and degree, have important implications for the way in which components of the migration syndrome respond to selection (Bradshaw, 1986).

Finally, traits that are subject to genotype × environment interactions have their phenotypic expression modulated by environmental cues and conditions. Modulation can also occur as a result of maternal effects.

### The processes by which a migration system functions

The processes that occur as a migration system functions give rise to fluctuations in the condition of the arena and determine the migrants' responses to these fluctuations. They maintain the components of the system within a range of states, enabling the population to survive indefinitely. How this self-maintenance is achieved, and why the system's components and processes take particular forms in particular environ-ments, are the primary questions that fundamental research on insect migration aims to answer.

We will now examine these processes, and the effects they have on each of the system components. Processes involving natural selection will be considered together in a final subsection.

### Processes involving the arena

Arena processes that influence migration can be divided into three categories: internal processes that do not involve the population but which alter the environment that it experiences; processes in which components of the arena environment affect the population; and processes in which the population affects components of the arena environment.

The internal processes are both physical and biological in nature. They include:

1 The planetary and atmospheric phenomena that give rise to climate, the seasons and weather;

2 The ecological processes that provide and regulate the abundance and phenologies of host plant, prey or other food resources, populations of natural enemies, biotic shelter sites, etc.;

3 Interactions between these that result in variation of habitat availability and quality.

An important characteristic of the effect of these processes on the arenas of populations of migratory species is that they induce large-amplitude variations of habitat quality in both space and time (Southwood, 1977). The environment experienced by both a migrant population itself and the biotic resources on which it depends appears to be determined largely by the internal arena processes (especially weather and climate) (Andrewartha & Birch, 1954). Because of their key role in the dynamics of populations, these must be fully integrated into any comprehensive model of a migration system.

The processes by which the arena environment directly affects the population are of six types:

1 Effects of environmental conditions on the phenotypic expression of migration-syndrome (and other) traits through genotype × environment interactions;

2 Effects of environmental conditions and resources, especially food and shelter, on the migrants' development, activity and reproduction;

3 Effects of unfavourable factors, e.g. extremes of temperature and rainfall, disease and natural enemies, causing mortality;

4 Effects of immediate and anticipatory cues that modulate the incidence and timing of migration and other key life-history events, such as dormancy and reproduction;

5 Effects of environmental stimuli influencing the migrants' behaviour between and during migrations, e.g. stimuli from resources influencing behaviour within habitat patches, stimuli influencing take-off, altitude of flight and orientation, descent and landing, etc.;

6 Effects on the outcome of migratory flights (e.g. due to wind direction), and so on the spatial distribution of the population.

These processes have a direct impact on reproductive success, mortality and the induction and outcome of migration. However, they also have indirect effects through their influence (e.g. by determining development rates or by providing environmental cues) on the incidence and timing of migration (and other life-history events), both of which have a major impact on fitness. They therefore affect both the demography of the population and its spatiotemporal development, i.e. they determine how the population trajectory develops. By imposing selection on members of the population, they alter the distributions of the values of the component traits of the migration syndrome, and the frequencies of the alleles at loci contributing to it, as the population advances along the strands of its trajectory. These effects are considered further in later sections.

The population affects the arena environment through processes such as depletion of resources, effects on predator and pathogen populations, or selection for host-plant resistance. Many of these processes alter the quality of the resources available to the migrants, and the severity of hazards they encounter. Empirical studies suggest that these influences, which are often density-dependent, have a negligible effect on migrant populations most of the time, but they can become key factors when densities are very high (Strong, 1984).

*Processes involving the population trajectory*

Both the demographic and spatiotemporal development of the population is represented in the model by the population trajectory. The trajectories of migratory species are characterised by their topological complexity (i.e. the frequency with which strands branch and fuse) and by their geographical extent (i.e. the magnitude of the displacements of their component strands). We will examine here the forms that these trajectories can take, and then see how they are shaped by processes involving the arena and the migration syndrome.

The strands of the trajectory, each representing a temporary subpopulation, advance along the time axis (vertical in Fig. 21.2c). These strands can branch, converge and fuse. Branching occurs when new temporary subpopulations are created, either by spatial separation during migration, or by development within a common area becoming asynchronous (e.g. by part of the subpopulation entering dormancy). Strands converge when the temporarily separated subpopulations they represent come, as a result of migration, to occupy a common area. However, convergent strands can fuse only if the subpopulations they represent reach the adult stage almost simultaneously, so that interbreeding occurs and the subpopulations lose their separate identities. Collocated strands retain their separate identities if interbreeding does not occur, e.g. because the subpopulations are developing asynchronously or (as in aphids after spring and summer migrations) reproducing asexually. The trajectory for a unitary population forms a fully connected reticulum (we use this term, rather than 'network', to emphasise its three-dimensional structure, with latitude, longitude and time axes – Fig. 21.2c): unconnected (but possibly intertwined) reticula would represent separate populations.

The population trajectories of migratory species can have many topologies (Fig. 21.3). Perhaps the simplest is an unbranched zig-zag trajectory (Fig. 21.3c) that represents a to-and-fro movement, e.g. between winter and summer breeding or dormancy sites. A variant of this is a helix-like trajectory (Fig. 21.3d) which represents movement in a 'circuit' around a series of sites. To-and-fro migration out of and back to a permanent breeding area, within which part of the population remains, produces a ladder-like trajectory (Fig. 21.3e). Spurs ('dead-end' strands) on a trajectory represent the extinction of subpopulations (Fig. 21.3b); this may occur soon after branching, or only after a number of subsequent generations. Finally, parts of the reticulum may be connected to the remainder only by inward-flowing strands; these represent subsidiary populations that periodically receive immigrants from the main population, but do not return any emigrants to it (Fig. 21.3f).

More complex topologies than the simple examples illustrated in Fig. 21.3, and substantial variation from year to year in both topology and intensity of the flux along the different strands, will be the norm, especially in discontinuous habitats (archipelagoes, ephemeral watercourses, irrigation regions, etc.) or in highly unpredictable environments (semi-arid areas with very variable rainfall, and areas that are climatologically marginal for the species' survival). In addition to variations in the total population size, different strands may dominate in different years, and strands that are spurs in some years may fuse with continuing strands in others. A common feature of all trajectories, however, is that there is some degree of 'inwards-turning' or 'spatial cycling'; i.e. the trajectory tends to reproduce itself as strands return repeatedly (but not necessarily at regular intervals) to regions of the arena where temporarily favourable habitats arise either seasonally or irregularly. In the pathway representation, this spatial cycling will manifest itself as 'loops': pathway strands that return to become superimposed on themselves after a number of generations (Fig. 21.2d).

The trajectory that the population actually follows will be determined in part by environmental factors. The processes by which the arena affects the population have already been identified. As habitat quality changes in space and time, subpopulations occupying different locations will develop, enter dormancy, migrate, reproduce or suffer mortality according to the conditions they encounter; in the trajectory, these changes will be reflected in the way the corresponding strands advance through sectors and stages, wax and wane in intensity, change location, or terminate. The distance and direction of location changes, and whether or not strands branch and diverge or converge and fuse, depends on the winds at the time of migration and how variable these are while the strand is advancing through the migration stage. Mortality during or following migration, as a result of, for example, unfavourable temperatures, offshore winds, drought, etc., will be a principal cause of strand termination. In all but the most non-seasonal of climates, these environmental influences will introduce some degree of annual periodicity into the trajectory, and synchronise any spatial cycling to the annual cycle of the seasons.

The second major determinant of trajectory shape, the insects' migration syndrome, will be considered in the next section.

Finally, we can note that subpopulations migrating down trajectory spurs, or joining persistent subsidiary populations (Fig. 21.3*f*), can make no further contribution to the development of the population. Spurs arise whenever migrants fail to locate habitat patches that are in a favourable condition or become trapped, e.g. by contrary winds or temperatures below the threshold for flight, in regions where conditions are deteriorating. This can happen in any part of the arena. However, some spurs arise when subpopulations migrate into regions where conditions are everywhere hostile, or from which there are never any opportunities for return movements back to the main (i.e. continuing) portion of the trajectory reticulum. It therefore seems useful to distinguish between 'core' and 'peripheral' parts of the population pathway: the core encompasses all continuing strands of the trajectory, while the periphery is occupied only by spurs (which may persist for a few generations if conditions and resources remain temporarily favourable) and subsidiary populations. Subpopulations that find temporarily excellent conditions within the periphery, and experience high reproductive success for a generation or two before eventually succumbing, are examples of what has been termed the 'Pied Piper' phenomenon (Rabb & Stinner, 1978; McNeil, 1987).

This distinction between core and periphery can be applied also to the arena, and it can be inferred that it is only the environmental conditions and resources of the core that are significant for the population's long-term survival. Probable examples of arena peripheries include northeastern China for *Nilaparvata lugens* (Chapter 3, this volume) and *Mythimna separata* (Chen *et al.*, 1989; but see Chapter 4, this volume), Japan for most if not all windborne migrants (see e.g. Chapters 3 and 6, this volume), and southeastern Australia for *Chortoicetes terminifera* (Farrow, 1975). An example of the periphery supporting a subsidiary population may be provided by *Autographa gamma* in Japan (O. Saito – personal communication).

*Processes involving the migration syndrome*

Four classes of process involving the syndrome can be recognised:

1 Processes by which the syndrome is passed from generation to generation;
2 Environmental effects that modulate the migrants' physiology and behaviour, e.g. through thresholds, rates, and cues;
3 Processes by which the syndromes of its members steer the population along the trajectory reticulum;
4 Modulation of the values of syndrome traits and their distributions as the population advances along the trajectory, by the action of contemporary natural selection.

The second class has been considered already, and the first and fourth classes will be treated in the next two subsections. We will describe here the third class of process.

An insect's migration syndrome, in combination with environmental conditions and cues, determines whether and when the insect will develop a capacity for migration, how long this capacity will be maintained, and in what circumstances migratory flight will be initiated and terminated. The syndromes of a subpopulation's members, together with physical factors such as wind and temperature, therefore determine the way in which the subpopulation's trajectory strand develops.

Because the members of each subpopulation are phenotypically diverse, they will vary in their responses to the conditions they encounter. For example, different values of traits determining the incidence or extent of migration, or different thresholds of response to the environmental cues modulating these traits, will influence the temporal and spatial distribution of the migrants' destinations. Thus, if migration is induced in some or all members of a subpopulation, variation in migration thresholds and migratory potential, and in windspeed and direction, will lead to reproduction taking place at a number of different locations and probably at different times, i.e. to branching and divergence of trajectory strands. Variation in other life-history traits (e.g. in development rates or in thresholds for diapause induction and termination), and in their responses to environmental conditions, may also contribute to temporal divergence between strands.

Whereas variation in some syndrome traits tends to cause trajectory strands to branch and diverge, other traits or mechanisms may have the opposite effect. Responses to resources in the same favourable habitat patch, during immediately post-migratory flights, may arrest individuals from different subpopulations (Gatehouse, 1995), or flying migrants may become concentrated by convergent wind systems (Drake & Farrow, 1989). Both mechanisms could reduce the geographical extent of strands and lead previously separated strands to converge at a common location. Temporal convergence between subpopulations that have entered diapause could occur as they resume development in response to improving environmental conditions (Danks, 1987). Strands may of course also sometimes converge, in both space and time, more or less by chance: the responses of subpopulations at separate locations to the differing conditions they experience before and during migration may result in them becoming synchronous and migrating to a common destination. Because the two mechanisms of spatial convergence described earlier both act on a relatively local scale (in comparison with the displacements typically achieved during specifically migratory flight), this more random process may be the predominant mechanism for bringing widely separated subpopulations together.

### Processes involving the genetic complex

The genetic complex determines how the migration-syndrome traits are passed from generation to generation, and how the syndrome responds to selection imposed by the environments through which the strands of the population trajectory pass. Five categories of process involving the genetic complex can be recognised:

1 Processes by which genes coding for migration-syndrome traits are transmitted from generation to generation, and variation in the trait values is maintained;
2 Processes modulating the phenotypic expression of these genes;
3 Processes by which gene frequencies at migration-syndrome loci are adjusted by contemporary natural selection;
4 Processes by which the genetic complex is itself modified by contemporary natural selection;
5 The process of genetic partitioning by migration.

The first, second and fifth of these categories are considered here, while the third and fourth are discussed in the next subsection.

The mode of inheritance of most traits contributing to the migration syndrome is polygenic (see above). Recombination of alleles at loci contributing to polygenic traits allows continuous variation in trait values among offspring, over a range that can extend beyond those of the parents; in large families, frequencies of phenotypic values of quantitative traits are taken to approximate to a normal distribution (Falconer, 1989). For migration-syndrome traits determining the timing of migration, or migratory potential, values associated with a successful outcome vary continually and often unpredictably with changing environmental conditions before and during migration. Therefore, an individual's fitness must depend on maintaining variation in these traits among its offspring. Polygenic inheritance with recombination achieves this, although other mechanisms, such as multiple mating (characteristic of many migratory noctuid moths), also contribute. The processes by which variation is maintained in clonal migrants such as aphids (see above) are unknown, but may be genetic (see above and Walker, 1986).

The expression of some traits contributing to the migration syndrome may depend on genotype × environment interactions. In facultative migrants, genes coding for traits associated with the process of migration, e.g. macroptery, takeoff and flight behaviour, are only expressed in response to specific environmental cues; in their absence, heritabilities of these traits are effectively reduced to zero, so that variation is protected from directional selection when migration is inappropriate (Roff, 1986b). Maternal effects have a similar outcome, temporarily reducing the heritability of traits to which they relate (Mousseau & Dingle, 1991). In other cases, genotype × environment interactions merely modulate the expression of traits; e.g. longer pre-reproductive periods may occur under shortening photoperiods, to allow for long-distance migration to escape the approaching winter at high latitudes, than are appropriate for shorter, summer migrations (e.g. Dingle, 1978; Han & Gatehouse, 1991).

Branching of trajectory strands inevitably results in some partitioning of migrants according to syndrome traits that influence their destinations. However, the groundspeed of windborne migrants is strongly influenced by stochastic variations in windspeed, so migratory potential may be only loosely related to the extent of displacement. Nevertheless, the outcome must be a degree of genetic partitioning, with the gene frequencies at these loci, and those contributing to correlated traits, varying between the new subpopulations. Similarly, the gene frequencies in new subpopulations formed by fusion of trajectory strands will reflect those of the subpopulations these strands represented.

*Processes involving contemporary natural selection*

As the population advances along its trajectory, natural selection acting on the migration-syndrome traits adjusts the distributions of their values and the frequencies of the alleles of the underlying genetic complex. The principal factors imposing this selection are success or failure in reaching favourable habitats, and the impact of the quality of those habitats on the fitness of the immigrants. The manner in which natural selection acts on syndrome traits will depend greatly on the form and characteristics of the arena. For example, an arena in which favourable habitats are widely separated by hostile regions (e.g. an archipelago) is likely to subject migratory potential to disruptive selection, as only very short or very long migrations can be successful. In migrants penetrating to high latitudes in summer, directional selection for high migratory potential can be expected, temporarily, during spring and autumn as current habitats deteriorate and/or favourable ones become available in different climatic zones (Chapters 1–3, this volume). However, for much of the time in most arenas, the distribution of new habitat patches in space and time is highly variable and often unpredictable, imposing constantly shifting selection for migratory potential and maintaining high levels of variation in traits influencing it.

When strands branch and migrants are partitioned according to the values of the migration-syndrome traits that steer them down one or other of the branches, the distributions of values of these traits (and others correlated with them) in the new subpopulations are likely to differ, with associated differences in gene frequencies at the affected loci. Independent development of these separate strands through different regions of the arena, where they are exposed to different selective pressures, is likely to lead to further phenotypic and genetic divergence. Evidence of geographic variation in some syndrome traits in several migratory species supports this expectation (Chapter 11, this volume). When separate strands converge and interbreeding between immigrants from the contributing subpopulations occurs, the range of genotypes among their offspring in the new subpopulation, and the range of phenotypic variation in syndrome traits, will be enhanced. Fusion of strands is therefore a further factor contributing to variation within subpopulations.

An important effect of natural selection acting on subpopulations along strands of the population trajectory is, therefore, to maintain variation in migration-syndrome traits, particularly those involved in initiating and implementing migratory behaviour. This variation is central to the functioning of the migration system, because it allows emigrants from these subpopulations to track short-lived and spatially restricted resources as they arise within the arena. As long as variability is maintained along the continuing trajectory reticulum, losses of migratory phenotypes along spurs arising from it will be sustainable.

A last category of process involving natural selection includes its action on the genetic complex itself. For example, the magnitude of genotype × environment interactions or of maternal effects, as well as the degree of genetic correlations, are all subject to selection.

## Summary and discussion

This overview has not aimed to develop or contribute new insights into the individual aspects of migration listed in the beginning of the chapter. Rather, its purpose has been to integrate consideration of all elements of a migration system – from geography and climate, through physiology and behaviour, to genetics and natural selection – within a unified conceptual framework. This extends Southwood's (1977) theoretical discussion of the ecological context for the evolution of migration and transposes it into a 'real' environment of a landscape with weather, resources, natural enemies, etc., to which the migrant population is specifically adapted, through its migration syndrome and the genetic complex underlying it. It is this holistic yet specific approach that we believe to be innovative.

A consistent feature of the arenas of migratory insects is that their habitats arise in different places at different times and persist for only limited periods. Migrants are able to track this continuously, and sometimes unpredictably, changing pattern of habitat patches and so can take advantage of the favourable conditions for development, reproduction and survival that temporary habitats offer. For windborne migrants, the arena is typically very large and may extend across two or more climatic zones. The wind is a key component of the arena for these species, as it provides the vehicle on which they can quickly achieve the long-distance displacements they need to make. In contrast, for migrants that fly within their flight boundary layer or for non-migrants, wind can only have a disruptive effect. Therefore, an essential feature of a migration arena is a pattern of airflows (or a degree of variability of windspeed and direction) that allows transport to occur into favourable habitats as they arise, irregularly or seasonally, in different locations.

A migrant population's adaptation to the spatiotemporal variability of its arena is evident in its population trajectory, which differs from those of non-migratory species by its relatively large geographical extent, the short-term occupancy of individual locations and by the presence of some degree of cycling which, in the pathway representation, may appear as a loop. Migration trajectories often (though not necessarily) have a reticulate topology with frequent branching and fusing of strands, representating the formation and merging of temporary subpopulations. These subpopulations rarely persist for more than a few generations, and so are too ephemeral to undergo the process of turnover (by colonisation and extinction) characteristic of the elements of metapopulations (Hanski & Gilpin, 1991). Their more complex spatial and temporal dynamics are therefore unlikely to be accommodated within the framework of current metapopulation models (Gliddon & Goudet, 1994; Harrison, 1994).

The migration syndrome differs in several important respects from the suites of traits influencing locomotion and demography in non-migrants. Morphological, behavioural, physiological and biochemical adaptations extend, often by many orders of magnitude, the migratory phenotypes' ranges of displacement and, by means of responses to environmental cues, associate the performance of migratory behaviour in facultative migrants with the availability and quality of their habitats. For some species, an important feature of the syndrome, absent in non-migrants (Dingle, Evans & Palmer, 1988; Gu & Danthanarayana, 1992), is the association of traits leading to displacement with demographic traits that increase the rate at which immigrants colonise their new habitats. In some long-distance migrants, a delay in the onset (or resumption) of reproduction, which represents a cost to fitness but provides a time interval during which migration can take place, may be offset by high early fecundity on arrival (Dingle, 1986b, 1991).

The functioning of the migration syndrome depends on the genetic complex which underlies it and associates, through a pattern of genetic correlations, many but not all of its component traits. Arguably the most significant features of the genetic complex are those that contribute to the maintenance of variation in traits influencing the induction and implementation of migratory flight. This variation, in thresholds of responses to environmental cues and in the components of migratory potential, ensures the ability of subpopulations to track changes in the spatial and temporal distribution of their habitats. Only by maintaining high levels of variation in these traits can they persist through the irregular and seasonal changes that occur in the physical and biotic environment of the arena.

The distribution of the phenotypic values of migration-syndrome traits, and the frequencies of genes at contributing loci, are continually adjusted by contemporary natural selection imposed by the conditions subpopulations encounter along the strands of the population trajectory. The distributions of these values therefore represent the adaptation of the population's migration syndrome to its particular arena. At any migration, migrants whose syndromes steer them down spurs of the trajectory are eliminated sooner or later. However, the variability within migration arenas is such that these same phenotypic values may keep other members of the population (or their descendants) within the continuing trajectory reticulum at a subsequent migration. The maintenance of variation in the migration-syndrome traits is therefore central to the functioning of the migration system. The frequent loss of migratory phenotypes down trajectory spurs is an inevitable consequence of this variability; however, it is evidently sustainable in persisting systems although natural selection can be expected to act to minimise it. The process of partitioning at trajectory branches, according to the phenotypic values of migration-syndrome traits, and subsequent genetic differentiation along the temporarily separated strands before they fuse again in the continuing trajectory reticulum, are features that have no counterpart in non-migratory populations.

We hope that the holistic approach to the description of insect migration adopted in the conceptual model presented here will provide a framework for future development of our understanding of the functioning of migration systems. We believe that explicit consideration of the properties of the arena (especially the spatiotemporal variability of resources, the availability of wind transport and the extent to which environmental changes are predictable) is a prerequisite for identifying and investigating the components of the migration syndrome and the genetic complex that underpins it, and for beginning to understand the selective forces that have shaped them. Progress in these areas will allow the development of improved predictive models of migration systems, and may eventually permit the spatial and temporal dynamics of migratory populations (i.e. their trajectories) to be forecast.

Although we have made no attempt to do so here, development of the model to allow quantitative representation of a migration system appears possible. The quantities required to describe all four components of the model are, at least in principle, measureable. The condition of both abiotic and biotic components of the arena can be determined through regular observations and surveys, including remote-sensing observations from satellites; similarly, the development of the trajectory can be monitored through regular surveys, with major displacements perhaps being detected with entomological radars. In both cases, the observations can be expressed quantitatively, in terms of habitat availability (and possibly quality), population densities, migration fluxes, etc. Geographic information systems (GIS) probably provide the most convenient means of storing, accessing and analysing the data. Such representations are already being developed, and the necessary datasets accumulated, for some economically important migration systems, in the expectation that this will allow improved forecasting of pest outbreaks and invasions (Tucker, 1994; Chapters 16, 19 and 20, this volume).

The migration syndromes of subpopulations could also be determined by field sampling, and can be represented quantitatively by statistics or probabilities for the incidence or values of the various traits. This information may also be of practical value for understanding, for example, the significance of changes in migratory potential at different times and locations for the origin and development of outbreaks of migratory pests (see e.g. Pedgley *et al.*, 1989; Tucker, 1994; Rose, Dewhurst & Page, 1995). Finally, the genetic complex underlying the migration syndrome is accessible to genetic analysis. It could, in principle, be represented quantitatively, for any subpopulation, by the frequencies of alleles at contributing loci (determined by techniques of molecular genetics), and estimates of the magnitude and sign of the genetic correlations between traits. Quantitative estimates of the genotypic variability for migration-syndrome traits in a subpopulation, together with knowledge of the mechanisms by which this variability is maintained, may make it possible to forecast its migratory movements in response to changing conditions. Similar analyses for the population as a whole may allow prediction of its capacity to adapt to changes in the spatial distribution of habitats caused by agricultural development or climate change. Such assessments of the robustness of migration systems in the face of changes in their arenas are potentially valuable to land-use planners, scientists developing pest management strategies, and conservationists concerned with protecting threatened migrant populations (see e.g. Brower & Malcolm, 1991).

Whether or not these expectations of progress are realised, we hope that this chapter will provide a firmer, more formalised conceptual basis for discussing insect migration than has hitherto been available. If this allows the development of testable hypotheses, more precise (e.g. quantitative) models, or alternative conceptual structures, it will have fulfilled our intended purpose.

## References

Andrewartha, H.G. (1957). The use of conceptual models in population ecology. *Cold Spring Harbour Symposia on Quantitative Biology*, **22**, 219–32.

Andrewartha, H.G. & Birch, L.C. (1954). *The Distribution and Abundance of Animals*. Chicago: University of Chicago Press.

Baker, R.R. (1978). *The Evolutionary Ecology of Animal Migration*. London: Hodder & Stoughton.

Bradshaw, W.E. (1986). Pervasive themes in insect life cycle strategies. In *The Evolution of Insect Life Cycles*, ed. F. Taylor & R. Karban, pp. 261–75. New York: Springer-Verlag.

Brower, L.P. & Malcolm, S.B. (1991). Animal migrations: endangered phenomena. *American Zoologist*, **31**, 265–76.

Charlesworth, B., Coyne, J.A. & Barton, N.H. (1987). The relative rates of evolution of sex chromosomes and autosomes. *American Naturalist*, **130**, 113–46.

Chen, R.-L., Bao, X.-Z., Drake, V.A., Farrow, R.A., Wang, S.-Y., Sun, Y.-J. & Zhai, B.-P. (1989). Radar observations of the spring migration into northeastern China of the oriental armyworm moth, *Mythimna separata*, and other insects. *Ecological Entomology*, **14**, 149–62.

Colvin, J. & Gatehouse, A.G. (1993). Migration and genetic regulation of the pre-reproductive period in the cotton-bollworm moth, *Helicoverpa armigera*. *Heredity*, **70**, 407–12.

Danks, H.V. (1987). *Insect Dormancy: an Ecological Perspective*. Ottawa: Biological Surveys of Canada Monograph 1.

Dingle, H. (1978). Migration and diapause in tropical, temperate and island milkweed bugs. In *Evolution of Insect Migration and Diapause*, ed. H. Dingle, pp. 254–76. New York: Springer-Verlag.

Dingle, H. (1980). Ecology and evolution of migration. In *Animal Migration, Orientation and Navigation*, ed. S.A. Gauthreaux, pp. 1–101. New York: Academic Press.

Dingle, H. (1984). Behaviour, genes and life histories: complex adaptations in uncertain environments. In *A New Ecology: Novel Approaches to Interactive Systems*, ed. P.W. Price, C.N. Slobodchikoff & W.S. Gaud, pp. 169–94. New York: John Wiley.

Dingle, H. (1985). Migration. In *Comprehensive Insect Physiology, Biochemistry and Pharmacology*, vol. 9, ed. G.A. Kerkut & L.I. Gilbert, pp. 375–415. Oxford: Pergamon Press.

Dingle, H. (1986*a*). Evolution and genetics of insect migration. In *Insect Flight: Dispersal and Migration*, ed. W. Danthanarayana, pp. 11–26. Berlin: Springer-Verlag.

Dingle, H. (1986*b*). Quantitative genetics of life history evolution in a migrant insect. In *Population Genetics and Evolution*, ed. G. de Jong, pp. 83–93. Berlin: Springer-Verlag.

Dingle, H. (1989). The evolution and significance of migratory flight. In *Insect Flight*, ed. G.J. Goldsworth & C.H. Wheeler, pp. 100–14. Boca Raton, Florida: CRC Press.

Dingle, H. (1991). Evolutionary genetics of animal migration. *American Zoologist*, **31**, 253–64.

Dingle, H., Evans, K.E. & Palmer, J.O. (1988). Responses to selection among life-history traits in a non-migratory population of milkweed bugs (*Oncopeltus fasciatus*). *Evolution*, **42**, 79–92.

Drake, V.A. (1994). The influence of weather and climate on agriculturally important insects: an Australian perspective. *Australian Journal of Agricultural Research*, **45**, 487–509.

Drake, V.A. & Farrow, R.A. (1988). The influence of atmospheric structure and motions on insect migration. *Annual Review of Entomology*, **33**, 183–210.

Drake, V.A. & Farrow, R.A. (1989). The 'aerial plankton' and atmospheric convergence. *Trends in Ecology and Evolution*, **4**, 381–5.

Falconer, D.S. (1989). *Introduction to Quantitative Genetics*, 3rd edn. London: Longman.

Farrow, R.A. (1975). Offshore migration and the collapse of outbreaks of the Australian plague locust (*Chortoicetes terminifera* Walk.) in southeast Australia. *Australian Journal of Zoology*, **23**, 569–95.

Farrow, R.A. (1990). Flight and migration in Acridoids. In *Biology of Grasshoppers*, ed. R.F. Chapman & A. Joern, pp. 227–314. New York: John Wiley.

Gatehouse, A.G. (1986). Migration in the African armyworm *Spodoptera exempta*: genetic determination of migratory capacity and a new synthesis. In *Insect Flight: Dispersal and Migration*, ed. W. Danthanarayana, pp. 128–44. Berlin: Springer-Verlag.

Gatehouse, A.G. (1987). Migration: a behavioural process with ecological consequences. *Antenna*, **11**, 10–12.

Gatehouse, A.G. (1989). Genes, environment, and insect flight. In *Insect Flight*, ed. G.J. Goldsworth & C.H. Wheeler, pp. 115–38. Boca Raton, Florida: CRC Press.

Gatehouse, A.G. (1995). Insect migration: variability and success in a capricious environment. *Researches in Population Ecology*, **36**, in press.

Gibo, D.L. (1986). Flight strategies of migrating monarch butterflies (*Danaus plexippus* L.) in southern Ontario. In *Insect Flight: Dispersal and Migration*, ed. W. Danthanarayana, pp. 172–84. Berlin: Springer-Verlag.

Gliddon, C.J. & Goudet, J. (1994). The genetic structure of metapopulations and conservation biology. In *Conservation Genetics*, ed. V. Loeschcke, J. Tomiuk & S.K. Jain, pp. 107–14. Basel: Birkhäuser Verlag.

Gu, H.-N. & Danthanarayana, W. (1992). Quantitative genetic analysis of dispersal in *Epiphyas postvittana*. II. Genetic covariations between flight capacity and life-history traits. *Heredity*, **68**, 61–9.

Gunn, A. & Gatehouse, A.G. (1993). The migration syndrome in the African armyworm moth, *Spodoptera exempta*: allocation of resources to flight and reproduction. *Physiological Entomology*, **18**, 149–59.

Han, E.-N. & Gatehouse, A.G. (1991). Effect of temperature and photoperiod on the calling behaviour of a migratory insect, the oriental armyworm *Mythimna separata*. *Physiological Entomology*, **16**, 419–27.

Hanski, I. & Gilpin, M. (1991). Metapopulation dynamics: brief history and conceptual domain. In *Metapopulation Dynamics: Empirical and Theoretical Investigations*, ed. M. Gilpin & I. Hanski, pp. 3–16. London: Academic Press.

Harrison, R.G. (1980). Dispersal polymorphisms in insects. *Annual Review of Ecology and Systematics*, **11**, 95–118.

Harrison, S. (1994). Metapopulations and conservation. In *Large-scale Ecology and Conservation Biology*, ed. P.J. Edwards, R.M. May & N.R. Webb, pp. 111–28. Oxford: Blackwell Scientific Publications.

Howard, W.E. (1960). Innate and environmental dispersal of individual invertebrates. *American Midland Naturalist*, **63**, 152–61.

Johnson, C.G. (1960). A basis for a general system of insect migration and dispersal by flight. *Nature*, **186**, 348–50.

Johnson, C.G. (1963). Physiological factors in insect migration by flight. *Nature*, **198**, 423–7.

Johnson, C.G. (1969). *Migration and Dispersal of Insects by Flight*. London: Methuen.

Kampmeier, G.E. (1994). The take-off and ascent component of aerial movement of biota: introductory remarks. In *Proceedings, 13th International Congress of Biometeorology, 12–18 September 1993, Calgary, Alberta, Canada*, pp. 988–92. Downsview, Ontario: Environment Canada.

Kennedy, J.S. (1961). A turning point in the study of insect migration. *Nature*, **189**, 785–91.

Kennedy, J.S. (1975). Insect dispersal. In *Insects, Science and Society*, ed. D. Pimentel, pp. 103–19. New York: Academic Press.

Kennedy, J.S. (1985). Migration, behavioural and ecological. In *Migration: Mechanisms and Adaptive Significance*, ed. M.A. Rankin, pp. 5–26. *Contributions in Marine Science*, vol. 27 (Suppl.). Port Aransas, Texas: Marine Science Institute, The University of Texas at Austin.

McNeil, J. N. (1987). The true armyworm, *Pseudaletia unipuncta*: a victim of the pied piper or a seasonal migrant? In *Recent Advances in Research on Tropical Entomology*, ed. M.F.B. Chaudhury, pp. 591–7. *Insect Science and its Application*, vol. 8, Special Issue. Nairobi: ICIPE Science Press.

Mousseau, T.A. & Dingle, H. (1991). Maternal effects in insect life histories. *Annual Review of Entomology*, **36**, 511–34.

Nicholls, N. (1991). The El Niño / Southern Oscillation and Australian vegetation. *Vegetatio*, **91**, 23–36.

Nottingham, S.F. & Hardie, J. (1989). Migratory and targeted flight in seasonal forms of the black bean aphid, *Aphis fabae*. *Physiological Entomology*, **14**, 451–8.

Nottingham, S.F., Hardie, J. & Tatchell, G.M. (1991). Flight behaviour of the bird cherry aphid, *Rhopalosiphum padi*. *Physiological Entomology*, **16**, 223–9.

Palmer, J.O. & Dingle, H. (1986). Direct and correlated responses to selection among life-history traits in milkweed bugs (*Oncopeltus fasciatus*). *Evolution*, **40**, 767–77.

Palmer, J.O. & Dingle, H. (1989). Responses to selection on flight behaviour in a migratory population of milkweed bug (*Oncopeltus fasciatus*). *Evolution*, **43**, 1805–8.

Pedgley, D.E., Page, W.W., Mushi, A., Odiyo, P., Amisi, J., Dewhurst, C.F., Dunstan, W.R., Fishpool, L.D.C., Harvey, A.W., Megenasa, T. & Rose, D.J.W. (1989). Onset and spread of an African armyworm upsurge. *Ecological Entomology*, **14**, 311–33.

Rabb, R.L. & Stinner, R.E. (1978). The role of insect dispersal and migration in population processes. In *Radar, Insect Population Ecology, and Pest Management*. NASA Conference Publication 2070, ed. C.R. Vaughn, W.W. Wolf & W. Klassen, pp. 3–16. Wallops Island, Virginia: NASA Wallops Flight Centre.

Roff, D.A. (1986a). The evolution of wing dimorphism in insects. *Evolution*, **40**, 1009–20.

Roff, D.A. (1986*b*). Evolution of wing polymorphism and its impact on life cycle adaptation in insects. In *The Evolution of Insect Life Cycles*, ed. F. Taylor & R. Karban, pp. 204–21. New York: Springer-Verlag.

Roff, D.A. & Fairbairn, D.J. (1991). Wing dimorphisms and the evolution of migratory polymorphisms among the Insecta. *American Zoologist*, **31**, 243–51.

Rose, D.J.W., Dewhurst, C.F. & Page, W.W. (1995). The bionomics of the African armyworm *Spodoptera exempta* in relation to its status as a migrant pest. *Integrated Pest Management Reviews*, **1**, in press.

Schaefer, G.W. (1976). Radar observations of insect flight. In *Insect Flight*, ed. R.C. Rainey, pp. 157–97. *Symposia of the Royal Entomological Society of London 7*. Oxford: Blackwell Scientific Publications.

Schmidt-Koenig, K. (1985). Migration strategies of monarch butterflies. In *Migration: Mechanisms and Adaptive Significance*, ed. M.A. Rankin, pp. 786–98. *Contributions in Marine Science*, vol. 27 (Suppl.). Port Aransas, Texas: Marine Science Institute, The University of Texas at Austin.

Shaw, M.P.J. (1970*a*). Effects of population density on alienicolae of *Aphis fabae* Scop. I. The effect of crowding on the production of alatae in the laboratory. *Annals of Applied Biology*, **65**, 191–6.

Shaw, M.P.J. (1970*b*). Effects of population density on alienicolae of *Aphis fabae* Scop. III. The effect of isolation on the development of form and behaviour of alatae in a laboratory clone. *Annals of Applied Biology*, **65**, 205–12.

Southwood, T.R.E. (1962). Migration of terrestrial arthropods in relation to habitat. *Biological Reviews*, **37**, 171–214.

Southwood, T.R.E. (1977). Habitat, the templet for ecological strategies? *Journal of Animal Ecology*, **46**, 337–65.

Southwood, T.R.E. (1981). Ecological aspects of insect migration. In *Animal Migration*, ed. D.J. Aidley, pp. 197–208. Cambridge: Cambridge University Press.

Strong, D.R. (1984). Density-vague ecology and liberal population regulation in insects. In *A New Ecology: Novel Approaches to Interactive Systems*, ed. P.W. Price, C.N. Slobodchikoff & W.S. Gaud, pp. 313–27. New York: John Wiley.

Taylor, L.R. (1958). Aphid dispersal and diurnal periodicity. *Proceedings of the Linnean Society of London*, **169**, 67–73.

Taylor, L.R. (1986). The four kinds of migration. In *Insect Flight: Dispersal and Migration*, ed. W. Danthanarayana, pp. 265–80. Berlin: Springer-Verlag.

Taylor, L.R. & Taylor, R.A.J. (1977). Aggregration, migration and population mechanics. *Nature*, **265**, 415–21.

Tucker, M.R. (1994). Inter- and intra-seasonal variation in outbreak distribution of the armyworm, *Spodoptera exempta* (Lepidoptera: Noctuidae), in eastern Africa. *Bulletin of Entomological Research*, **84**, 275–87.

Walker, T.J. (1985). Butterfly migration in the boundary layer. In *Migration: Mechanisms and Adaptive Significance*, ed. M.A. Rankin, pp. 704–23. *Contributions in Marine Science*, vol. 27 (Suppl.). Port Aransas, Texas: Marine Science Institute, The University of Texas at Austin.

Walker, T.J. (1986). Stochastic polyphenism: coping with uncertainty. *Florida Entomologist*, **69**, 46–62.

# Index

459